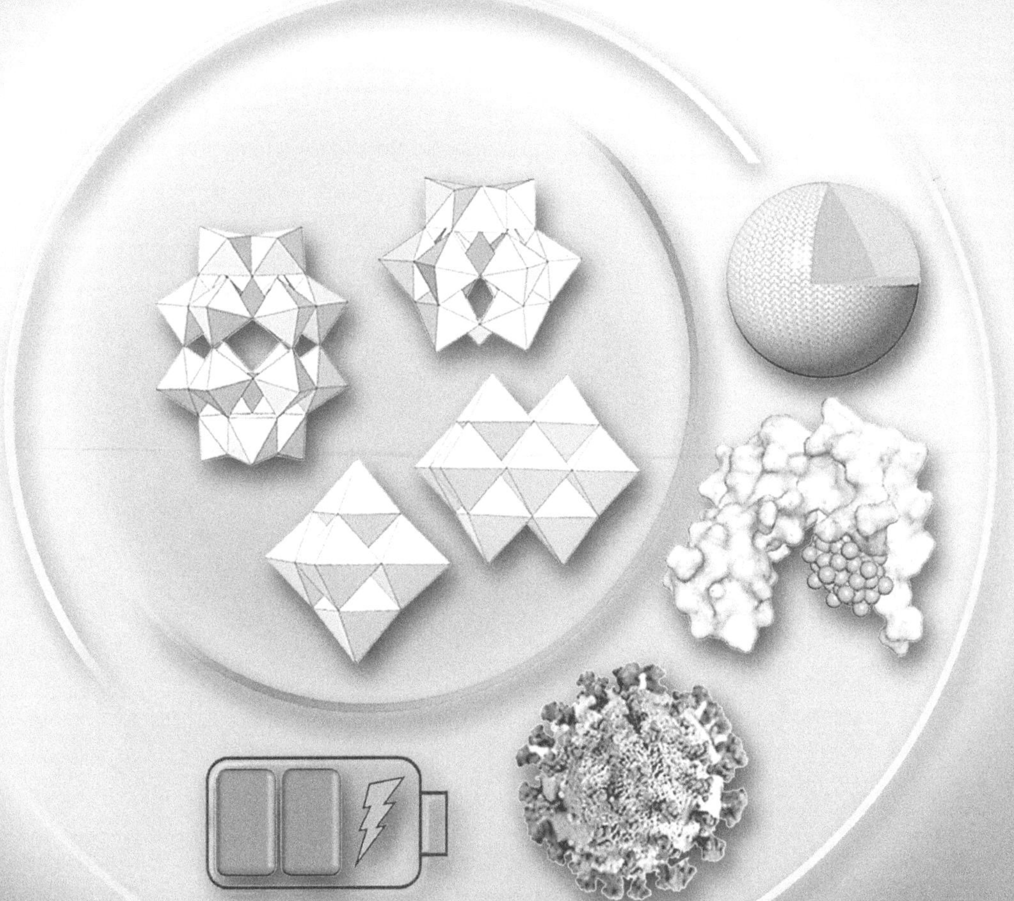

Polyoxometalates

Polyoxometalates

Advances, Properties, and Applications

edited by

Leire Ruiz Rubio | José Luis Vilas Vilela
Beñat Artetxe | Juan Manuel Gutiérrez-Zorrilla

JENNY STANFORD
PUBLISHING

Published by

Jenny Stanford Publishing Pte. Ltd.
101 Thomson Road
#06-01, United Square
Singapore 307591

Email: editorial@jennystanford.com
Web: www.jennystanford.com

British Library Cataloguing-in-Publication Data
A catalogue record for this book is available from the British Library.

Polyoxometalates: Advances, Properties, and Applications

ISBN 978-981-4968-14-0 (Hardcover)
ISBN 978-1-003-27744-6 (eBook)

Contents

Preface

The design of molecular materials represents one of the current hot topics within the field of materials science because their well-defined structure allows structure–property relationships to be determined, and thus, tailored physical properties can be fine-tuned for required practical applications. In this sense, we are proud to introduce the growing family of polyoxometalate (POM) clusters to the reader, which constitute a unique class of compounds due to their rich solution equilibria and their unique compositional, electronic, reactive, and structural diversity. This book reviews anionic metal-oxide cluster chemistry from fundamental aspects (structure, properties, self-assembly processes, derivatization) to functional materials that incorporate these molecular units, as well as their applications in fields of current socio-economic interest such as catalysis, electronics, and biomedicine.

Chapter 1 provides an overview of the general principles and properties of POMs. The predominant species in aqueous solutions of traditional iso- and hetero-POMs are discussed, together with the main structural archetypes for each particular group. This piece could serve as a compilation of general opening remarks for the rest of the chapters included in this themed book.

One of the most intriguing properties of POM macroions is featured in **Chapter 2**. The special solution behavior displayed by these anionic clusters is different from that shown by smaller simple ions or larger colloids, and thus, it allows their controlled self-assembly in dilute solutions.

Besides fundamentals, traditional strategies for the functionalization of POM platforms are also presented in the introductory chapter, which include the organically derivatized clusters. The latter approach is nicely employed for the rational design of POM-peptide hybrid materials reviewed in **Chapter 3**. Different routes for the preparation of covalent hybrids are discussed initially, whereas their inherent properties such as stereochemistry and folding are described later on. The second part is devoted to ionic compounds and their interesting applications.

One of the major drawbacks for the incorporation of POM anions into functional materials originates from their crystalline solid nature, which makes them hard to process. **Chapter 4** summarizes the intensive work carried out in the incorporation of POM anions into organic polymers.

The second part of this book focuses on some of the most fascinating applications exhibited by POM-based systems. Mechanistic insights of some selected heterogeneous catalysts are provided in **Chapter 5**. The influence on the catalytic activity of both POM species and active porous supports are the main aspects to be commented and analyzed in detail within this chapter.

Moreover, the study and application of molecular building blocks for the fabrication of electronic components represents an exciting field in which POM chemistry has a key role to play. Beyond classical bulk ferromagnets and semiconductors, metal-oxo clusters can act as nano-molecular computing elements for the fabrication of data storage materials. This field is comprehensively analyzed in **Chapter 6**.

Among the large-scale alternatives to fossil fuels, electrochemical energy storage and conversion systems are the most developed technologies. POMs have been recently used to replace the redox active species of the popular all-vanadium redox-flow systems with very promising results. All these studies are gathered in **Chapter 7**.

Chapter 8 covers recent advances in POMs with anticancer, antibacterial, and antiviral activity with a particular interest on mechanistic insights. Although their activity is often higher than some approved drugs, their mechanism of action is still not well understood.

Finally, we would like to thank all the authors for their highly valuable scientific contributions, effort and patience all along the publication process, specially taking into account the ongoing pandemic, which has affected not only our lives but also the initially programmed schedule for this book. All the support from our research groups (LabQuiMac, Dpt. Physical Chemistry and

POM group, Dpt. Organic and Inorganic Chemistry) and Universidad del País Vasco UPV/EHU is highly appreciated. Last but not least, a special acknowledgement to Jenny Stanford Publishing for making this journey much easier and providing us with a certain degree of flexibility. It has been indeed a prolific and rewarding experience.

Dr Leire Ruiz Rubio
Dr José Luis Vilas Vilela
Dr Beñat Artetxe
Dr Juan Manuel Gutiérrez-Zorrilla

Chapter 1

General Principles and Structural Chemistry of Polyoxometalates

Estibaliz Ruiz-Bilbao, Leticia Fernández-Navarro, Beñat Artetxe, and Juan M. Gutiérrez-Zorrilla

Departamento de Química Orgánica e Inorgánica, Facultad de Ciencia y Tecnología, Universidad del País Vasco, UPV/EHU, P.O. Box 644, 48080 Bilbao, Spain

benat.artetxe@ehu.eus

A general overview of the main properties and structural principles of the growiing field of polyoxometalates (POMs) is provided in this introductory chapter. The most common structural archetypes of these anionic metal-oxo clusters are discussed, as well as the predominant species in aqueous solutions which cover from traditional iso- and hetero-vanadates, molybdates, and tungstates to less conventional anions including giant molybdates, uranium peroxoclusters, and noble-metal POMs. In addition, classical derivatization methods are described in the second part, which consist in the incorporation of transition-metal and/or lanthanide centers in the inorganic cluster skeleton and the organic functionalization of POMs. This piece could serve as a compilation of opening remarks for the rest of the chapters included in this themed-book.

Polyoxometalates: Advances, Properties, and Applications
Edited by Leire Ruiz Rubio, José Luis Vilas Vilela, Beñat Artetxe, and Juan Manuel Gutiérrez-Zorrilla

ISBN 978-981-4968-14-0 (Hardcover), 978-1-003-27744-6 (eBook)
www.jennystanford.com

1.1 Introduction

Polyoxometalates (POMs) represent a well-known family of anionic metal–oxygen nanoclusters that constitute a unique class of compounds not only for their diverse structural complexity, but also for their compositional, electronic, and reactive versatility [1, 2]. These features make POMs useful for a wide range of potential applications in the fields of current interest related to technology, health, energy, or environment, such as material science [3], catalysis [4], magnetism [5], and biomedicine [6]. It is commonly thought that POM chemistry started in the 19th century, when Jöns Jacob Berzelius reported back in 1826 the synthesis of the ammonium salt of the $[PMo_{12}O_{40}]^{3-}$ anion which was prepared through the addition of an excess of ammonium molybdate to a solution of phosphoric acid [7]. However, Juan José and Fausto Elhuyar (1783) had already been able to prepare a bitter, spicy, and yellow salt almost 40 years before which was later identified as the $(NH_4)_{10}[H_2W_{12}O_{42}]\cdot10H_2O$ salt [8]. In this context, it is worth mentioning that the Elhuyar brothers were the discoverers of the element W (tungsten or wolfram). Later on, making use of the analytical work by Jean Charles Galissard de Marignac in 1864 who establishes the 12:1 composition for the tungstosilicic acid [9], Alfred Werner and Linus C. Pauling developed different potential structural models for POMs [10, 11]. The clathrate-like model proposed by Pauling attracted crystallographers to solve this challenging system. In 1933, the British scientist James F. Keggin, who was working in Lawrence Bragg's group at the University of Manchester, determined the crystal structure of the $H_3[PMo_{12}O_{40}]\cdot5H_2O$ acid by using powder X-ray diffraction. This event represents the first structural evidence of a POM cluster and, hence, this topology became referred to as the Keggin structure after him [12].

Although hundreds of POMs and their physico-chemical properties had been reported by the middle of the 20th century, there was not such a deep study of their formation, degradation, and interconversion equilibria in solution [13]. The systematic work carried out by the French school contributed to shed light in this regard and established some general schemes for the synthesis of POMs [14]. Based on Michael T. Pope's seminal work in the 1980s [1, 2], POM chemistry has expanded very quickly in the last decades as exemplified by the vast number of patents, books, special thematic issues, and works published in some of the most prestigious journals within the inorganic chemistry [15, 16]. Recently, polyoxotungstates played a key role in the X-ray diffraction structural elucidation of ribosomes that leads to the 2009 Nobel Prize in Chemistry (Venkatraman Ramakrishnan, Thomas A. Steitz, and Ada E. Yonath) [17]. These clusters can be used as co-crystallizing agents for large biomolecules which not only allows a more efficient crystallization process, but it also facilitates the structural determination due to their high symmetry and heavy atom content [18]. Therefore, POM chemistry can be considered as one of the current hot topics of inorganic chemistry at present, which encompasses all the areas related to solution, solid state, and supramolecular chemistries including, (i) d/f metal containing derivatives with interesting photo/catalytic (special mention for water oxidation catalysts), magnetic, electronic, and optical properties; (ii) organic derivatization of POMs for their easy incorporation into functional materials; (iii) giant Mo-based molecular systems; (iv) POM-based open porous frameworks; (v) solution behavior of macroions which involves their self-assembly processes in solution; (vi) theoretical and computational perspectives; and (vii) interaction between biomolecules/cells and POM species. This way, this chapter aims to give an overview of the fundamental principles of POM chemistry with strong focus on the structural aspects. The last section describes the most explored strategies for the functionalization of POM platforms, comprising (i) 3d/4f metal-substituted POMs and (ii) organic derivatization of metal-oxo clusters.

1.2 Structural Principles

POMs are composed of early transition metals (M, so-called *addenda* metals) from groups 5 and 6, usually in their highest oxidation states; traditionally V, Mo, or W and less frequently Nb or Ta [19]. These metal centers are excellent precursors of polynuclear anions because they display a suitable combination of charge and ionic radius (Fig. 1.1A), as well as empty and accessible d orbitals available for electronic back-donation in the formation of π M–O bonds. In the last few years, examples of noble metal POMs have also been reported [20], together with some other related POM families, such as peroxo-actinyl species [21].

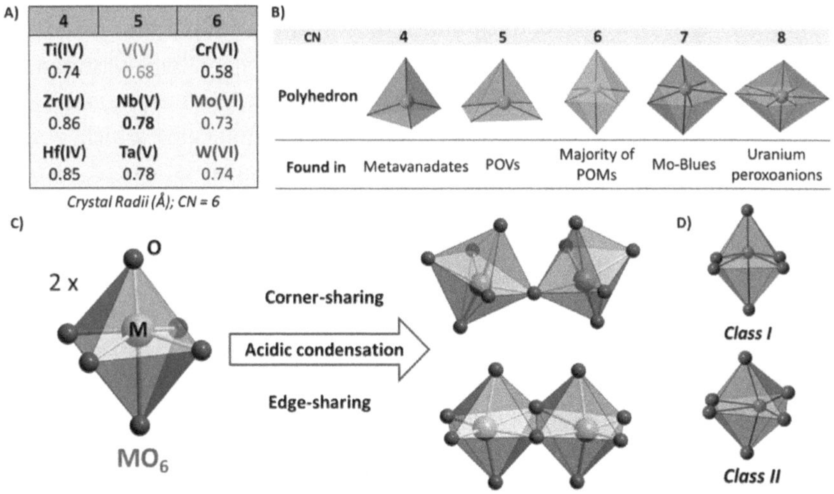

Figure 1.1 (A) Shannon and Prewitt crystal radii for metals from groups 4–6 in their highest oxidation states; (B) most usual coordination number (CN) and geometries of metal centers in POMs; (C) schematic representation of the two common linkage modes between octahedral $\{MO_6\}$ units; (D) ball-and-stick and polyhedral representations of the fundamental $\{MO_6\}$ units for *class I* and *class II* POMs.

From a structural point of view, POMs are formed by the acidic condensation of a variable number of MO_x polyhedra. The coordination number of *addenda* metals ranges from 4 to 7, but the octahedral MO_6 is the most common geometry by far. Some Mo^{VI} and W^{VI} centers can also occupy a seven-coordinated $\{O{=}M(O_6)\}$ polyhedron in POMs, and this has led to the development of several macro-sized giant-architectures [22]. Higher coordination numbers are usually observed in uranium peroxoanions (Fig. 1.1B). All these units can condense by sharing corners or edges (exceptionally faces), in such a way that each polyhedron must show a maximum of two unshared terminal oxido ligands (Fig. 1.1C) to minimize the *trans* effect of the terminal M–O bonds and avoid the dissociation of the cluster according to Lipscomb's principle [23]. Corner sharing offers flexibility, while edge sharing confers certain degree of rigidity on the cluster formed. This oligomerization is not infinite; due to the polarization of the peripheral tungsten centers toward the outer oxygen shell, vacant and energetically accessible d orbitals from the metal ion allow for the formation of terminal M=O double bonds, preventing the subsequent formation of bridges with additional polyhedra. Each metal center located in an MO_6 environment is displaced toward terminal oxygen atoms as a result of the participation of d orbitals in the π M–O bond. Thus, two types of polarization can be distinguished: toward a single terminal oxygen atom (*class I*) or toward two terminal oxygen atoms located in relative *cis* configuration (*class II*) (Fig. 1.1D). This fact leads to polyanions with very different electronic properties. The non-bonding Lowest Unoccupied Molecular Orbitals (LUMO) in members of *class I* make them more easily and reversibly reduced by chemical, photochemical, or electrochemical methods. This is the case of mixed-valence "blue" species based on Mo^{V}/Mo^{VI} moieties, which can be easily and reversibly reduced with no significant alterations in the POM skeleton. In contrast, reduction of *class II* clusters is more difficult and shows irreversible nature [24].

1.3 General Properties

The size, shape, composition, redox potential, acid strength, charge density, and functionality of a cluster can be controlled through synthetic POM chemistry to a great extent. In spite of this wide structural and compositional diversity, most POMs exhibit some common features [25]:

(i) **Physical properties:** They are discrete species with high charge, relatively big size, and high symmetry. POMs are thermally stable and resistant to oxidizing agents (air, water) and often, they maintain their structure in aqueous solution. Some species, such as heteropolyacids, can display extremely high acidity.

(ii) **Synthesis and formation equilibria**: The preparation of POM clusters covers a wide number of synthetic approaches that go from simple one-pot procedures to more sophisticated flow-system approaches or 3D printed reactionwares [26]. The simplest synthesis method implies an aqueous solution of $[MO_n]^{m-}$ anions which involves complex self-assembly processes *via* acidic condensation equilibria of $\{MO_6\}$ units that are still not-well understood. Afterwards, clusters need to be precipitated by adding suitable counterions (e.g., alkali metals, organic alkylammonium cations). POM species containing alkaline counterions are usually water soluble, whereas organic counterions confer high solubility in organic solvents to the polyanions [27].

The rich formation equilibria of POMs is very dependent on the concentration of the reactants, the sequence in which different reagents are added, pH, ionic strength, presence of extra ligands, counterions employed, temperature, or pressure (hydrothermal synthesis, microwave assisted reactions), among others. Moreover, the vast majority of studies developed in aqueous solution have shown that, for a given pH value, there are many POM species that coexist. For that reason, the species that are isolated do not need to be the predominant ones in solution, but the less soluble in combination with the corresponding counterion.

(iii) **Redox properties:** POMs can undergo multielectronic redox processes without any structural change or degradation. Electrons added to POMs are often delocalized over several metal atoms, and this facilitates fast electron transfers. It is worth mentioning that polyoxomolybdates are much easier to reduce than their tungstate analogs. Electron transfer by POMs is often coupled to cation or proton transfer. Thus, usually the net charge of the polyoxoanions does not change upon reduction, because edge-sharing oxygen atoms can be easily protonated [28]. This concept, often found in biological systems, avoids the presence of highly charged species, and results in increased stability, which is in the origin of their highly stable redox cycles when used in electrochemical systems. Therefore, this ability for the reversible electron/proton storage paves their way for their potential use as sustainable catalysts, electrochemical energy storage systems, molecular memory devices, molecular transistors, or single-molecule spintronics.

(iv) POMs can be used as **inorganic ligands** due to their capacity to anchor cations, anions, or neutral molecules to their surface oxygen atoms. Furthermore, some POM structures are derived from larger parent cluster by removing one or more *addenda* metal atoms from the shell. These defect structures are known as lacunary species. Their main feature is that lacunary polyanions can act as multidentate ligands toward different electrophiles (such as 3d and 4f metals) through the oxygen atoms delimiting their vacant sites.

(v) The high solubility that some POM clusters show in organic solvents facilitates their **organic derivatization**. The most usual functionalization approaches involve the replacement of shell oxygen atoms with those belonging to O- and N-donor ligands, such as alkoxo, diazenido, or imido units or the incorporation of p-block organoderivatives including organophosphoryl, -silyl, -germyl or -stanyl residues to the vacancies of lacunary POM units. The resulting hybrid derivatives have been identified as a key factor for the clusters to be suitably incorporated into functional materials for advanced applications [29].

1.4 Classification

Traditionally, POMs are classified according to their composition into iso- and hetero-polyoxometalates. Members of the former group exclusively contain *addenda* metals (M) and oxygen atoms acting as bridging or terminal oxido ligands (sometimes also as hydroxo groups, especially when bridging), whereas those from the latter show additional elements known as heteroatoms (X). Heteroatoms usually occupy central positions of the clusters and their nature does not show any restriction, as many of the elements of the periodic table can play this role. These especially include a variety of p-block elements and transition metals which exhibit coordination numbers ranging from 4 to 12. In general terms, isopolyanions display strongly basic surface oxygen atoms and they are less stable than heteropolyanions. This section compiles the most archetypical structures of classical iso- and hetero-polyoxometalates, excluding those containing mixed-*addenda* metal centers. The intriguing giant Mo-based system and less usual clusters, such as noble metal POMs and uranium peroxoanions are briefly described as well.

1.4.1 Isopolyoxometalates (isoPOMs)

The structures of most isoPOMs are derived from the $M_{10}O_{28}$ decametalate anion [30, 31] *via* elimination of octahedral units. This parent POM cluster is formed by ten edge-sharing MO_6 octahedra arranged in ideal D_{2d} symmetry. Examples of most representative isoPOMs include: (i) the β-M_8O_{26} octametalate with ideal D_{2h} symmetry [32], which derives from the decametalate cluster through the removal of two opposite units from the central level of six edge-sharing octahedra; (ii) the M_7O_{24} heptametalate (parametalate-A) [33] with C_{2v} symmetry, which can be best described as a bent arrangement of seven edge-sharing MO_6 octahedra; and (iii) the Lindqvist-type M_6O_{19} hexametalate with an ideal O_h symmetry that can be achieved through the removal of two octahedra from the central level and additional two from the upper and lower levels from the decametalate cluster (Fig. 1.2). Unlike the case of vanadium for which the isolation of pristine Lindqvist-type POMs remains elusive, hexamolybdates [34] and –tungstates [35] with a general octahedral symmetry can be easily obtained as alkyl-ammonium salts.

Isopolyoxovanadates are interesting species mainly due to the versatile redox properties of vanadium which displays up to four different relatively stable oxidation states in water solution (+2, +3, +4, +5) with associated easily distinguishable colors. When it comes to fully oxidized species, the tetrahedral orthovanadate $[VO_4]^{3-}$ oxoanion is only stable in very alkaline solutions, but tends to form pyrovanadate $[V_2O_7]^{4-}$ dimers for high vanadium concentrations [36]. These tetrahedral units further condense by sharing corners to afford oligomeric species like *cyclo*-tetravanadates $[V_4O_{12}]^{4-}$ [37] or polymeric metavanadate $(VO_3)_n^{n-}$ chains at neutral to moderately basic pH.

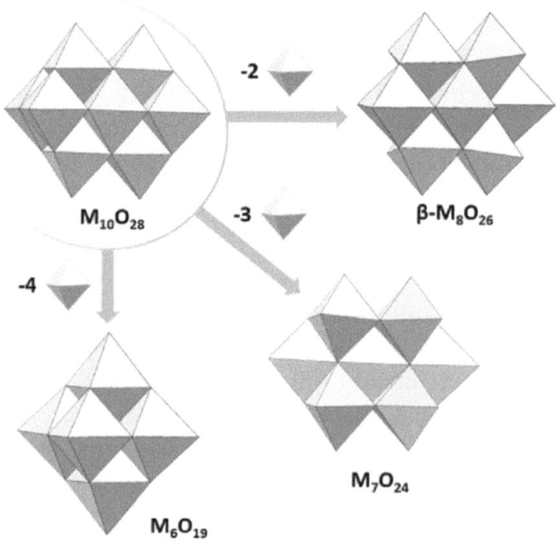

Figure 1.2 Polyhedral representation of some of the most usual isoPOM structures.

The spontaneous ring-opening of the cyclic species and subsequent polymerization into the *catena*-form [38] represents one of the first examples of a solid-state phase transition involving POM species for which the initial and final stages were structurally characterized by single-crystal X-ray diffraction [39]. At moderately acidic pH, the coordination number of vanadium increases from 4 to 6 and the orange-colored $[H_nV_{10}O_{28}]^{(6-n)-}$ decavanadate cluster becomes the predominant species in solution (Fig. 1.3) [40]. This anion can display several extents of protonation and represents the most studied polyoxovanadate due to its interesting biological activity [41]. Oligomers with different nuclearities, such as 3, 5, and 12 [42], have only been isolated in organic media and all of them exhibit corner-sharing $\{VO_4\}$ units. Additional meta-oxo units can also condense to well-known parent structures as exemplified by the $[V_{13}O_{34}]^{3-}$ anion derived from decavanadate species [43] and the bicapped Keggin-type structure of the $[V_{15}O_{42}]^{9-}$ cluster [44], in which the tetrahedral $\{VO_4\}$ unit occupies the inner heteroatomic site and 12 shell $\{VO_6\}$ fragments, together with two square pyramidal $\{VO_5\}$ capping-units form the outer shell. Besides, several mixed-valent (V^V/V^{IV} and V^{IV}/V^{III}), fully-reduced (V^{IV}) and totally reduced (V^{III}) polyoxovanadates have also been isolated using chemical or photochemical reduction routes, which usually display square pyramidal metal geometries. The spherical $[V^{IV}{}_{18}O_{42}]^{12-}$ shell constitutes one of the most popular examples of such reduced species due to the ability to incorporate small anionic or neutral guests in its inner void as a function of the pH [45].

In contrast, solution equilibria of molybdates are much more complex because their low kinetic stability prevents them from displaying well-defined pH-dependent stability ranges [46]. In fact, structures exhibiting reduced molybdenum centers usually consist in large species different from those of the parent fully oxidized clusters. In basic aqueous solutions of Mo(VI), only the molybdate $[MoO_4]^{2-}$ ion is stable, whereas neutral to moderately acidic pH affords the heptamolybdate (or paramolybdate-A) $[Mo_7O_{24}]^{6-}$ cluster. It should be mentioned the remarkable anticancer activity displayed by the isopropyl ammonium salt and derived photolytes of the latter species [47]. Further acidification leads to the formation of the well-known β-isomer of the octamolybdate $[Mo_8O_{26}]^{4-}$ anion, the structural diversity of which affords up to eight additional isomers (some of them are interconvertible [48]), namely, α, γ, δ, ε, ζ, η, θ, and ι that can be individually isolated depending on the synthetic conditions (solvent, counterions, synthetic method). From a structural point of view, all of them display some four- or five-coordinated molybdenum centers. Larger aggregates are only found at very acidic pH (< 2); this is the case of the $[Mo_{36}O_{112}(H_2O)_{18}]^{8-}$ anion [49] which was identified as the predominant species under these conditions (Fig. 1.3). In these large assemblies, some Mo centers occupy a seven-coordinated polyhedron. These building-blocks are very usual among the family of partially reduced, blue, macro-sized anions that represent some of the most spectacular architectures within the inorganic chemistry. These examples will be discussed separately in Section 1.4.3.

Despite their higher resistance to reduction and slower kinetics compared to their Mo counterparts, the basic and slightly acid aqueous solutions of tungstates are dominated by similar species to those found for molybdates; that is, the tungstate $[WO_4]^{2-}$ oxoanion and heptatungstate (paratungstate-A) $[W_7O_{24}]^{6-}$ polyanion. The latter is rapidly formed at pH = 5–7, but it slowly converts into paratungstate-B $[H_2W_{12}O_{42}]^{10-}$ composed of two W_3O_{13}- and two W_3O_{14}-like trimers arranged alternately [50]. These two anions coexist in an equilibrium shifted toward heptatungstate which is rapidly formed, but solids isolated from these solutions usually contain the $[H_2W_{12}O_{42}]^{10-}$ species because it is thermodynamically more stable and its salts are usually less soluble [51]. Defective structures closely related to the heptatungstate anion have also been obtained from water solutions [52]. The *Ψ*-metatungstate $[W_{24}O_{72}(OH)_{12}]^{12-}$ represents the predominant species in the pH = 3–4 range and it is composed of a central ring of six corner-sharing octahedra to which six W_3O_{13} groups containing one trigonal bipyramidal unit are linked [53]. It must be highlighted that this cluster establishes equilibrium with another metatungstate derivative, the $[H_2W_{12}O_{40}]^{6-}$ anion which displays the well-known tungsten–oxygen shell of the Keggin-type structure but for the absence of any heteroatom [54]. Further assemblies of monovacant $\{W_{11}O_{39}\}$ metatungstate species linked by octahedral $\{WO_6\}$ units resulted in some of the largest isopolyoxotungstates prepared to date in which all the 36 W centers are hexa-coordinated [55, 56]. As depicted in Fig. 1.3, condensation of two monovacant hexatungstate units affords the $[W_{10}O_{32}]^{4-}$ decatungstate which is the main species in an acidic aqueous solution (pH = 1–2) [57]. Some other interesting clusters include, (i) the family of $[H_4W_{19}O_{62}]^{6-}$ anions formed by a $\{W_{18}\}$ Wells–Dawson-like cage that encapsulates a $\{WO_6\}$ moiety [58], and (ii) giant isopolyoxotungstates exhibiting a common $\{(M)M_5\}$ moiety containing a central M atom in a pentagonal bipyramidal environment [59]. Even though, their size (<100 W centers) is still small in comparison to giant isopolyoxomolybdates (>300 Mo centers).

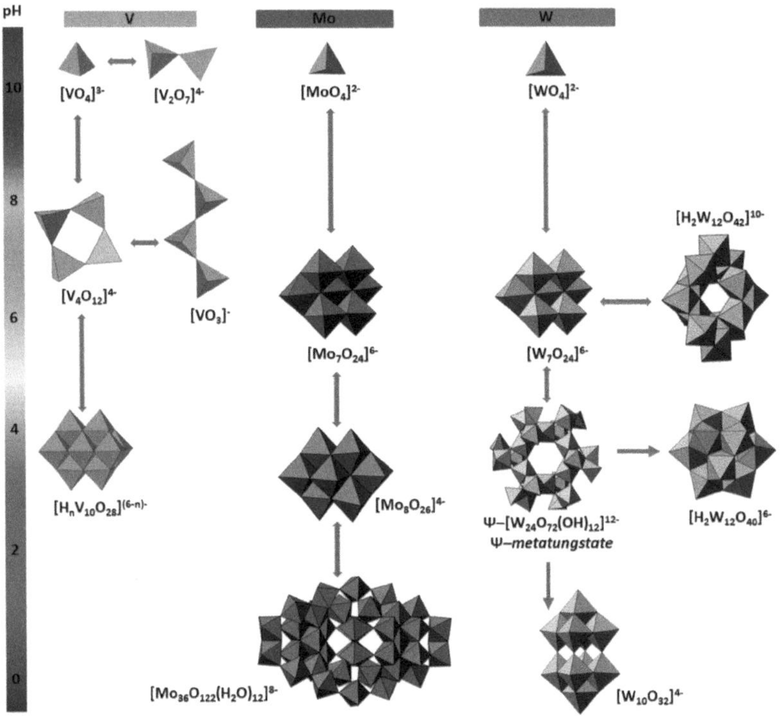

Figure 1.3 Structure of the predominant isoPOMs (V, Mo, and W) in aqueous solution as a function of pH. Color code: $\{VO_x\}$, orange polyhedra; $\{MoO_x\}$, blue polyhedra; $\{WO_x\}$, gray polyhedra (x = 4–7).

Regarding less usual *addenda* metals, the chemistry of niobates and tantalates was firstly limited to both solution and solid-state studies carried out for Lindqvist-type anions [60, 61]. These works, together with those carried out for decaniobate species [62], revealed that in contrast to V, Mo, and W; (i) aqueous synthesis need to be carried out at basic pH values; (ii) fully oxidized metal ions are too small to remain as mononuclear units in solution; (iii) the trend is inversed when comparing the solubility of their alkaline metal salts; that is, the solubility for Nb- and Ta-based POMs increases as follows: Li < Na < K < Rb/Cs; (iv) the presence of counterions determines the predominant species that are formed in aqueous solution; and (v) they can show different extents of protonation above pH 8. In the last two decades, a number of reports devoted to isopolyoxoniobates have grown considerably, whereas polyoxotantalates still represent an incipient field. It is of great relevance the fact that the first structural report of the decaniobate anion [63] was carried out 30 years before that of the tantalum analog [64]. Besides these two archetypical examples, related derivatives include: (i) the heptaniobate $[Nb_7O_{22}]^{9-}$ anion which is observed in water solution. Its structure could be rationalized as a hexaniobate core with an additional $\{NbO_6\}$ octahedron condensed through face-sharing; (ii) the $[Nb_{20}O_{54}]^{8-}$ dimer formed by the condensation of two decaniobate units [65]; and (iii) the $\{Nb_{24}O_{72}\}$ and $\{Nb_{32}O_{96}\}$ clusters constituted by 3 and 4 hexaniobate units arranged *via* face-sharing in a central 6 and 8 membered niobate rings, respectively (Fig. 1.4a,b) [66]. Analogous to that described for giant Mo and W clusters, molecular species showing 27 and 31 Nb centers which display pentagonal $\{Nb(Nb)_5\}$ building-blocks (Fig. 1.4c) [67] and the assembly of the former units in dimers (Nb_{52}), trimers (Nb_{81}), or tetramers (Nb_{114}) affords some of the largest polyoxoniobates known to date [68]. The record belongs to the giant framework $[Nb_{288}O_{768}(OH)_{48}(CO_3)_{12}]^{180-}$ which exhibits a size comparable to that of Mo-blues [69].

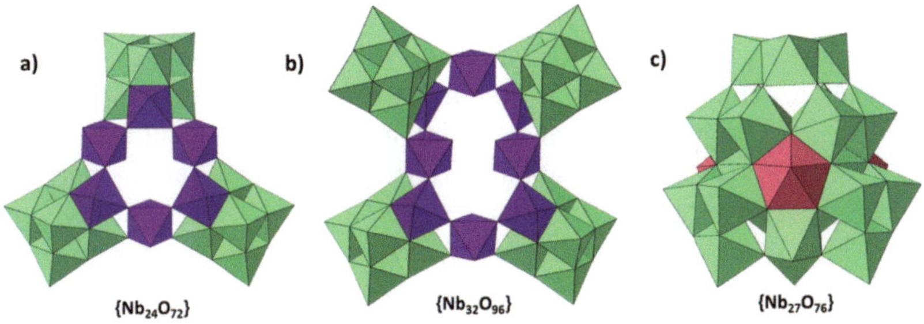

Figure 1.4 Molecular structure of (a) $[H_{21}Nb_{24}O_{72}]^{3-}$, (b) $[H_{28}Nb_{32}O_{96}]^{4-}$ (Color code: $\{Nb_6O_{19}\}$, light green; bridging $\{NbO_6\}$ groups, purple) and (c) $[HNb_{27}O_{76}]^{16-}$ anions (Color code: $\{NbO_6\}$ octahedra in dark green; $\{NbO_5\}$ centers, pink).

1.4.2 Heteropolyoxometalates (heteroPOMs)

As a result of their high solution stability, heteropolyoxometalates are the most explored group within the POM chemistry. Through the removal of octahedra, addition of new polyhedra and relative rotation of some building blocks, the vast majority of heteroPOM structures can be originated from the following three parent clusters with high symmetry; the $\{XM_{12}O_{40}\}$ Keggin structure, the $\{XM_{12}O_{38}\}$ cluster, and the Dexter–Silverton structure $\{XM_{12}O_{42}\}$.

Acidic condensation of MO_6 units ($M = W^{VI}$, Mo^{VI}) in the presence of tetrahedral $[XO_4]^{n-}$ oxoanions (usually X = B^{III}, Si^{IV}, Ge^{IV}, P^{V}, As^{V}, but also first row transition metals like Fe^{III}, Co^{II}, etc.) results in the $\{\alpha\text{-}XM_{12}O_{40}\}$ Keggin-type structure. The α-Keggin anion is constituted by four $\{M_3O_{13}\}$ trimers formed by three edge-sharing $\{MO_6\}$ octahedra that are linked to each other and to the central tetrahedron through corner-sharing in an ideal T_d symmetry. Regarding group 5 metals, the $[XNb_{12}O_{40}]^{16-}$ heteroniobates (X = Si, Ge) have been recently prepared as isolated clusters [70], whereas related examples for vanadium only include mixed-*addenda* metal V/Mo or V/W clusters and the bicapped α-Keggin-type $[PV_{14}O_{42}]^{9-}$ [71]. Keggin anions can be reduced with no significant structural changes. This is the case of the $[PMo_{12}O_{40}]^{3-}$ "electron sponge" which can incorporate up to 24 electrons after a super-reduction process that involves the formation of metal–metal bonds [72]. The protonation of bridging oxygen atoms is usually associated with reduction processes, and hence, Keggin-type POMs can act as both proton and electron reservoirs simultaneously. These properties, together with their high solution and thermal stability, make them very useful as acid- and redox-catalysts, including photo- and electro-catalytic reactions [73].

Keggin-type heteropolyanions have five structural isomers, known as Baker–Figgis isomers. These result from the 60° rotation of one (β), two (γ), three (δ), or four (ε) M_3O_{13} trimers from the parent α-Keggin anion (Fig. 1.5). The α and β isomers are the most stable from the five, because they do not show any edge-sharing linkages between different M_3O_{13} trimers. The α, β, and γ isomers of Keggin-type tungstosilicates can be readily synthesized and have been successfully characterized both in solution and solid state [74]. Nevertheless, the δ isomer structure has only been observed for the aluminum-based $[(AlO_4)Al_{12}(OH)_{24}(H_2O)_{12}]^{7+}$ cationic cluster [75], and the $[H_2W_4V_8(VO_4)O_{33}(C_6H_{13}NO_3)]^{5-}$

anion which is stabilized with a tripodal organic ligand [76]. Reports containing ε-Keggin-type building block as independent cluster are limited to a few examples, such as the lanthanide-stabilized $[\varepsilon\text{-}PMo_{12}O_{36}(OH)_4\{Ln(H_2O)_4\}_4]^{5+}$ phosphomolybdate series (Ln = La to Sm) [77] and the aluminum $[MO_4Al_{12}(OH)_{24}(H_2O)_{12}]^{n+}$ (M = Al, Ga, Ge) polyoxocations.

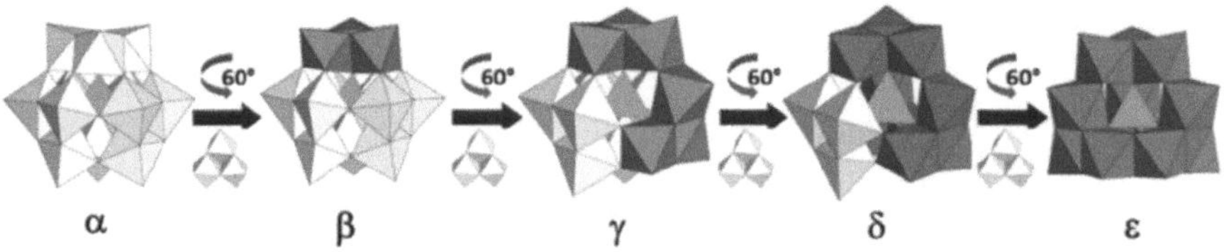

Figure 1.5 Baker–Figgis isomers of the Keggin-type structure. Color code: $\{MO_6\}$, gray; $\{XO_4\}$, green. Rotated trimers are highlighted in dark.

Some POM structures are derived from larger parent clusters by removing one or more *addenda* metal atoms from their shell. This process results in defect structures with vacant sites known as lacunary species. Different lacunary species can be obtained from the parent plenary $\{\alpha\text{-}XM_{12}O_{40}\}$ Keggin anion. From the α-isomer mono-$\{XM_{11}O_{39}\}$ and tri-vacant $\{XM_9O_{34}\}$, species can be generated upon elimination of one or three octahedra, respectively. The $\{\alpha\text{-}XM_9O_{34}\}$ anion shows two different isomers named as A-α or B-α depending on whether the $\{WO_6\}$ octahedra eliminated belongs to a $\{M_3O_{13}\}$ edge-sharing trimer or to a $\{M_3O_{15}\}$ corner-sharing triad, respectively. Similar trilacunary species can be isolated for the β-isomer. However, the lower symmetry of the $\{\beta\text{-}XM_{12}O_{40}\}$ allows for three different monolacunary species to be formed depending on the position of the vacant site. The vacant position in the β_1 derivative is located in the triad opposite to the 60° rotated trimer, whereas for the β_2 and the β_3 forms it lies at the central belt and the rotated trimer, respectively. Up to date, only the dilacunary $\{\gamma\text{-}XM_{10}O_{36}\}$ species has been isolated from the γ isomer, in which two edge-sharing octahedra from the two rotated trimers are removed. Chemical equilibria of Keggin-type tungstosilicates in water solution (Fig. 1.6) illustrate the complexity of the pH-dependent interconversion pathways between plenary and lacunary POMs.

All the 10 species represented can be prepared and isolated as alkaline salts from mixtures of silicate and tungstate sources depending on the pH, by following the procedures reported by Téze and Hervé [74]. Furthermore, for some specific heteroatoms such as As^{III}, Sb^{III}, Te^{IV}, or Se^{IV}, the presence of a lone pair of electrons precludes the full closure of the Keggin shell, so that only B-XW_9O_{33} trivacant species can be formed [78].

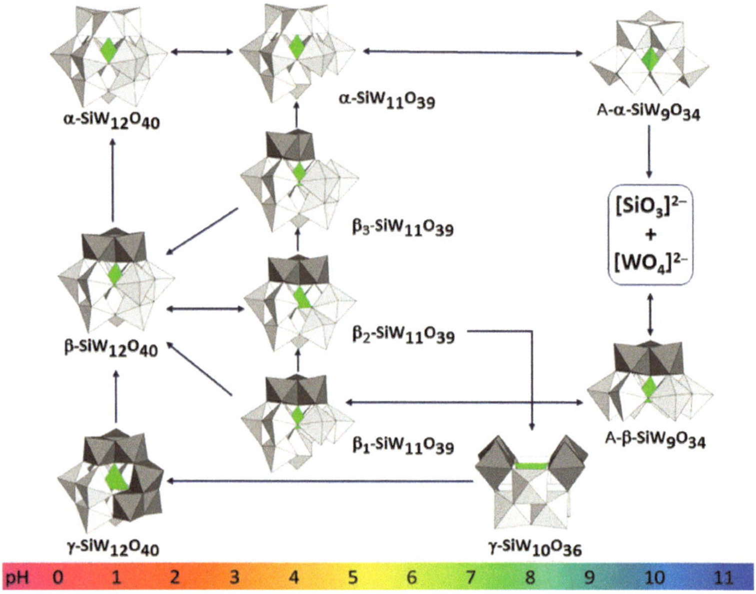

Figure 1.6 Scheme of the chemical equilibria between different plenary and lacunary Keggin-type tungstosilicates.

In some other cases, lacunary species can undergo further condensation reactions in water, leading to new types of heteroPOMs (Fig. 1.7). For instance, one of the most remarkable examples is given by the $X_2M_{18}O_{62}$ Wells–Dawson cluster which is constituted by two A-α-XM_9O_{34} (X = P^V, As^V, S^{VI}) trilacunary units sharing corners through the belt $\{MO_6\}$ octahedra in an ideal D_{3h} symmetry [79]. In contrast, if the trivacant Keggin-type blocks are combined in staggered conformation, the less stable α^* isomer is formed. The Baker–Figgis-like β isomer comprises Keggin-type A-α and A-β trilacunary building blocks, whereas the γ form contains two β building blocks [80]. Conversely, the γ^*

species represents the unique isolation of such type of cluster in solid state [81]. Two types of monolacunary species can be obtained from the parent plenary α form, so called α_1 and α_2 depending on whether the removed octahedron belongs to a belt or cap positions, respectively. Trilacunary $\{\alpha\text{-}X_2M_{15}O_{56}\}$ units can be readily prepared by the removal of a cap trimer, whereas polylacunary species like the hexavacant $\{\alpha\text{-}X_2M_{12}O_{48}\}$ are obtained after the removal of four belt $\{MO_6\}$ units and one additional center from each cap [82]. The latter building block is

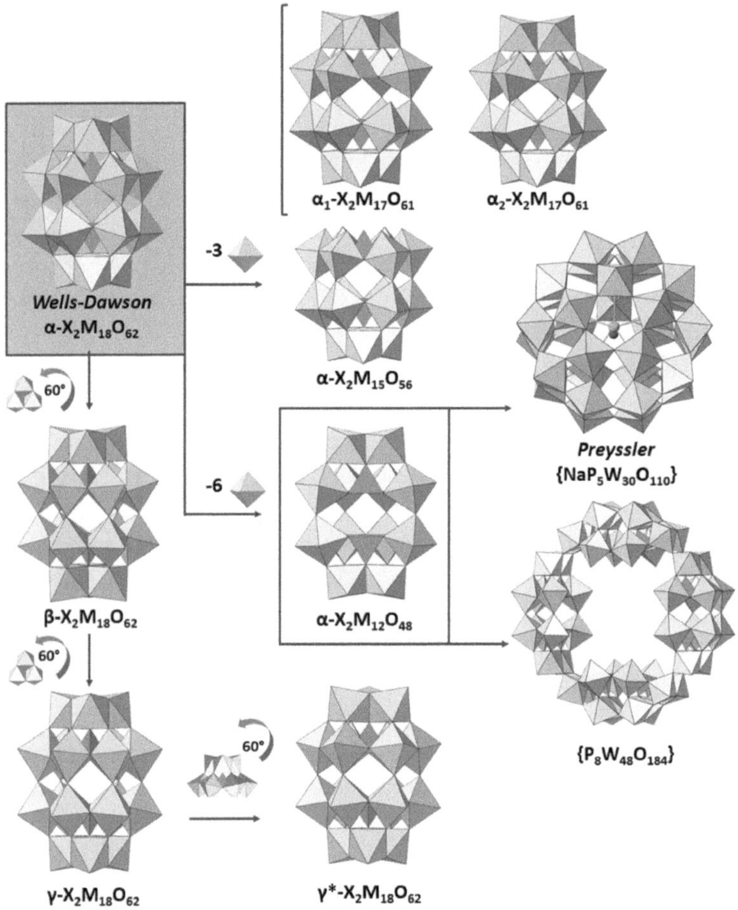

Figure 1.7 Representation of the Baker–Figgis isomers of the Wells–Dawson POM and the most representative lacunary derivatives, together with the related Preyssler and $\{P_8W_{48}O_{184}\}$ archetypes.

found in the $[P_8W_{48}O_{184}]^{40-}$ macrocycle [83] and Preyssler-type $[NaP_5W_{30}O_{110}]^{14-}$ anions [84]. Besides Keggin and Wells–Dawson type species, it is worth mentioning that the smaller Strandberg-type $X_2M_5O_{23}$ anions [85] are also predominant species that can be found at neutral pH in the aqueous equilibria of the M/X combination mentioned above in this paragraph. These clusters consist of a ring formed by five edge-sharing octahedra, except for two of them which are linked to each other by corner sharing. Two tetrahedral heteroatoms cap this ring from both top and bottom sides.

The second parent cluster displays a general formula $\{XM_{12}O_{38}\}$ and it is formed by 12 edge-sharing $\{MO_6\}$ units that comprise a central $\{XO_6\}$ heteroatom in an arrangement with general O_h symmetry. This structure has not been observed yet as an independent polyanion, but related bicapped derivatives can be found in the mineral Sherwoodite [86]. Removal of three octahedra generates the chiral XM_9O_{32} cluster (Waugh-type POM) [87] with ideal D_3 symmetry, which has only been observed for a particular combination of M/X (M = Mo, X = Mn^{IV}, Co^{IV}, Ni^{IV}) [88].

In addition, the removal of two opposed $\{M_3O_{13}\}$ trimers from the parent cluster results in the planar $\{\alpha\text{-}XM_6O_{24}\}$ heterohexametalate commonly known as Anderson–Evans polyanion [89]. This cluster can be best described as a ring of six edge-sharing octahedra which encapsulates a central $\{XO_6\}$ unit in an ideal D_{3d} symmetry. Moreover, this anion shows a less common "bent" β-isomer of C_{2v} symmetry which is structurally similar to that of the heptametalate clusters, but with a heteroatom in its central position [90]. The structure of the $[\alpha\text{-}Te^{VI}Mo_6O_{24}]^{6-}$ anion was first proposed by Anderson [91] back in 1937 and it was confirmed by X-ray diffraction experiments by Evans 10 years later [92]. Since that breakthrough, dozens of structures displaying this archetype have been reported which contain both Mo or W acting as *addenda* metal, as well as many of the first-row transition metals and p-block elements in their highest oxidation states as heteroatoms (M = Mo, X = first row transition metals, Rh^{III}, Pd^{IV}, Pt^{IV}, Al^{III}, Ga^{III}, Sb^{V}, Te^{VI}, I^{VII}; and M = W, X = Mn^{II}, Mn^{IV}, Ni^{II}, Ir^{IV}, Pt^{IV}, Sb^{V}, Te^{VI}). Traditionally, Anderson–Evans type POMs can be classified into two groups: first, the non-protonated

A-type with central heteroatoms in high oxidation states and second, protonated *B*-type with heteroatoms in low oxidation states. The six triply shared oxygen atoms (μ_3–O) that connect the heteroatom and two consecutive *addenda* atoms are basic and, thus, they can be easily protonated. This allows condensation reactions with trisalkoxo moieties leading to the organic functionalization of inorganic clusters in a procedure that will be detailed in the last section of this chapter. This feature together with its small size, planar shape, presence of heavy atoms, and solution stability make Anderson–Evans polyoxotungstates ideal additives for protein crystallography [93]. Different related structural examples include: (i) those in which the central heteroatom is replaced by one XO_4 tetrahedral unit (X = V, As, Te) on each side of the planar ring [94] and (ii) the Evans–Showell-type $[Co_2Mo_{10}O_{34}(OH)_4]^{6-}$ heteropolyanion formed by the perpendicular interpenetration of two Anderson clusters (Fig. 1.8) [95].

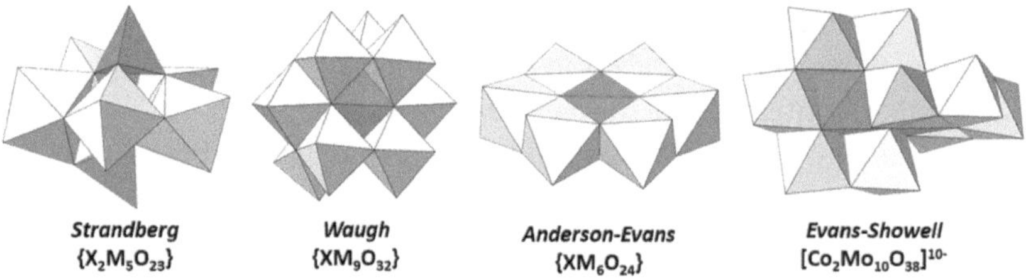

Figure 1.8 Polyhedral representation of the Strandberg, Waugh, Anderson-Evans and Evans-Showell type heteroPOMs (from left to right).

Finally, a reduced number of examples of heteropoly-oxomolybdates in which the X position is occupied by tetravalent 4f/5f cations have only been reported for the $XM_{12}O_{42}$ Dexter–Silverton cluster [96]. Formally, they are constituted by six pairs of face-shared dimers in an I_h shape comprising a central heteroatom.

1.4.3 Unconventional POMs

Beyond the traditional POM archetypes, some families of metal-oxo clusters display unusual features in terms of size, complexity, and composition. Although they can be well divided into iso- and heteroPOM groups according to classical criteria, we decided to discuss them separately. This section includes a brief comment on giant polyoxomolybdates showing more than 100 Mo atoms and sizes comparable to those of proteins, together with actinyl peroxometalates, exemplified by uranium clusters and the growing family of noble-metal POMs (such as palladium, platinum, and gold oxo-frameworks).

1.4.3.1 Giant molybdenum clusters

Partial reduction of acidified molybdate solutions allows the generation of a variety of building blocks that can be linked together to create very complex and large molecular systems *via* stepwise self-assembly processes [97]. Such considerable structural and chemical diversity arises from the combination of favorable kinetics, thermodynamics, and structural factors, which makes the so-called molybdenum "blue" and "brown" giant cluster families unique within the field of inorganic chemistry [23]. Mo-blues are defined by the fact that they contain mixed-valence Mo^{V}/Mo^{VI} *addenda* metals and have delocalized electrons capable of intervalence charge transfer from Mo^{V} to Mo^{VI} facilitated by the π-orbitals of the bridging oxo ligands. This electronic interaction is responsible for their characteristic intense blue color and accounts for their reversible redox nature. In contrast, Mo-browns are further reduced polyanions and have electrons localized between reduced Mo^{V} centers as Mo–Mo bonds which contribute to the brown color of these clusters, and make them less sensitive to redox processes than their blue counterparts [98]. All these large species have pentagonal $\{(Mo)Mo_5\}$ moieties as a common structural feature, containing a central Mo atom in a pentagonal bipyramidal $\{MoO_7\}$ environment surrounded by five edge-sharing $\{MoO_6\}$ octahedra. The linkage among $\{(Mo)Mo_5\}$ synthonic units can be established by $\{Mo_8\}$ building blocks, edge-(Mo-brown) or corner (Mo-blue) sharing $\{Mo_2\}$ units and single $\{Mo\}$ centers.

Scheele described the first reproducible experiment involving Mo-blue species in 1793 [99]. However, the structure of any related systems remained unknown until the 20th century, when Müller and coworkers succeeded in the crystallization of the $\{Mo_{154}\}=[Mo_{154}O_{462}H_{14}(H_2O)_{70}]^{14-}$ (Fig. 1.9b) wheel-shaped anion [100]. Since then, larger macroclusters, such as $\{Mo_{176}\}$, have been synthesized [101], which is able to undergo molecular growth by the addition of two $\{Mo_{36}\}$ caps to its inner cavity to lead to $\{Mo_{248}\}$ species [102]. Besides the "big wheels," "Keplerates" (big balls) such as the $\{Mo_{132}\}$ cluster (Fig. 1.9a) represent a second structural family for which a number of metal substituted $\{Mo_{72}M_{30}\}$ (M = V, Cr, Fe) forms have been reported [103]. The potential of these reduced molybdenum systems to form large assemblies is so high that the $\{Mo_{368}\}$ "hedgehog" or "blue lemon" structure $[H_{16}Mo_{368}O_{1032}(H_2O)_{240}(SO_4)_{48}]^{48-}$ still represents the largest (ca. 5 nm) POM obtained to date [104] (Fig. 1.9c).

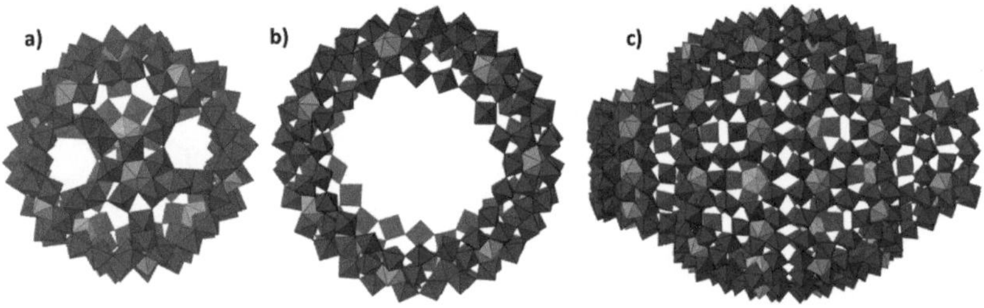

Figure 1.9 Molecular structures of (a) $\{Mo_{132}\}$ Keplerate, (b) $\{Mo_{154}\}$ wheel, and (c) $\{Mo_{368}\}$ blue lemon. Mo(V) atoms with pentagonal bipyramid geometry are depicted in cyan.

The pentagonal bipyramidal $\{MO_7\}$ unit has also been observed for tungsten-based giant structures like Keplerate-type anions [105] or even as part of a "defect" pentagonal $\{(W)W_4\}$ core in large hexameric selenotungstates [106]. However, the lower lability and much slower assembly processes of tungstates compared to molybdates have made the search of mixed-valence tungsten analogs an almost unexplored field.

1.4.3.2 Uranium clusters

Recent discoveries in POM chemistry have challenged the paradigms of what we refer now as a polyoxometalate. In this context, peroxo actinide compounds should be highlighted as they exhibit certain structural key similarities to POM architectures despite the fact that they are certainly not classical metal-oxo clusters. Although clusters containing Th, Np, and Pu can be found in the literature, uranium-based systems represent by far the most studied subclass [107]. The shell of these macroions is composed of hexavalent U cations in a hexagonal bipyramidal coordination geometry bridged together *via* peroxide (and sometimes hydroxyl) ligands. Furthermore, the passivating nature of their two *trans* terminal "triple" bonds often leads to hollow species that can host cationic guests in their inner site. For that reason, the templating effect of internal and external cations on the formation of these types of systems has been widely studied [108]. Since the publication of the first actinyl peroxide nanospheres ($[[(UO_2)(OH)(O_2)]_x^{x-}$, x = 24, 32; and $[(UO_2)_{28}(O_2)_{42}]^{28-}$) by Burns and collaborators [109], dozens of related systems have been successfully synthesized in basic aqueous solutions.

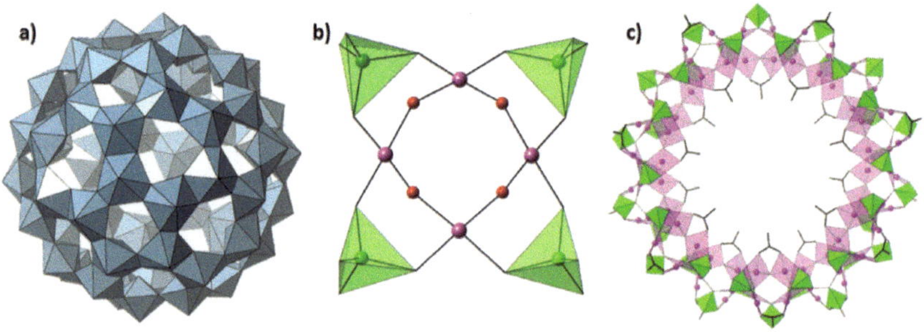

Figure 1.10 Molecular structures of (a) U_{60}; (b) $[Au^{III}O_4(As^{V}O_4)_4]^{8-}$ anion; and (c) Pd_{84} wheel. Color code: {UO_8}, blue polyhedra; {PdO_x}, pink polyhedra; {PO_4} and {AsO_4}, green polyhedra; Au, pink ball; O, red ball; C, black stick.

According to their morphology and composition, these can be classified as (i) hollow cages displaying penta- and hexagonal faces, such as the $[UO_2(O_2)(OH)]_{60}^{60-}$ (U_{60}, Fig. 1.10a) that could be considered the structural analog of the C_{60} buckminsterfullerene; (ii) cage clusters containing topological squares like the U_{40} anion; (iii) open-bowl-shaped architectures as exhibited by the U_{16} unit; (iv) crown-shaped rings like the U_{32} derivative; and (v) hybrid clusters which contain oxalate or pyrophosphate moieties as bridging units [21].

1.4.3.3 Noble metal containing POMs

Although significant effort has been devoted to the incorporation of catalytically active noble metals (such as Ru, Os, Rh, Pd, Pt, Ag, and Au) into the inorganic skeleton of POM anions [20], structures in which noble metals act as *addenda* atoms still represent an incipient family within the chemistry of these anionic metal-oxo clusters. Anderson–Evans archetypes containing noble metals in heteroatom position [110] or lacunary polyoxotungstates that play their role as inorganic ligands toward noble metal cations [111, 112] are the most representative examples of the former group. On the other hand, only POMs containing Pd, Pt, and Au as *addenda* metals have been discovered up to date. The polyoxo-noble-metalate field started in 2004, when Pley and Wickleder prepared the ammonium salt of the polyoxoplatinate $[Pt^{III}_{12}O_8(SO_4)_{12}]^{4-}$ [113]. Members of the polyoxoaurate family are limited to the $[Au^{III}O_4(As^{V}O_4)_4]^{8-}$ anion (Fig. 1.10b) [114] which exhibits a tetrameric structure of four Au^{III} square-planar ions linked by four oxo and four arsenate bridging ligands, as well as its selenite analog [115]. Without any doubt, polyoxopalladates(II) represent the most significant subset of this class. Condensation of square planar palladate anions in the presence of external heterogroups (e.g., AsO_4^{3-}, PO_4^{3-}, and SeO_4^{2-}) leads to a large variety of structures that ranges from cube- (Pd_{13}), star- (Pd_{15}), bowl- (Pd_7), dumbbell- (Pd_{22}), and giant wheel-shaped (Pd_{84}) (Fig. 1.10c) architectures [116] to open shell forms able to serve as molecular hosts for small cationic guests [117].

1.5 Functionalization of POMs

Besides the aforementioned structural diversity of POMs, the incorporation of additional functionalities into these molecular architectures considerably broadens the catalog of available metal-oxo frameworks. This derivatization can be easily achieved *via* conventional synthetic routes and it usually leads to compounds with properties that unfunctionalized clusters cannot exhibit. Two main strategies have been traditionally used for this purpose; the incorporation of 3d and/or 4f metal centers and the organic derivatization of inorganic POM clusters.

In regard to the first approach, lacunary polyoxotungstate fragments are effective in stabilizing arrays of d- and f-metal centers, because they can act as inorganic ligands toward electrophilic transition metal and lanthanide ions. This ability has resulted in an enormous catalog of metal-substituted POM series, which usually show additional properties brought by the presence metal atoms, such as magnetism, Lewis acidity, luminescence, etc. The chemistry of these clusters is dominated by polyoxotungstates, in such a way that a large number of species ranging from simple entities to giant structures have proved to result from the incorporation of 3d or 4f metals. In the last decade, several heterometallic POMs containing 3d and 4f centers simultaneously have been synthetized [118], but this family of compounds will not be reviewed in this chapter.

With respect to the second strategy, POMs are ideal candidates to be used as active molecular entities in functional materials due to their inherent features (e.g., high solution and thermal stability, high acidity, and versatile redox properties). Their organic functionalization represents a key factor for these clusters to be suitably incorporated into matrixes like polymers or carbon nanotubes, and also to interact with diverse surfaces (e.g., silica, silicon, gold, alumina, and graphite) or metallic nanoparticles [119], because the resulting organic–inorganic hybrids avoid the processing drawbacks that unmodified POMs can display. This strategy might pave their way for the fabrication of new multifunctional devices (e.g., immobilized catalysts, photoactive systems for energy production, and electron storage systems); however, it usually requires elaborated organic functions that can only be achieved *via* multistep synthetic work on preformed hybrid POM platforms [120].

1.5.1 3d Metal Containing POMs

The combination of 3d metals with lacunary clusters confers remarkable magnetic and catalytic properties on POM-based systems. Tungsten frameworks ensure the adequate isolation of magnetic clusters, which allows the modelization of magnetic properties for a given molecular compound. Furthermore, the potential of POM-based first-row transition metal containing catalysts have addressed problems of current interest, such as energetic challenge. Well-known and simple structures containing earth-abundant metals can act as efficient water oxidation catalysts for the production of renewable fuels, avoiding the use of costly noble-metal catalysts [121]. Several topologies and nuclearities of 3d metal containing polyoxotungstates (Fig. 1.11) can be obtained following synthetic strategies that range from simple solution approaches to more extreme solvothermal routes.

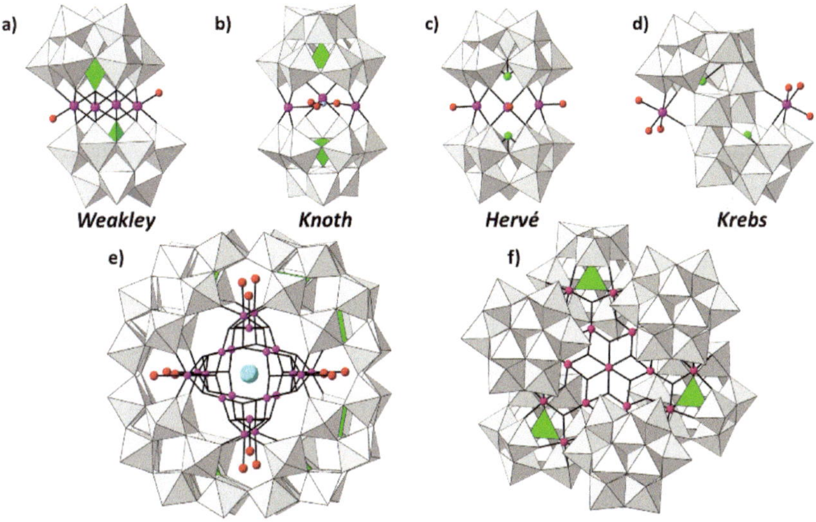

Figure 1.11 Archetypes of transition-metal substituted sandwich-type polyoxometalates, together with molecular structures of some multi-nuclear 3d metal (depicted in pink) containing species: (a) $[Co_4(H_2O)_2(GeW_9O_{34})_2]^{12-}$ (*Weakley*); (b) $[\{Cu_3(\mu\text{-}NO_3)\}(PW_9O_{34})_2]^{13-}$ (*Knoth*); (c) $[\{Cu(H_2O)\}_3(BiW_9O_{33})_2]^{12-}$ (*Hervé*); (d) $[\{Co(H_2O)_3\}_2(WO_2)_2(SbW_9O_{33})_2]^{10-}$ (*Krebs*); (e) $[Cu_{20}Cl(OH)_{24}(H_2O)_{12}\text{-}(P_8W_{48}O_{184})]^{25-}$; (f) $[Mn_{19}(OH)_{12}(SiW_{10}O_{37})_6]^{34-}$.

The most simple examples are given by lacunary Keggin [122] and Wells–Dawson [123] species in which 3d metal centers occupy vacant positions to recover the skeleton of parent plenary clusters. In some specific cases, the number of incorporated transition metals exceeds the number of vacancies [124].

Nevertheless, these monomeric structures are scarcely found in literature when reacting trilacunary anions with 3d metals and dimeric sandwich-type POMs are obtained instead, which comprise metal centers in central belt positions. Typically, trilacunary Keggin or Wells–Dawson anions react with 3d metal centers to afford four archetypical sandwich-type species. Weakley-type architectures with the general formula $[M_4(H_2O)_2(B\text{-}\alpha\text{-}XW_9O_{34})_2]^{n-}$ (X = Si, Ge, P, As^V) or $[M_4(H_2O)_2(\alpha\text{-}X_2W_{15}O_{56})_2]^{n-}$ (X = P, As^V) trap a $M_4O_{14}(H_2O)_2$ (M = Mn to Zn) core [125]. The outer $M(H_2O)$ groups from their central, rhomblike moieties of four edge-sharing MO_6 octahedra are relatively labile, as indicated by the isolation of defect di- and trisubstituted derivatives [126], which can effectively act as lacunary species and incorporate different 3d metal ions (M′) in the vacancies to form mixed $M'_2M_2O_{16}$ or $M'M_3O_{16}$ clusters [127]. All these POM clusters have exhibited fascinating magnetic, electrochemical and catalytic properties. Among the latter field, the $[Co_4(H_2O)_2(PW_9O_{34})_2]^{10-}$ anion stands out as the first efficient POM-based water oxidation catalyst [128]. In some specific cases, 3d metals can also occupy heteroatomic positions leading to clusters with the general formula, $[M_4(H_2O)_2(B\text{-}\alpha\text{-}M'W_9O_{34})_2]^{n-}$ [129] and the family of Tourné-type $[M_3W(H_2O)_2(B\text{-}\alpha\text{-}MW_9O_{34})_2]^{12-}$ POMs [130]. Exchange of the external $M(H_2O)$ groups by different transition-metal ions has also been studied for these families.

A-type trilacunary fragments can encapsulate a triangle of three first row transition metal cations, which complete their octahedral spheres with two coordination water molecules, to produce $[M_3(H_2O)_6(A\text{-}\alpha\text{-}XW_9O_{34})_2]^{12-}$ (X = traditionally P, but occasionally Si) Knoth-type sandwich POMs [131]. Immobilization of the iron(III) derivative in cationic SiO_2 nanoparticles have shown to be a suitable route to prepare a highly efficient heterogeneous catalyst for aerobic oxidations [132]. These reactions usually lead to the concomitant incorporation of anions, such as carbonate or nitrate that bridge triangularly arranged metal centers, especially for larger ions (e.g., Y, Zr, Hf, and lanthanides) [133]. In addition, even if no evidence for the formation of Wells–Dawson type tungstosilicate or -germanate analogs has been reported to date, two $[A\text{-}\alpha\text{-}XW_9O_{34}]^{10-}$ anions can link to each other *via* two pairs of corner sharing octahedral to afford the so-called open Wells–Dawson $[\alpha\text{-}X_2W_{18}O_{66}]^{16-}$ anion [134]. This cluster exhibits a central pocket which can accommodate up to six additional 3d metal centers [135].

On the other hand, the presence of a lone-pair of electrons in the heteroatom precludes the closure of the Keggin shell and, thus, only *B*-type trilacunary species can be formed. In this sense, the 3d metal directed assembly of two $\{B\text{-}\alpha\text{-}XW_9O_{33}\}$ units results in Hervé-type sandwich-species with the general formula $[(M(H_2O))_3(B\text{-}\alpha\text{-}XW_9O_{33})_2]^{n-}$, in which the three square-pyramidal metals in belt position exhibit one terminal water molecule each. Since the report of the As^{III}/Cu^{II} system [136], anions with different combinations of heteroatom/3d metal have been synthesized (X = Sb^{III}, Bi^{III}, Se^{IV}, Te^{IV}, and M = Cu^{II}, but also X = As^{III}, Sb^{III}, and M = Mn^{II}, Co^{II}, Ni^{II}, Zn^{II}, VO^{2+}) with relevant magnetic properties and spin-frustration effect [137]. Related examples include those in which one $WO(H_2O)$ group replaces one of the belt metal centers and results in a lower degree of transition-metal substitution; and tetra- and hexa-substituted derivatives with higher content of 3d metals. The $[(MCl)_6(XW_9O_{33})_2]^{12-}$ (X = As^{III}, Sb^{III} and M = Mn^{II}, Cu^{II}) anions from the latter group were prepared in organic media and exhibited field-induced single-molecular magnet behavior [138].

Association of two $\{B\text{-}\beta\text{-}XW_9O_{33}\}$ units leads to the Krebs-type [83] $[\{WO_2(OH)\}_2(WO_2)_2(SbW_9O_{33})_2]^{12-}$ anion in which trilacunary fragments are connected *via* two inner *cis*-$\{WO_2\}$ and outer *fac*-$\{WO_2(OH)\}$ octahedra. The latter groups are labile and can be easily exchanged for $\{M(H_2O)_3\}$ moieties, resulting in a large family of 3d metal disubstituted species with general formula $[\{M(H_2O)_3\}_2(WO_2)_2(XW_9O_{33})_2]^{n-}$, where M = Zn, Ni, Co^{II}, Fe^{III} or Mn^{II}, and X accounts for Sb^{III} or Bi^{III} [139]. Very recently, the elusive Cu^{II} derivative was isolated as well [140]. Inner positions have also been fully or partially exchanged for $\{M(H_2O)_2\}$ groups in several tri- and tetra-substituted analogs, especially for certain heteroatoms (As^{III}, Te^{IV}, Se^{IV}) or trivalent metals and some of them have been employed as efficient Lewis acid catalyst in oxidation reactions [141].

Besides the archetypical, dimeric sandwich type anions, the huge variety of 3d metal substituted POMs, that can accommodate up to 40 3d metal centers, has been regularly reviewed [142]. Large species are usually formed by the self-assembly of lacunary fragments in which 3d metal centers occupy all the vacant positions to display the skeleton of plenary Keggin-type clusters [143]. Occasionally, this association is directed by small alkaline cations, anions (e.g., sulfate, phosphate, and carbonate) or additional transition metal or tungstate centers [144]. Another family of 3d metal containing POMs includes those examples in which preformed multinuclear metal clusters are stabilized by lacunary POM fragments. This is the case of the $[Mn_{19}(OH)_{12}(SiW_{10}O_{37})_6]^{34-}$ anion [145] in which the central magnetic core is reminiscent to Gatteschi's $[Mn_{12}(CH_3COO)_{16}(H_2O)_4O_{12}]$ single-molecule magnet [146]. Some other works report unprecedented polylacunary Keggin fragments coordinated to transition metal centers. It is worth highlighting that tetra-, penta-, and hexa-vacant species cannot be isolated in the absence of these 3d metals which act as stabilizing agents [147]. Opposed to these fragments, the hexalacunary Wells–Dawson-type $[H_2P_2W_{12}O_{48}]^{12-}$ anion does not need any stabilization agent in its synthesis and, therefore, it can be easily isolated as a potassium salt. This anion [148], together with related dimeric, trimeric $[P_6W_{39}O_{147}(H_2O)_3]^{30-}$, and tetrameric $[H_7P_8W_{48}O_{184}]^{33-}$ [149] crown-shaped assemblies can be used to trap high-nuclearity 3d metal clusters.

1.5.2 4f Metal Containing POMs

Owing to their large size, oxophilic character, and high coordination numbers (usually higher than 7), lanthanides have been successfully incorporated into lacunary polyoxotungstate fragments leading to complex architectures and a large variety of assemblies. Extensive work has been carried out for several years in this field due to the interesting structural and physicochemical properties that lanthanide containing POMs might show, including their role as Lewis acid catalysts, magnetism or luminescence [150]. From a magnetic point of view, the intrinsic large magnetic anisotropy together with large ground-state spin values displayed by rare-earth metal ions under certain ligand fields, accompanied by the rigidity and insulating ability of POMs as ligands (which hinder possible alterations in the ligand field or intermolecular magnetic interactions from taking place) have been proved to result in molecular systems showing slow relaxation of the magnetization. Different 4f containing POMs have been reported to behave as effective single-ion or single-molecule magnets (SIMs or SMMs) or spin qubits in recent years; this may pave their way for being applied in high density memory-storage/sensor devices, spintronics or quantum computing [5]. In addition, it is well known that POMs can act as multidentate antenna ligands because the photoexcitation of the O→M (M = Nb, Mo, W) ligand to metal charge transfer (LMCT) bands sensitize the luminescent 4f metal centers as a result of the cited intramolecular energy transfer. Owing to their multidentate character, they can also block coordination sites against aqua ligands responsible for emission quenching. These facts can lead to sharp and monochromatic emission bands and relatively long luminescence lifetimes for Sm, Eu, Tb, and Dy species in the visible region or for Nd, Er, and Yb in the near infrared [151].

Peacock and Weakley described the first family of lanthanide-containing POMs in 1971, namely, the series $[Ln(W_5O_{18})_2]^{9-}$ (Ln^{III} = La–Sm, Ho, Yb) [152]. These species consist of two monolacunary Lindqvist type POMs encapsulating a 4f metal ion in an eight-coordinated square-antiprismatic $LnO_4(O_4')$ fashion (Fig. 1.12a). Since then, all the sandwich-type clusters formed by a lanthanide ion trapped between two monolacunary fragments have been known as Peacock–Weakley type anions after them. Besides decatungstate-based sandwich-type species, structural characterization of only a few other types of lanthanide containing isoPOMs has been reported to date [153]. Nevertheless, a wide range of Peacock–Weakley-type POMs containing Keggin or Wells–Dawson monolacunary units have been reported, including the $\{Ln(\alpha\text{-}XW_{11}O_{39})_2\}$ (X = Si^{IV}, P^V, As^V) [154], $\{Ln(\beta_2\text{-}SiW_{11}O_{39})_2\}$ [155] or $\{Ln(\alpha\text{-}P_2W_{17}O_{61})_2\}$ ($\alpha = \alpha_1, \alpha_2$) [156] families (Fig. 1.12b). Although lanthanide substituted monomeric species were identified in solution in related early works, self-assembled derivatives forming chains or dimers [157] and showing accessible coordination water molecules, have also been isolated in the solid state.

In addition to Peacock–Weakley species, some other lanthanide-containing dimeric POMs are known. These are based on di- [158] or tri-lacunary Keggin or Wells–Dawson [159] units and include Knoth-type POMs incorporating 4f metals [131]. Only a few examples of trimeric lanthanopolyoxotungstates are found in the literature [160], whereas tetrameric species represent a large group where almost all members display dilacunary Keggin-type POMs or trilacunary units with lone-pair containing heteroatoms. Francesconi's $[(Eu_2PW_{10}O_{38})_4(W_3O_{14})]^{30-}$ anion [161] could be considered as a representative example of the second group. Higher nuclearity has also been observed in crown-shaped $\{K\subset\{Eu(H_2O)_2(\alpha\text{-}AsW_9O_{33}\}_6\}$ hexameric assemblies [162] and the largest lanthanide-containing tungstobismuthate reported to date [163].

The larger size of 4f ions compared to 3d metals usually prevents their full incorporation in lacunary frameworks as *addenda* atoms, and therefore, additional sites are available in the coordination sphere of the lanthanide centers for further derivatization. Even if the largest polyoxotungstate architectures known in terms of the number of tungsten atoms, that is the $[Mn_{40}P_{32}W_{224}O_{888}]^{144-}$ anion [164], has been obtained by employing first row transition metal linkers, the combination of lanthanide atoms with dilacunary $\{GeW_{10}O_{38}\}$ or trilacunary $\{B\text{-}As^{III}W_9O_{33}\}$ units has also lead to several gigantic metal-oxo frameworks. In 1997, the $[As_{12}Ce_{16}(H_2O)_{36}W_{148}O_{524}]^{76-}$ anion was synthesized by Pope and co-workers (Fig. 1.12c) and it constituted the largest polyoxotungstate known until 2010. This anion incorporates 12 $\{B\text{-}\alpha\text{-}AsW_9O_{33}\}$ subunits, 4 monolacunary Lindqvist-type $\{W_5O_{18}\}$ subunits, and 16 Ce centers linked by extra tungstate groups in a disk-shaped assembly with ideal D_{2d} symmetry [165]. Similar gigantic POMs, the $[Ce_{20}Ge_{10}W_{100}O_{376}(OH)_4(H_2O)_{30}]^{56-}$ and $[Gd_8As_{12}W_{124}O_{432}(H_2O)_{22}]^{60-}$ can be both described as dimeric entities composed of two halves related by an inversion center. The pentameric half of the former dumbbell-shaped POM comprises five $\{\beta\text{-}Ce_2GeW_{10}O_{38}\}$ subunits linked to each other by the coordination sphere of Ce atoms [166], whereas hexameric halves in the latter POM are composed of six $\{AsW_9O_{33}\}$ and two $\{Gd_2W_4\}$ building blocks [167]. The last member of the family of giant lanthanide-containing POMs is the crown-shaped $[K_7Ce_{24}Ge_{12}W_{120}O_{444}(OH)_{12}(H_2O)_{64}]^{52-}$ POM published in 2010 by Reinoso *et al.* [168] which displays a central hexameric core of alternating lanthanide disubstituted $\beta(1,5)/\beta(1,8)$ Keggin-type subunits, to which six additional $\gamma(1,5)$ blocks are linked (Fig. 1.12d). Related works involve the use different lanthanide ions, alkaline cation directed assemblies, interesting solution behavior [169], as well as magnetic and luminescent studies [170].

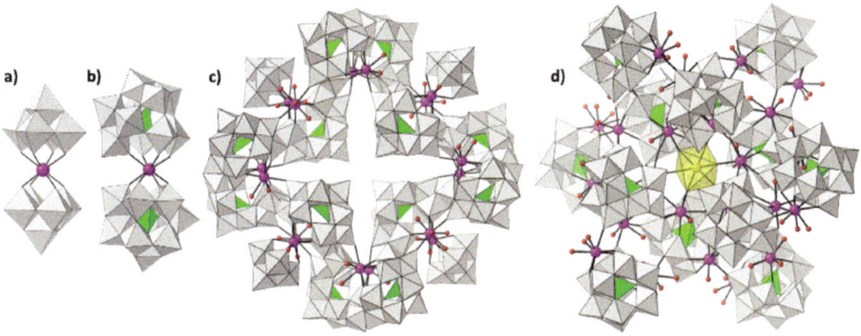

Figure 1.12 Representative structures of lanthanide containing polyoxometalates. (a) $[Ln(W_5O_{18})_2]^{9-}$; (b) $\{Ln(\alpha\text{-}XW_{11}O_{39})_2\}$; (c) $[As_{12}Ce_{16}(H_2O)_{36}W_{148}O_{524}]^{76-}$; (d) $[K_7Ce_{24}Ge_{12}W_{120}O_{444}(OH)_{12}(H_2O)_{64}]^{52-}$. Color code: Ln, pink ball; $\{WO_6\}$, gray polyhedron; $\{XO_3\}$ and $\{XO_4\}$, green polyhedron; water molecules, red ball; $\{K(H_2O)_6\}$, yellow polyhedron.

1.5.3 Organic Functionalization

Two main strategies can be followed to incorporate custom-designed organic moieties into POM systems. The first strategy consists in preparing the desired organic molecule, followed by its reaction with polyoxoanions. In the second strategy, the organic function exhibits a reactive pendant group, which is first incorporated into the cluster. The resulting POM is then used as platform for further post-functionalization. Two different classes of POM-based organic–inorganic hybrids can be distinguished attending to the type of interaction between the two components. The first class gathers all the systems in which electrostatic interactions, hydrogen bonds, or Van der Waals forces are established, whereas in the second class, the organic and inorganic moieties are linked *via* strong covalent bonds. In this sense, different synthetic methods have been developed for the covalent attachment of organic groups to inorganic POM skeletons and hence, well-defined, solution-stable hybrid POMs can be obtained [171]. Some of these examples are gathered in Fig. 1.13.

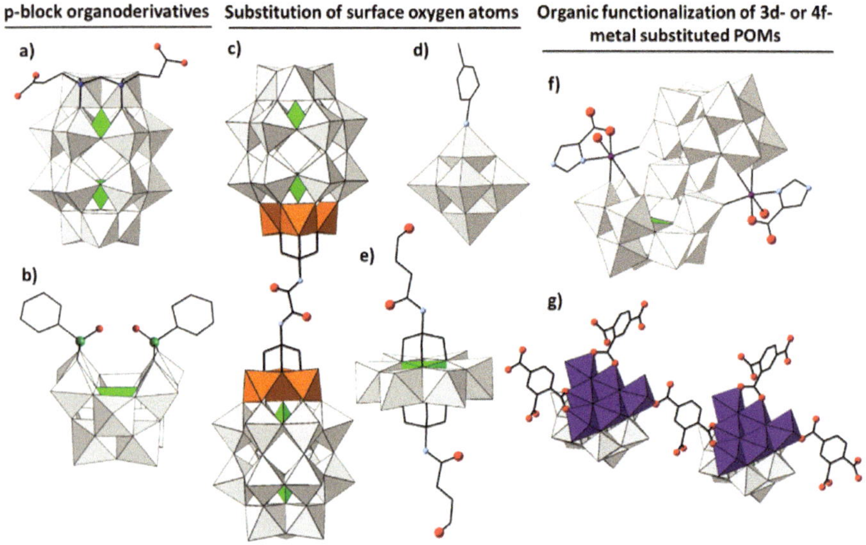

Figure 1.13 Representative structures of POMs bearing organic functions. (a) $[\alpha_2\text{-}P_2W_{17}O_{61}\{(HOOC(CH_2)_2Si)_2O\}]^{6-}$; (b) $[\gamma\text{-}SiW_{10}O_{36}(PhPO)_2]^{4-}$ (Ph = phenyl); (c) $[\{P_2V_3W_{15}O_{59}(OCH_2)_3CNHCO\}_2]^{12-}$; (d) $[Mo_5O_{18}(MoNC_6H_4CH_3)]^{2-}$; (e) $[MnMo_6O_{18}(O(CH_2)_3NH\text{-}Gly\text{-}OH)_2]^{3-}$ (Gly = glycine); (f) $[\{Co(imc)(H_2O)\}_2(WO_2)_2(B\text{-}\beta\text{-}SbW_9O_{33})_2]^{12-}$ (imc = imidazolcarboxylate); (g) $\{[Ni_6(OH)_3(H_2O)_5(PW_9O_{34})](1,2,4\text{-}Hbtc)$ (btc = benzenetricarboxylate). Color code: $\{MO_6\}$ (M = Mo, W), light gray polyhedra; $\{VO_6\}$, orange polyhedra; $\{XO_n\}$ (X = Si, P, Sb, Mn), light green polyhedra; Si from organosilyl group, dark blue ball; P from organophosphoryl moiety, dark green ball; N, light blue ball; Co, pink ball; Ni, purple polyhedra; O, red ball; C, black stick. H atoms are omitted for clarity.

1.5.3.1 p-block organoderivatives

One of the most extensively explored routes consist in combining lacunary polyoxotungstates with p-block organoderivatives, predominantly organosilyl, -phosphoryl and -stannyl moieties, but also -germyl, -arsenyl and -stibyl groups. Since the first description of a series of organosilyl derivatives $[SiW_{11}O_{39}\{O(SiR)_2\}]^{4-}$ (R = C_2H_5, C_6H_5, $NC(CH_2)_3$, C_3H_5) [172] in 1979, a large variety of organic functions have been incorporated into monolacunary Keggin or Wells–Dawson-type structures. The resulting molecular hybrid POMs can contain up to two organosilyl groups or one dimeric μ-oxo-bridged $(RSi)_2O$ unit per vacant site [173]. Moreover, from two to four organo-groups can be anchored in the larger lacunae of di- or tri-lacunary Keggin units [174]. Some of these molecular hybrids have been linked to metallic nanoparticles, connected to carbon nanotubes, or incorporated into silicon surfaces for molecular electronics [175]. In contrast, only a few works describe the functionalization of POMs with organogermyl moieties [176]. Similarly, organophosphonate groups have been widely used for the functionalization of Strandberg-type polyoxomolybdates, lacunary polyoxotungstates derived from the Keggin, and Wells–Dawson structures or diverse polyoxovanadates [177, 178]. The resulting hybrid clusters have shown optical activity or molecular capsule shape among other features [179, 180].

Organotins constitute one of the most studied groups because they show some noticeable advantages like the high stability of the Sn–C bond in aqueous media. Knoth's group first studied the reactivity of commercial mono-organotin groups with vacant heteropolyoxotungstates [181] and a plethora of complex molecules were inserted later on [182, 183]. Over the past decade, the group of Kortz has been one of the main contributors to the chemistry of organotin containing POMs. Dimethyltin ($SnMe_2$) moieties were employed in combination with polylacunary POM frameworks leading to outstanding structures such as the dodecameric $[\{SnMe_2(H_2O)\}_{24}\{SnMe_2\}_{12}(A\text{-}XW_9O_{34})_{12}]^{36-}$ (X = P, As) ball-shaped anion [184]. This group is also responsible for the handful of reports on organo-antimony POM derivatives [185].

1.5.3.2 Substitution of surface oxygen atoms

Another strategy to covalently attach organic groups to the POM skeleton involves the replacement of shell O atoms with O- or N-donor ligands, as exemplified by trisalkoxo capped Anderson–Evans clusters, Lindqvist-type polyoxovanadates or $[H_4P_2V_3W_{15}O_{62}]^{5-}$ anions (V_3-Wells–Dawson) as well as by organoimido/diazenido derivatives of Lindqvist-type molybdates. During the last decade, an impressive work has been carried out in the organic functionalization of Anderson–Evans anions with trisalkoxide ligands [91]. Due to the high alkaline character of the μ_3–O oxygen atoms of these clusters, they can easily be replaced in such a way that the trisalkoxo ligand caps the two opposite triangular faces of the central $\{XO_6\}$ octahedron in an idealized D_{3d} symmetry (δ isomer) [186]. The symmetric and double-sided functionalization was first reported for trivalent transition metal containing polyoxomolybdates (X^{III} = Fe^{III}, Mn^{III}, Cr^{III}). This approach allowed the incorporation of a vast number of organic functionalities (e.g., alkyl chains, amines, alcohols, metal complexes, and organic dyes) that could be further derivatized by diverse post-functionalization reactions [187]. Some authors have also been able to construct metal–organic frameworks [188] and supramolecular hydrogels [189] using such kind of hybrid POMs as linkers. More recently, asymmetric and single-sided functionalizations have been achieved [190]. In contrast, ligand substitution can occasionally take place over the triangular cavities of the Anderson–Evans structure by replacing two μ_3–O oxygen atoms from the $\{XO_6\}$ octahedron and one μ_2–O from the $\{M_6O_{18}\}$ ring. In this case, the less frequent χ isomer with ideal C_{2h} symmetry is obtained [191]. The proton-controlled δ to χ isomerization [192] has allowed the synthesis of mixed δ-χ forms, as well [193]. It is also worth highlighting the recent reports on the organic derivatization of the bent β isomer of the Anderson–Evans anion [194] together with the scarce successful attempts for the functionalization of polyoxotungstates [195].

Preparation of tris-alkoxo functionalized species is also well-described in the case of mixed-*addenda* V_3-Wells–Dawson anions [196]. The use of bridging ligands has resulted in hybrids with very diverse conformation (dumbbell-, triangular-, dendrimer-shape), which can self-assemble into vesicles in aqueous solutions [197]. Furthermore, this structure allows not only the direct derivatization with trisalkoxo ligand, but also the covalent attachment of small organic molecules showing amide and diol groups [198]. Trisalkoxo type ligands have also been used to prepare Lindqvist-type hexavanadates difunctionalized in *trans* fashion [199]. In contrast, only one example of the *cis*-functionalization has been published to our knowledge [200]. Regarding the N-donor ligands, since the first organoimido derivative of the Lindqvist-type molybdate [201], this family of hybrid POMs has experienced a vast growing. In this context, a wide range of mono- and di-organoimido derivatives have been synthesized [202] following to the efficient synthetic process described by Peng and coworkers [203]. A recent study has demonstrated that not only terminal oxo ligands, but also bridging O atoms can be replaced with these ligands [204]. In contrast to organoimido derivatives, functionalization with organidiazenido-type molecules has been comparatively less studied [205].

1.5.3.3 Organic functionalization of 3d or 4f metal substituted POMs

Derivatization of pre-formed metal-containing POM precursors involves the replacement of labile solvent molecules (e.g., acetone, acetate anions) coordinated to the exposed metal centers. This strategy seems to be the less used so far, due to the difficulty of finding solution stable POM precursors with accessible metal centers. Besides the substitution of coordination water molecules by monodentate organic ligands in 3d metal monosubstituted Keggin-type anions or Weakley- and Hervé-type sandwich POMs [206], the reactivity of Krebs-type tungstoantimonates toward planar and aromatic N,O chelating ligands has been systematically studied, because this anion exhibit up to three accessible and labile water molecules per external belt metal center [207]. A second route implies the reaction of metal complexes bearing organic ligands with lacunary polyoxotungstates. This approach has allowed the preparation of several chiral compounds displaying potential applications in asymmetric catalysis, material science, and applied biology, among others [208]. Some other authors have opted for one-pot procedures in which the simultaneous combination of trilacunary Keggin-type units, an excess of 3d metal source and carboxylate ligands resulted in multi-transition-metal containing species with coordinated organic molecules [209]. Hydrothermal techniques were efficiently used by Yang and coworkers for the formation of extended and porous POlyoxometalate-Metal-Organic-Framework (POMOF) networks in which $\{Ni_6PW_9O_{34}\}$ building blocks were interconnected by bridging ligands [210]. In regard to 4f metal substituted POMs, the oxophylic nature of the lanthanide centers makes carboxylic acids suitable ligands for such functionalization, but the negative charge of both carboxylate groups and POM anions together with the steric hindrance that a given POM could produce make it more challenging. These drawbacks have been overcome by using small chelating ligands in a large excess [211], or even positively charged POMs, such as the tetra-lanthanide-capped and mixed-valence ε-Keggin-type phosphomolybdate [212].

1.6 Conclusion

This chapter nicely illustrates the compositional and structural diversity of this outstanding class of metal-oxo clusters. Moreover, the main classical derivatization methods, that is, the incorporation of transition-metal and/or lanthanide centers in their skeleton and the organic functionalization of POMs, considerably extend the catalog of available molecular species and allow the fine tuning of the desired electronic or redox properties for a given application (e.g., catalysis, medicine, magnetism, and photochemistry). Since the first pioneering studies by synthetic and structural chemists, the field has grown very quickly taking advantage from the rapid development of crystallography and single-crystal X-ray diffraction and currently, it has become the focus of interdisciplinary research from areas, such as physics, biology, biomedicine, materials science, and theoretical chemistry. Although hundreds of POM-based compounds have been reported to date, mechanisms governing their self-assembly processes in solution are not still well understood. Thorough control of the molecular assembly could allow the rational design of POMs with required features. POMs excel as ideal candidates to be incorporated into multifunctional, molecular, and composite materials with potential applications in fields related to energy, electronics, and environmental science. Indeed, present research is devoted to the fabrication of POM-based devices and their practical applications. Therefore, this piece could serve as a general introduction for the rest of the chapters included in this themed-book.

Acknowledgments

This work was funded by Eusko Jaurlaritza/Gobierno Vasco (EJ/GV), grants IT1291-19, PIBA2018-59 and ELKARTEK bG20 (KK-2020/00008), and Ministerio de Ciencia, Innovación y Universidades (grant MAT2017-89553-P). E.R.B. and L.F.N. are indebted to EJ/GV for their pre-doctoral fellowships (PRE_2019_1_0106 and PRE_2018_1_0143).

References

1. Pope, M.T. (1983). *Heteropoly and Isopoly Oxometalates*, Springer-Verlag, Berlin, Germany.
2. Pope, M.T., and Yamase, T. (eds.) (2002). *Polyoxometalate Chemistry for Nanocomposite Design*, Kluwer, Dordrecht, The Netherlands.
3. Ammam, M. (2013). Polyoxometalates: formation, structures, principal properties, main deposition methods and application in sensing, *J. Mater. Chem. A*, **1**, pp. 6291–6312.
4. Wang, S.S. and Yang, G.-Y. (2015). Recent advances in polyoxometalate-catalyzed reactions, *Chem. Rev.*, **115**, pp. 4893–4962.
5. Clemente-Juan, J.M., Coronado, E. and Gaita-Ariño, A. (2012). Magnetic polyoxometalates: from molecular magnetism to molecular spintronics and quantum computing, *Chem. Soc. Rev.*, **41**, pp. 7464–7478.
6. Bijelic, A., Aureliano, M., and Rompel, A. (2018). The antibacterial activity of polyoxometalates: structures, antibiotic effects and future perspectives, *Chem. Commun.*, **54**, pp. 1153–1169.

7. Berzelius, J.J. (1826). Beitrag zur näheren Kenntniss des Molybdäns, *Poggend. Ann. Phys. Chem.*, **6**, pp. 369–392.
8. De Luyart, F. and De Luyart, J.J. (1783). *Extractos de las Juntas Generales celebradas por la Real Sociedad Bascongada de Amigos del País*, Bergara, Spain, p. 46.
9. De Marignac, M.C. (1864). Untersuchungen über die Kieselwolframsäuren, *Justus Liebigs Ann. Chem.*, **132**, pp. 25–31.
10. Werner A. (1907). Untersuchungen über anorganische Konstitutions- und Konfigurations-Fragen, *Ber. Dtsch. Chem. Ges.*, **40**, pp. 15–69.
11. Pauling, L. (1929). The molecular structure of the tungstosilicates and related compounds, *J. Am. Chem. Soc.*, **51**, pp. 2868–2880.
12. Keggin, J.F. (1933). Structure of the molecule of 12-phosphotungstic acid, *Nature*, **131**, pp. 908–909.
13. Baker L.C.W. and Glick, D.C. (1998). Present general status of understanding of heteropoly electrolytes and a tracing of some major highlights in the history of their elucidation, *Chem. Rev.*, **98**, pp. 3–50.
14. Souchay, P. (1943). Study of the tungstic heteropoly acids-conditions of existence and properties of the various ions contained in tungstate solutions, *Ann. Chim. Appl.*, **18**, pp. 61–72.
15. Gumerova, N.I. and Rompel, A. (2018). Synthesis, structures and applications of electron-rich polyoxometalates, *Nat. Rev. Chem.*, **2**, pp. 1–20.
16. Galán-Mascarós, J.R. and Kortz, U. (eds.) (2018). *Acta Crystallogr. Sec. C*, **74**, special issue 11.
17. Schluenzen, S., Tocilj, A., Zarivach, R., Harms, J., Glueh-Mann, M., Janell, D., Bashan, A., Bartels, H., Agmon, I., Franceschi, F. and Yonath, A. (2000). Structure of functionally activated small ribosomal subunit at 3.3 Å resolution, *Cell*, **102**, pp. 615–623.
18. Bijelic, A. and Rompel, A. (2015). The use of polyoxometalates in protein crystallography—an attempt to widen a well-known bottleneck, *Coord. Chem. Rev.*, **299**, pp. 22–38.
19. Nyman, M. (2011). Polyoxoniobate chemistry in the 21st century, *Dalton Trans.*, **40**, pp. 8049–8059.
20. Izarova, N.V., Pope, M.T. and Kortz, U. (2012). Noble metals in polyoxometalates, *Angew. Chem. Int. Ed.*, **51**, pp. 9492–9510.
21. Nyman, M. and Burns, P.C. (2012). A comprehensive comparison of transition-metal and actinyl polyoxometalates, *Chem. Soc. Rev.*, **41**, pp. 7354–7367.

22. Müller, A. and Gouzerh, P. (2012). From linking of metal-oxide building blocks in a dynamic library to giant clusters with unique properties and towards adaptive chemistry, *Chem. Soc. Rev.*, **41**, pp. 7431–7463.

23. Lipscomb, W.N. (1965). Paratungstate ion, *Inorg. Chem.*, **4**, pp. 132–134.

24. López, X., Carbó, J.J., Bo, C. and Poblet, J.M. (2012). Structure, properties and reactivity of polyoxometalates: a theoretical perspective, *Chem. Soc. Rev.*, **41**, pp. 7537–7571.

25. Pope, M.T. (1994). *Polyoxoanions*, King, R.B. (ed.) "Encyclopaedia of Inorganic Chemistry", John Wiley & Sons, Chichester, UK, pp. 3361–3371.

26. Richmond, C.J., Miras, H.N., Ruiz de la Oliva, A., Zhang, H., Sans, V., Paramonov, L., Makatsoris, C., Inglis, R., Brechin, E.K., Long, D.-L. and Cronin, L. (2012). A flow-system array for the discovery and scale up of inorganic clusters, *Nature Chem.*, **4**, pp. 1037–1043.

27. Misra, A., Kozma, K., Streb, C. and Nyman, M. (2019). Beyond charge balance: counter-cations in polyoxometalate chemistry, *Angew. Chem. Int. Ed.*, **59**, pp. 596–612.

28. López, X., Bo, C. and Poblet, J.M. (2002). Electronic properties of polyoxometalates: electron and proton affinity of mixed-*addenda* Keggin and Wells–Dawson anions, *J. Am. Chem. Soc.*, **124**, pp. 12574–12582.

29. Proust, A., Matt, B., Villanneau, R., Guillemot, G., Gouzerh, P. and Izzet, G. (2012). Functionalization and post-functionalization: a step towards polyoxometalate-based materials, *Chem. Soc. Rev.*, **41**, pp. 7605–7622.

30. Wéry, A.S.J., Gutiérrez-Zorrilla, J.M., Luque, A., Román, P. and Martínez-Ripoll, M. (1996). Influence of protonation on crystal packing and thermal behaviour of tert-butylammonium decavanadates, *Polyhedron*, **15**, pp. 4555–4564.

31. Fuchs, J., Hartl, H., Schiller, W. and Gerlach, U. (1976). Die Kristallstruktur des Tributylammoniumdekawolframats $[(C_4H_9)_2NH]_4W_{10}O_{23}$, *Acta Crystallogr. Sec. B*, **32**, pp. 740–749.

32. Gutiérrez-Zorrilla, J.M., Yamase, T. and Sugeta, M. (1994). Tetrakis-(isopropylammonium) β-octamolybdate(VI), *Acta Crystallogr. Sec. C*, **50**, pp. 196–198.

33. Román, P., Wéry, A.S.J., Luque, A. and Gutiérrez-Zorrilla, J.M. (1994). Hexakis(tert-butylammonium) heptamolybdate(VI)–water (1/7), *Acta Crystallogr. Sec. C*, **50**, pp. 1031–1034.

34. Rheingold, A.L., White, C.B., Haggerty, B.S. and Maatta, E.A. (1993). Bis(tetrabutylammonium) nonadecaoxohexamolybdenum(VI): a second polymorph, *Acta Crystallogr. Sec. C*, **49**, pp. 756–758.

35. Fuchs, J., Freiwald, W. and Hartl, H. (1978). Neubestimmung der Kristallstruktur von Tetrabutylammonium hexawolframat, *Acta Crystallogr. Sec. B*, **34**, pp. 1764–1770.

36. Monakhov, K.Y., Bensch, W. and Kögerler, P. (2015). Semimetal-functionalised polyoxovanadates, *Chem. Soc. Rev.*, **44**, pp. 8443–8483.

37. Román, P., Wéry, A.S.J., Luque, A. and Gutiérrez-Zorrilla, J.M. (1993). Observation of a novel cyclic tetrametavanadate anion isolated from aqueous solution, *Inorg. Chem.*, **32**, pp. 775–776.

38. Wéry, A.S.J., Gutiérrez-Zorrilla, J.M., Luque, A., Ugalde, M. and Román, P. (1996). Phase transitions in metavanadates. Polymerization of tetrakis(tert-butylammonium)-cyclo-tetrametavanadate, *Chem. Mater.*, **8**, pp. 408–413.

39. Reinoso, S., Artetxe, B., San Felices, L. and Gutiérrez-Zorrilla, J.M. (2016). *Polyoxometalates: Properties, Structure and Synthesis*, Roberts, A. P. (ed.), Chapter 6 "Single-crystal-to-single-crystal transformations in stimuli-responsive compounds based on polyoxometalate clusters", Nova Science Publishers, Hauppauge, NY, USA, pp. 143–212.

40. Aureliano, M. and Crans, D.C. (2009). Decavanadate ($V_{10}O_{28}^{6-}$) and oxovanadates: oxometalates with many biological activities, *J. Inorg. Biochem.*, **103**, pp. 536–546.

41. Aureliano, M. (2011). Recent perspectives into biochemistry of decavanadate, *World J. Biol. Chem.*, **2**, pp. 215–225.

42. Hamilton, E.E., Fanwick, P.E. and Wilker, J.J. (2002). The elusive vanadate $(V_3O_9)^{3-}$: isolation, crystal structure, and nonaqueous solution behavior, *J. Am. Chem. Soc.*, **124**, pp. 78–82.

43. Hou, D., Hagen, K.D. and C.L. Hill (1992). Tridecavanadate, $[V_{13}O_{34}]^{3-}$, a new high-potential isopolyvanadate, *J. Am. Chem. Soc.*, **114**, pp. 5864–5866.

44. Hou, D., Hagen, K.D. and Hill, C.L. (1993). Pentadecavanadate, $V_{15}O_{42}^{9-}$, a new highly condensed fully oxidized isopolyvanadate with kinetic stability in water, *J. Chem. Soc., Chem. Commun.*, pp. 426–428.

45. Johnson, G.K. and Schlemper, E.O. (1978). Existence and structure of the molecular ion 18-vanadate (IV), *J. Am. Chem. Soc.*, **100**, pp. 3645–3646.

46. Maksimovskaya, R.I. and Maksimov, G.M. (2007). ^{95}Mo and ^{17}O NMR studies of aqueous molybdate solutions, *Inorg. Chem.*, **46**, pp. 3688–3695.

47. Ogata, A., Ishikawa, E., Morishita, Y., Mitsui, S., Yamashita, A., Hasumi, K., Takamoto, S., Yamase, T. and Eriguchi, M. (2008). Antitumour effect of polyoxomolybdates: induction of apoptotic cell death and autophagy in in vitro and in vivo models, *Br. J. Cancer*, **98**, pp. 399–409.

48. Bridgeman, A.J. (2002). The electronic structure and stability of the isomers of octamolybdate, *J. Phys. Chem. A*, **106**, pp. 12151–12160.

49. Krebs, B., Stiller, S., Tytko, K.H. and Mehmke, J. (1991). Structure and bonding in the high molecular weight isopolymolybdate ion, $[Mo_{36}O_{112}(H_2O)_{16}]^{8-}$. The crystal structure of $Na_8[Mo_{36}O_{112}(H_2O)_{16}]\cdot 58\ H_2O$, *Eur. J. Solid State Inorg. Chem.*, **28**, pp. 883–903.

50. Duncan, J.F. and Kepert, D.L. (1962). Polyanion equilibria in aqueous solution. Part II. A thermodynamic study of the paratungstate A anion, *J. Chem. Soc.*, pp. 205–214.

51. Fan, L.-Y., Cao, J. and Hu, C.-W. (2015). What can electrospray mass spectrometry of paratungstates in an equilibrating mixture tell us?, *RSC Adv.*, **5**, pp. 83377–83382.

52. Fuchs, J., Palm, R. and Hartl, H. (1996). $K_7HW_5O_{19}\cdot 10H_2O$—a novel isopolyoxotungstate (VI), *Angew. Chem. Int. Ed. Engl.*, **35**, pp. 2651–2653.

53. Brüdgam, I., Fuchs, J., Hartl, H. and Palm, R. (1998). Two new isopolyoxotungstates(VI) with the empirical composition $Cs_2W_2O_7\cdot 2H_2O$ and $Na_2W_2O_7\cdot H_2O$: an icosatetratungstate and a polymeric compound, *Angew. Chem. Int. Ed. Engl.*, **37**, pp. 2668–2671.

54. Freedman, M.L. (1959). The tungstic acids, *J. Am. Chem. Soc.*, **81**, pp. 3834–3839.

55. Long, D.L., Abbas, H., Kögerler, P. and Cronin, L. (2004). A high-nuclearity "celtic-ring" isopolyoxotungstate, $[H_{12}W_{36}O_{120}]^{12-}$, that captures trace potassium ions, *J. Am. Chem. Soc.*, **126**, pp. 13880–13881.

56. Miras, H.N., Yan, J., Long, D.L. and Cronin, L. (2008). Structural evolution of "S"-shaped $[H_4W_{22}O_{74}]^{12-}$ and "§"-shaped $[H_{10}W_{34}O_{116}]^{18-}$ isopolyoxotungstate clusters, *Angew. Chem. Int. Ed.*, **47**, pp. 8420–8423.

57. Hasting, J.J. and Howarth, O.W. (1992). A ^{183}W, ^{1}H and ^{17}O nuclear magnetic resonance study of aqueous isopolytungstates, *J. Chem. Soc., Dalton Trans.*, pp. 209–215.

58. Long, D.L., Kögerler, P., Parenty, A.D.C., Fielden, J. and Cronin, L. (2006). Discovery of a family of isopolyoxotungstates $[H_4W_{19}O_{62}]^{6-}$ encapsulating a $\{WO_6\}$ moiety within a $\{W_{18}\}$ Dawson-like cluster cage, *Angew. Chem. Int. Ed.*, **45**, pp. 4798–4803.

59. Zhan, C.-H., Winter, R.S., Zheng, Q., Yan, J., Cameron, J.M., Long, D.L. and Cronin, L. (2015). Assembly of tungsten-oxide-based pentagonal motifs in solution leads to nanoscale $\{W_{48}\}$, $\{W_{56}\}$, and $\{W_{92}\}$ polyoxometalate clusters, *Angew. Chem. Int. Ed.*, **54**, pp. 14308–14312.

60. Lindqvist, I. (1952). The structure of the hexaniobate ion in $(Na_2O)_7(Nb_2O_5)_6(H_2O)_{32}$, *Ark. Kemi*, **5**, pp. 247–250.

61. Lindqvist, I. and Aronsson, B. (1953). The structure of the hexatantalate ion in $(K_2O)_4(Ta_2O_5)_3(H_2O)_{16}$, *Ark. Kemi*, **7**, pp. 49–52.

62. Sures, D., Segado, M., Bo, C. and Nyman, M. (2018). Alkali-driven disassembly and reassembly of molecular niobium oxide in water, *J. Am. Chem. Soc.*, **140**, pp. 10803–10813.

63. Graeber, E.J. and Morosin, B. (1977). The molecular configuration of the decaniobate ion $(Nb_{17}O_{28}{}^{6-})$, *Acta Crystallogr. Sec. B*, **33**, pp. 2137–2143.

64. Matsumoto, M., Ozawa, Y., Yagasaki, A. and Zhe, Y. (2013). Decatantalate—the last member of the group 5 decametalate family, *Inorg. Chem.*, **52**, pp. 7825–7827.

65. Maekawa, M., Ozawa, Y. and Yagasaki, A. (2006). Icosaniobate: a new member of the isoniobate family, *Inorg. Chem.*, **45**, pp. 9608–9609.

66. Huang, P., Qin, C., Su, Z.-M., Xing, Y., Wang, X.-L., Shao, K.-Z., Lan, Y.-Q. and Wang, E.-B. (2012). Self-assembly and photocatalytic properties of polyoxoniobates: $\{Nb_{24}O_{72}\}$, $\{Nb_{32}O_{96}\}$, and $\{K_{12}Nb_{96}O_{288}\}$ clusters, *J. Am. Chem. Soc.*, **134**, pp. 14004–14010.

67. Tsunashima, R., Long, D.L., Miras, H.N. Gabb, D., Pradeep, C.P. and Cronin, L. (2010). The construction of high-nuclearity isopolyoxoniobates with pentagonal building blocks: $[HNb_{27}O_{76}]^{16-}$ and $[H_{10}Nb_{31}O_{93}(CO_3)]^{23-}$, *Angew. Chem. Int. Ed.*, **49**, pp. 113–116.

68. Jin, L., Zhu, Z.-K., Wu, Y.-L., Qi, Y.-J., Li, X.-X. and Zheng, S.-T. (2017). Record high-nuclearity polyoxoniobates: discrete nanoclusters $\{Nb_{114}\}$, $\{Nb_{81}\}$, and $\{Nb_{52}\}$, and extended frameworks based on $\{Cu_3Nb_{78}\}$ and $\{Cu_4Nb_{78}\}$, *Angew. Chem. Int. Ed.*, **56**, pp. 16288–16292.

69. Wu, Y.-L., Li, X.-X., Qi, Y.-J., Yu, H., Jin, L. and Zheng, S.-T. (2018). $\{Nb_{288}O_{768}(OH)_{48}(CO_3)_{12}\}$: a macromolecular polyoxometalate with

close to 300 niobium atoms, *Angew. Chem. Int. Ed.*, **57**, pp. 8572–8576.

70. Nyman, M., Bonhomme, F., Alam, T.M., Parise, J.B. and Vaughan, G.M.B. (2004). $[SiNb_{12}O_{40}]^{16-}$ and $[GeNb_{12}O_{40}]^{16-}$: highly charged Keggin ions with sticky surfaces, *Angew. Chem. Int. Ed.*, **43**, pp. 2787–2792.

71. Selling, A., Andersson, I., Pettersson, L., Schramm, C.M., Downey, S.L. and Grate, J.H. (1994). Multicomponent polyanions. 47: the aqueous vanadophosphate system, *Inorg. Chem.*, **33**, pp. 3141–3150.

72. Wang, H., Hamanaka, S., Nishimoto, Y., Irle, S., Yokoyama, T., Yoshikawa, H. and Awaga, K. (2012). In operando X-ray absorption fine structure studies of polyoxometalate molecular cluster batteries: polyoxometalates as electron sponges, *J. Am. Chem. Soc.*, **134**, pp. 4918–4924.

73. Kozhevnikov, I.V (2002). *Catalysis by Polyoxometalates*, vol 2, John Wiley & Sons Inc., New York, USA.

74. Tézé, A. and Hervé, G. (1990). *Inorganic Syntheses* **27**, Finke, R.G. and Lyon, D.K. (eds.) Chapter 16. α-, β- and γ-dodecatungstosilicic acids: isomers and related lacunary compounds, John Wiley & Sons, New York, pp. 85–96.

75. Casey, W.H. (2006). Large aqueous aluminum hydroxide molecules, *Chem. Rev.*, **106**, pp. 1–16.

76. Sartzi, H., Miras, H.N., Vilà-Nadal, L., Long, D.-L. and Cronin, L. (2015). Trapping the δ isomer of the polyoxometalate-based Keggin cluster with a tripodal ligand, *Angew. Chem. Int. Ed.*, **54**, pp. 15488–15492.

77. Mialane, P., Dolbecq, A., Lisnard, L., Mallard, A., Marrot, J. and Sécheresse, F. (2002). $[\varepsilon\text{-}PMo_{12}O_{36}(OH)_4\{La(H_2O)_4\}_4]^{5+}$: the first ε-$PMo_{12}O_{40}$ Keggin ion and its association with the two-electron-reduced α-$PMo_{12}O_{40}$ isomer, *Angew. Chem. Int. Ed.*, **41**, pp. 2398–2401.

78. Bösing, M., Loose, I., Pohlmann, H. and Krebs, B. (1997). New strategies for the generation of large heteropolymetalate clusters: the β-*B*-SbW_9, fragment as a multifunctional unit, *Chem. Eur. J.*, **3**, pp. 1232–1237.

79. Dawson, B. (1953). The structure of the 9(18)-heteropoly anion in potassium 9(18)-tungstophosphate, $K_6(P_2W_{18}O_{62})\cdot 14H_2O$, *Acta Crystallogr.*, **6**, pp. 113–126.

80. Contant, R. and Thouvenot, R. (1993). A reinvestigation of isomerism in the Dawson structure: syntheses and ^{183}W NMR

structural characterization of three new polyoxotungstates $[X_2W_{18}O_{62}]^{6-}$ (X = P^V, As^V), *Inorg. Chim. Acta*, **212**, pp. 41–50.

81. Zhang, F.-Q., Guan, W., Yan, L.-K., Zhang, Y.-T., Xu, M.-T., Hayfron-Benjamin, E. and Su, Z.-M. (2011). On the origin of the relative stability of Well–Dawson isomers: a DFT study of r-, β-, γ-, r*-, β*-, and γ*-$[(PO_4)_2W_{18}O_{54}]^{6-}$ anions, *Inorg. Chem.* **50**, pp. 4967–4977.

82. Contant, R. and Ciabrini, J. P. (1981). Stereospecific preparations of new n-molybdo-(18-n)-tungsto-2-phosphates and related "defect" compounds (n = 2, 4 or 5), *J. Inorg. Nucl. Chem.*, **43**, pp. 1525–1528.

83. Contant, R. and Tézé, A. (1985). A new crown heteropolyanion $K_{28}Li_5H_7P_8W_{48}O_{184}\cdot 92H_2O$: synthesis, structure, and properties, *Inorg. Chem.*, **24**, pp. 4610–4614.

84. Alizadeh, M.H., Harmalker, S.P., Jeannin, Y., Martin-Frere, J. and Pope, M.T. (1985). A heteropolyanion with fivefold molecular symmetry that contains a nonlabile encapsulated sodium ion. The structure and chemistry of $[NaP_5W_{30}O_{110}]^{14-}$, *J. Am. Chem. Soc.*, **107**, pp. 2662–2669.

85. Strandberg, R. (1973). Multicomponent polyanions. IV. The molecular and crystal structure of $Na_6Mo_5P_2O_{23}.(H_2O)_{13}$, a compound containing sodium-coordinated pentamolybdodiphosphate anions, *Acta Chem. Scand.*, **27**, pp. 1004–1018.

86. Evans, H.T. and Konnert, J.A. (1978). The crystal chemistry of sherwoodite, a calcium 14 vanadoaluminate-heteropoly complex, *Am. Mineral.*, **63**, pp. 863–868.

87. Waugh, J.L.T., Shoemaker, D.P. and Pauling, L. (1954). On the structure of the heteropoly anion in ammonium 9-molybdomanganate, $(NH_4)_6MnMo_9O_{32}(H_2O)_8$, *Acta Crystallogr.*, **7**, pp. 438–441.

88. Chen, Y., Zhang, C., Yang, C., Zhang, J., Zheng, K., Fang, Q. and Li, G. (2017). A Waugh type $[CoMo_9O_{32}]^{6-}$ cluster with atomically dispersed Co^{IV} originates from Anderson type $[CoMo_6O_{24}]^{3-}$ for photocatalytic oxygen molecule activation, *Nanoscale*, **9**, pp. 15332–15339.

89. Blazevic, A. and Rompel, A. (2016). The Anderson–Evans polyoxometalate: from inorganic building blocks *via* hybrid organic–inorganic structures to tomorrows "Bio-POM", *Coord. Chem. Rev.*, **307**, pp. 42–64.

90. Lee, U. and Sasaki, Y. (1984). Isomerism of the hexamolybdo-platinate (IV) polyanion. Crystal structures of $K_{3.5}[\alpha\text{-}H_{4.5}PtMo_6O_{24}]\cdot 3H_2O$ and $(NH_4)_4[\beta\text{-}H_4PtMo_6O_{24}]\cdot 1.5H_2O$, *Chem. Lett.*, **13**, pp. 1297–1300.

91. Anderson, J.S. (1937). Constitution of the Poly-acids, *Nature*, **140**, p. 850.

92. Evans, H.T. Jr. (1948). The crystal structures of ammonium and potassium molybdotellurates, *J. Am. Chem. Soc.*, **70**, pp. 1291–1292.

93. Bijelic, A. and Rompel, A. (2017). Ten good reasons for the use of the tellurium-centered Anderson–Evans polyoxotungstate in protein crystallography, *Acc. Chem. Res.*, **50**, pp. 1441–1448.

94. Vidyavathy, B. and Vidyasagar, K. (1998). Low-temperature syntheses and characterization of novel layered tellurites, $A_2Mo_3TeO_{12}$ (A = NH_4, Cs), and "zero-dimensional" tellurites, $A_4Mo_6Te_2O_{24}\cdot 6H_2O$ (A = Rb, K), *Inorg. Chem.*, **37**, pp. 4764–4774.

95. Evans, H.T. and Showell, J.S. (1969). Molecular structure of the decamolybdodicobaltate(III) ion, *J. Am. Chem. Soc.*, **91**, pp. 6881–6882.

96. Dexter, D.D. and Silverton, J.V. (1968). A new structural type for heteropoly anions. The crystal structure of $(NH_4)_2H_6(CeMo_{12}O_{42})\cdot 12H_2O$, *J. Am. Chem. Soc.*, **90**, pp. 3589–3590.

97. Schäffer, C., Todea, A.M., Gouzerh, P. and Müller, A. (2012). Spontaneous self-assembly of a giant spherical metal-oxide Keplerate: addition of one building block induces "immediate" formation of the complementary one from a constitutional dynamic library, *Chem. Commun.*, **48**, pp. 350–352.

98. Piepgrass, K. and Pope, M.T. (1987). Heteropoly "Browns" as *Class I* Mixed Valence (W(IV,VI)) Complexes. Tungsten-183 NMR of W(IV) Trimers, *J. Am. Chem. Soc.*, **109**, pp. 1586–1587.

99. Scheele, C.W. (1793). Sämtliche Physische und Chemische Werke. Hermbstädt D.S.F. (ed.), Martin Sändig oHG, Niederwalluf/Wiesbaden, Germany, pp. 185–200 (reprint 1971).

100. Müller, A., Krickemeyer, E., Meyer, J., Bögge, H., Peters, F., Plass, W., Diemann, E., Dillinger, S., Nonnenbruch, F., Randerath, M. and Menke, C. (1995). $[Mo_{154}(NO)_{14}O_{420}(OH)_{28}(H_2O)_{70}]^{(25\pm5)-}$: a water-soluble big wheel with more than 700 atoms and a relative molecular mass of about 24000, *Angew. Chem. Int. Ed. Eng*, **34**, pp. 2122–2124.

101. Müller, A., Krickemeyer, E., Bögge, H., Schmidtmann, M., Beugholt, C., Kögerler, P. and Lu, C. (1998). Formation of a ring-shaped reduced "metal oxide" with the simple composition $[(MoO_3)_{176}(H_2O)_{80}H_{32}]$, *Angew. Chem. Int. Ed.*, **37**, pp. 1220–1223.

102. Müller, A., Shah, S.Q.N.; Bögge, H. and Schmidtmann, M. (1999). Molecular growth from a Mo_{176} to a Mo_{248} cluster, *Nature*, **397**, pp. 48–50.

103. Müller, A., Sarkar, S., Shah, S.Q.N., Bögge, H., Schmidtmann, M., Sarkar, S., Kögerler, P., Hauptfleisch, B., Traitwein, A.X. and Schünemann, V. (1999). Archimedean synthesis and magic numbers: "Sizing" giant molybdenum-oxide-based molecular spheres of the Keplerate type, *Angew. Chem. Int. Ed.*, **38**, pp. 3238–3241.

104. Müller, A., Beckmann, E., Bögge, H., Schmidtmann, M. and Dress, A. (2002). Inorganic chemistry goes protein size: a Mo_{368} nano-hedgehog initiating nanochemistry by symmetry breaking, *Angew. Chem. Int. Ed.*, **41**, pp. 1162–1167.

105. Scäffer, C., Merca, A., Bögge, H., Todea, A.M., Kistler, M.L., Liu, T., Thouvenot, R., Gouzerh, P. and Müller, A. (2009). Unprecedented and differently applicable pentagonal units in a dynamic library: a Keplerate of the type $\{(W)W_5\}_{12}\{Mo_2\}_{30}$, *Angew. Chem. Int. Ed.*, **48**, pp. 149–153.

106. Gao, J., Yan, J., Beeg, S., Long, D.-L. and Cronin, L. (2013). One-pot versus sequential reactions in the self-assembly of gigantic nanoscale polyoxotungstates, *J. Am. Chem. Soc.*, **135**, pp. 1796–1805.

107. Qiu, J. and Burns, P.C. (2013). Clusters of actinides with oxide, peroxide, or hydroxide bridges, *Chem. Rev.*, **113**, pp. 1097–1120.

108. Miro, P., Pierrefixe, S., Gicquel, M., Gil, A. and Bo, C. (2010). On the origin of the cation templated self-assembly of uranyl-peroxide nanoclusters, *J. Am. Chem. Soc.*, **132**, pp. 17787–17794.

109. Burns, P.C., Kubatko, K.-A., Sigmon, G., Fryer, B.J., Gagnon, J.E., Antonio, M.R. and Soderholm, L. (2005). Actinyl peroxide nanospheres, *Angew. Chem. Int. Ed.*, **44**, pp. 2135–2139.

110. Day, V.W., Goloboy, J.C. and Klemperer, W.G. (2009). Synthesis and solid state structures of the hydrogen-bonded hexamolybdoplatinate(IV) tetramer $[(PtMo_6O_{24})_4H_{23}]^{9-}$ and the hexamolybdoplatinate(IV) trimers $[(PtMo_6O_{24})_3H_{16}]^{8-}$ and $[(PtMo_6O_{24})_3H_{14}]^{10-}$, *Eur. J. Inorg. Chem.*, pp. 5079–5087.

111. Kikukawa, Y., Kuroda, Y., Yamaguchi, K. and Mizuno, N. (2012). Diamond-shaped $[Ag_4]^{4+}$ cluster encapsulated by silicotungstate ligands: synthesis and catalysis of hydrolytic oxidation of silanes, *Angew. Chem. Int. Ed.*, **51**, pp. 2434–2437.

112. Geletti, Y.V., Besson, C., Hou, Y., Yin, Q., Musaev, D.G., Quiñonero, D., Cao, R., Hardcastle, K.I., Proust, A., Kögerler, P. and Hill, C.L. (2009).

Structural, physicochemical, and reactivity properties of an all-inorganic, highly active tetraruthenium homogeneous catalyst for water oxidation, *J. Am. Chem. Soc.*, **131**, pp. 17360–17370.

113. Pley, M. and Wickleder, M.S. (2004). The cluster ion $[Pt_{12}O_8(SO_4)_{12}]^{4-}$, *Angew. Chem. Int. Ed.*, **43**, pp. 4168–4170.

114. Izarova, N.V., Vankova, N., Heine, T., Ngo Biboum, R., Keita, B., Nadjo, L. and Kortz, U. (2010). Polyoxometalates made of gold: the polyoxoaurate $[Au^{III}_4As^{V}_4O_{20}]^{8-}$, *Angew. Chem. Int. Ed.*, **49**, pp. 1886–1889.

115. Xiang, Y., Izarova, N.V., Schinle, F., Hampe, O., Keita, B. and Kortz, U. (2012). The selenite-capped polyoxo-4-aurate(III), $[Au^{III}_4O_4(Se^{IV}O_3)_4]^{4-}$, *Chem Commun.*, **48**, pp. 9849–9851.

116. Yang, P. and Kortz, U. (2018). Discovery and evolution of polyoxopalladates, *Acc. Chem. Res.*, **51**, pp. 1599–1608.

117. Yang, P.; Xiang, Y.; Lin, Z.; Bassil, B. S.; Cao, J.; Fan, L.; Fan, Y.; Li, M.-X.; Jiménez-Lozano, P.; Carbó, J.J.; Poblet, J.M. and Kortz, U. (2014). Alkaline earth guests in polyoxopalladate chemistry: from nanocube to nanostar *via* an open-shell structure, *Angew. Chem. Int. Ed.*, **53**, pp. 11974–11978.

118. Das, V., Kaushik, R. and Hussain, F. (2020). Heterometallic 3d 4f polyoxometalates: an emerging field with structural diversity to multiple applications, *Coord. Chem. Rev.*, **413**, pp. 213271.

119. Izzet, G., Volatron, F. and Proust, A. (2017). Tailor–made covalent organic-inorganic polyoxometalate hybrids: versatile platforms for the elaboration of functional molecular architectures, *Chem. Rec.*, **17**, pp. 250–266.

120. Anyushin, A., Kondinski, A. and Parac-Vogt, T.N. (2020). Hybrid polyoxometalates as post-functionalization platforms: from fundamentals to emerging applications, *Chem. Soc. Rev.*, **49**, pp. 382–432.

121. Blasco-Ahicart, M., Soriano-López, J., Carbó, J.J., Poblet, J.M. and Galán-Mascarós, J.R. (2018). Polyoxometalate electrocatalysts based on earth-abundant metals for efficient water oxidation in acidic media, *Nature Chem.*, **10**, pp. 24–30.

122. Al-Oweini, R., Bassil, B.S., Friedl, J., Kottisch. V., Ibrahim, M., Asano, M., Keita, B., Novitchi, G., Lan, Y., Powell, A., Stimming, U. and Kortz, U. (2014). Synthesis and characterization of multinuclear manganese-containing tungstosilicates, *Inorg. Chem.*, **53**, pp. 5663–5673.

123. Li, N., Zhang, S., Zhao, J., Pan, Q., Xing, B., Li, J., Jiang, S. and Qu, W. (2020). Synthesis of two monomeric tri-substituted γ*-Dawson-type polyoxotungstates, *Eur. J. Inorg. Chem.*, pp. 690–694.

124. Kortz, U., Tézé, A. and Hervé, G. (1999). A cubane-substituted polyoxoanion: structure and magnetic properties of $Cs_2[H_2PW_9Ni_4O_{34}(OH)_3(H_2O)_6]\cdot 5H_2O$, *Inorg. Chem.*, **38**, pp. 2038–2042.

125. Weakley, T.J.R. and Finke, R.G. (1990). Single-crystal X-ray structures of the polyoxotungstate salts $K_{8.3}Na_{1.7}[Cu_4(H_2O)_2(PW_9O_{34})_2]\cdot 24H_2O$ and $Na_{14}Cu[Cu_4(H_2O)_2(P_2W_{15}O_{56})_2]\cdot 53H_2O$, *Inorg. Chem.*, **29**, pp. 1235–1241.

126. Clemente-Juan, J.M., Coronado, E., Gaita-Ariño, A., Giménez-Saiz, C., Güdel, H.-U., Sieber, A., Bircher, R. and Mutka, H. (2005). Magnetic polyoxometalates: anisotropic exchange interactions in the Co_3^{II} moiety of $[(NaOH_2)Co_3(H_2O)(P_2W_{15}O_{56})_2]^{17-}$, *Inorg. Chem.*, **44**, pp. 3389–3395.

127. Mbomekalle, I.M., Mialane, P., Dolbecq, A., Marrot, J., Sécheresse, F., Berthet, P., Keita, B. and Nadjo, L. (2009). Rational synthesis, structure, magnetism and electrochemistry of mixed iron–nickel-containing Wells–Dawson-fragment-based sandwich-type polyoxometalates, *Eur. J. Inorg. Chem.*, pp. 5194–5204.

128. Yin, Q., Tan, J.M., Besson, C., Geletti, Y.V., Musaev, D.G., Kuznetsov, A.E., Luo, Z., Hardcastle, K.I. and Hill, C.L. (2010). A fast soluble carbon-free molecular water oxidation catalyst based on abundant metals, *Science*, **328**, pp. 342–345.

129. Barats, D., Leitus, G., Popovitz-Biro, R., Shimon, L.J.W. and Neumann, R. (2008). A stable "end-on" iron(III)–hydroperoxo complex in water derived from a multi-iron(II)-substituted polyoxometalate and molecular oxygen, *Angew. Chem. Int. Ed.*, **120**, pp. 10056–10060.

130. Tourné, C.M., Tourné, G.F. and Zonnevijlle, F. (1991). Chiral polytungstometalates $[WM_3(H_2O)_2(XW_9O_{34})_2]^{12-}$ (X = M = Zn or Co^{II}) and their M-substituted derivatives. Syntheses, chemical, structural and spectroscopic study of some D,L sodium and potassium salts, *J. Chem.* Soc. *Dalton Trans.*, pp. 143–155.

131. Knoth, W.H., Domaille, P.J. and Harlow, R.L. (1986). Heteropolyanions of the types $M_3(W_9PO_{34})_2^{12-}$ and MM′M″ $(W_9PO_{34})_2^{12-}$: novel coordination of nitrate and nitrite, *Inorg. Chem.*, **25**, pp. 1577–1584.

132. Okun, N.M., Anderson, T.M. and Hill, C.L. (2003). $[(Fe^{III}(OH_2)_2)_3(A\text{-}\alpha\text{-}PW_9O_{34})_2]^{9-}$ on cationic silica nanoparticles, a new type of material and efficient heterogeneous catalyst for aerobic oxidations, *J. Am. Chem. Soc.*, **125**, pp. 3194–3195.

133. Fang, X., Anderson, T.M., Neiwert, W.A. and Hill, C.L. (2003). Yttrium polyoxometalates. synthesis and characterization of a carbonate-encapsulated sandwich-type complex, *Inorg. Chem.*, **42**, pp. 8600–8602.

134. Laronze, N., Marrot, J. and Hervé, G. (2003). Synthesis, molecular structure and chemical properties of a new tungstosilicate with an open Wells–Dawson structure, α-$[Si_2W_{18}O_{66}]^{16-}$, *Chem. Commun.*, pp. 2360–2361.

135. Guo, J., Zhang, D., Chen, L., Song, Y., Zhu, D. and Xu, Y. (2013). Syntheses, structures and magnetic properties of two unprecedented hybrid compounds constructed from open Wells–Dawson anions and high-nuclear transition metal clusters, *Dalton Trans.*, **42**, pp. 8454–8459.

136. Robert, F., Leyrie, M. and Hervé, G. (1982). Structure of potassium diaquatricuprooctadecatungstodiarsenate(III)(12-) undecahydrate, *Acta Crystallogr.*, **B38**, pp. 358–362.

137. Kortz, U., Al-Kassem, N., Savelieff, M.G., Al Kadi, N.A. and Sadakane, M. (2001). Synthesis and characterization of copper-, zinc-, manganese-, and cobalt-substituted dimeric heteropolyanions, $[(\alpha\text{-}XW_9O_{33})_2M_3(H_2O)_3]^{n-}$ ($n = 12$, X = As^{III}, Sb^{III}, M = Cu^{2+}, Zn^{2+}; n = 10, X = Se^{IV}, Te^{IV}, M = Cu^{2+}) and $[(\alpha\text{-}AsW_9O_{33})_2WO(H_2O)M_2(H_2O)_2]^{10-}$ (M = Zn^{2+}, Mn^{2+}, Co^{2+}), *Inorg. Chem.*, **40**, pp. 4742–4749.

138. Yamase, T., Ishikawa, H., Abe, H., Fukaya, K., Nojiri, H. and Takeuchi, H. (2012) Molecular magnetism of M_6 hexagon ring in D_{3d} symmetric $[(MCl)_6(XW_9O_{33})_2]^{12-}$ (M = Cu^{II} and Mn^{II}, X = Sb^{III} and As^{III}), *Inorg. Chem.*, **51**, pp. 4606–4619.

139. Loose, I., Droste, E., Bösing, M., Pohlmann, H., Dickman, M.H., Rosu, C., Pope, T. and Krebs, B. (1999). Heteropolymetalate clusters of the subvalent main group elements Bi^{III} and Sb^{III}, *Inorg. Chem.*, **38**, pp. 2688–2694.

140. Von Allmen, K.D., Grundmann, H., Linden, A. and Patzke, G.R. (2017). Synthesis and characterization of 0D–3D copper-containing tungstobismuthates obtained from the lacunary precursor Na_9 $[B\text{-}\alpha\text{-}BiW_9O_{33}]$, *Inorg. Chem.*, **56**, pp. 327–335.

141. Carraro, M., Bassil, B.S., Sorarù, A., Berardi, S., Suchopar, A., Kortz, U. and Bonchio, M. (2013). A Lewis acid catalytic core sandwiched by inorganic polyoxoanion caps: selective H_2O_2-based oxidations with $[Al^{III}_4(H_2O)_{10}(\beta\text{-}XW_9O_{33}H)_2]^{6-}$ (X = As^{III}, Sb^{III}), *Chem. Commun.*, **49**, pp. 7914–7916.

142. Oms, O., Dolbecq, A. and Mialane, P. (2012). Diversity in structures and properties of 3d incorporating polyoxotungstates, *Chem. Soc. Rev.*, **41**, pp. 7497–7536.

143. Ritchie, C., Ferguson, A., Nojiri, H., Miras, H.N., Song, Y.-F., Long, D.-L., Bulkholder, E., Murrie, M., Kögerler, P., Brechin, E.K. and Cronin, L. (2008). Polyoxometalate-mediated self-assembly of single-molecule magnets: $\{[XW_9O_{34}]_2[Mn^{III}_4Mn^{II}_2O_4(H_2O)_4]\}^{12-}$, *Angew. Chem. Int. Ed.*, **47**, pp. 5609–5612.

144. Ibrahim, M., Lan, Y., Bassil, B.S., Xiang, Y., Suchopar, A., Powell, A.K. and Kortz, U. (2011). Hexadecacobalt(II)-containing polyoxometalate-based single-molecule magnet, *Angew. Chem. Int. Ed.*, **50**, pp. 4708–4711.

145. Bassil, B.S., Ibrahim, M., Al-Oweini, R., Asano, M., Wang, Z., van Tol, J., Dalal, N.S., Choi, K.-Y., Bibourn, R.N., Keita, B., Nadjo, L. and Kortz, U. (2011). A planar $\{Mn_{19}(OH)_{12}\}^{26+}$ unit incorporated in a 60-tungsto-6-silicate polyanion, *Angew. Chem. Int. Ed.*, **50**, pp. 5961–5964.

146. Sessoli, R., Gatteschi, D., Caneschi, A. and Novak, M.A. (1993). Magnetic bistability in a metal-ion cluster, *Nature*, **365**, pp. 141–143.

147. Mitchell, S.G., Molina, P.I., Khanra, S., Miras, H.N., Prescimone, A., Cooper, G.J.T., Winter, R.S., Brechin, E.K. Long, D.-L., Cogdell, R.J. and Cronin, L. (2011). A mixed-valence manganese cubane trapped by inequivalent trilacunary polyoxometalate ligands, *Angew. Chem. Int. Ed.*, **50**, pp. 9154–9157.

148. Yang, Z., Wang, Y., Liang, Z., Lu, J., Ma, P., Niu, J. and Wang, J. (2019). An unprecedented $[\{Fe_5O_5(OH)_2(OAc)_2\}_2\{W_2O_2(OH)\}]$ cluster sandwiched in the tetravacant tungstophosphate, *Dalton Trans.*, **48**, pp. 16857–16860.

149. Mal, S.S. and Kortz, U. (2005). The wheel-shaped Cu_{20} tungstophosphate $[Cu_{20}Cl(OH)_{24}(H_2O)_{12}(P_8W_{48}O_{184})]^{25-}$ ion, *Angew. Chem. Int. Ed.*, **44**, pp. 3777–3780.

150. Boskovic, C. (2017) Rare earth polyoxometalates, *Acc. Chem. Res.*, **50**, pp. 2205–2214.

151. Yamase, T. (2009). Chapter 243 luminescence of polyoxometall-olanthanoates and photochemical nano-ring formation, *Handb. Phys. Chem. Rare Earths*, **39**, pp. 297–356.

152. Peacock, R.D. and Weakley, T.J.R. (1971). Heteropolytungstate complexes of the lanthanide elements. Part I. Preparation and reactions, *J. Chem. Soc. A*, pp. 1836–1839.

153. Li, H., Yang, W., Wang, X., Chen, L., Ma, J., Zheng, L. and Zhao, J. (2016). Self-assembly of a family of isopolytungstates induced by the synergistic effect of the nature of lanthanoids and the pH variation in the reaction process: syntheses, structures, and properties, *Cryst. Growth Des.*, **16**, pp. 108–120.

154. Iijima, J., Ishikawa, E., Nakamura, Y. and Naruke, H. (2010). Synthesis and structural investigation of sandwich polyoxotungstates containing cerium (III/IV) and mono-lacunary Keggin tungstophosphate units, *Inorg. Chim. Acta*, **363**, pp. 1500–1506.

155. Bassil, B.S., Dickman, M.H., von der Kammer, B. and Kortz, U. (2007). The monolanthanide-containing silicotungstates $[Ln(\beta_2\text{-}SiW_{11}O_{39})_2]^{13-}$ (Ln = La, Ce, Sm, Eu, Gd, Tb, Yb, Lu): a synthetic and structural investigation, *Inorg. Chem.*, **46**, pp. 2452–2458.

156. Luo, Q., Howell, R.C., Dankova, M., Bartis, J., Williams, C.W., DeW. Horrocks, W., Jr., Young, V.G., Jr., Rheingold, A.L., Francesconi, L.C. and Antonio, M.R. (2001). Coordination of rare-earth elements in complexes with monovacant Wells–Dawson polyoxoanions, *Inorg. Chem.*, **40**, 1894–1901.

157. Sadakane, M., Dickman, M.H. and Pope, M.T. (2000). Controlled assembly of polyoxometalate chains from lacunary building blocks and lanthanide-cation linkers, *Angew. Chem. Int. Ed.*, **39**, pp. 2914–2916.

158. Artetxe, B., Reinoso, S., San Felices, L., Lezama, L., Gutiérrez-Zorrilla, J.M., García, J.A., Galán-Mascarós, J. R., Haider, A., Kortz U. and Vicent, C. (2014). Cation-directed dimeric versus tetrameric assemblies of lanthanide-stabilized dilacunary Keggin tungstogermanates, *Chem. Eur. J.*, **20**, pp. 12144–12156.

159. Fang, X., Anderson, T.M., Benelli, C. and Hill, C.L. (2005). Polyoxometalate-supported Y– and Yb^{III}–hydroxo/oxo clusters from carbonate-assisted hydrolysis, *Chem. Eur. J.*, **11**, pp. 712–718.

160. Ma, P., Wan, R., Wang, Y., Hu, F., Zhang, D., Niu, J. and Wang, J. (2016). Coordination-driven self-assembly of a 2D graphite-like framework constructed from high-nuclear Ce10 cluster encapsulated polyoxotungstates, *Inorg. Chem.*, **55**, pp. 918–924.

161. Howell, R.C., Perez, F.G., Jain, S., DeW. Horrocks, W., Jr., Rheingold, A.L. and Francesconi, L.C. (2001). A new type of heteropolyoxometalates formed from lacunary polyoxotungstate ions and Europium or Yttrium cations, *Angew. Chem. Int. Ed.*, **40**, pp. 4031–4034.

162. Fukaya, K. and Yamase, T. (2003). Alkali-metal-controlled self-assembly of crown-shaped ring complexes of lanthanide/

$[\alpha\text{-AsW}_9O_{33}]^{9-}$: $[K\subset\{Eu(H_2O)_2(\alpha\text{-AsW}_9O_{33})\}_6]^{35-}$ and $Cs\subset\{Eu(H_2O)_2(\alpha\text{-AsW}_9O_{33})\}_4]^{23-}$, *Angew. Chem. Int. Ed.*, **42**, pp. 654–658.

163. Liu, J.C., Han, Q., Chen, L.-J., Zhao, J.-W., Streb, C. and Song, Y.-F. (2018). Aggregation of giant cerium-bismuth tungstate clusters into a 3D porous framework with high proton conductivity, *Angew. Chem. Int. Ed.*, **57**, pp. 8416–8420.

164. Fang, X., Kögerler, P., Furukawa, Y., Speldrich, M. and Luban, M. (2011). Molecular growth of a core-shell polyoxometalate, *Angew. Chem. Int. Ed.*, **50**, pp. 5212–5216.

165. Wassermann, K., Dickman, M.H. and Pope, M.T. (1997). Self-assembly of supramolecular polyoxometalates: the compact, water-soluble heteropolytungstate anion $[As^{III}_{12}Ce^{III}_{16}(H_2O)_{36}W_{148}O_{524}]^{76-}$, *Angew. Chem. Int. Ed.*, **36**, pp. 1445–1448.

166. Bassil, B.S., Dickman, M.H., Römer, I., von der Kammer, B. and Kortz, U. (2007). The tungstogermanate $[Ce_{20}Ge_{10}W_{100}O_{376}(OH)_4(H_2O)_{30}]^{56-}$: a polyoxometalate containing 20 cerium(III) atoms, *Angew. Chem. Int. Ed.*, **46**, pp. 6192–6195.

167. Hussain, F., Conrad, F. and Patzke, G.R. (2009). A gadolinium-bridged polytungstoarsenate(III) nanocluster: $[Gd_8As_{12}W_{124}O_{432}(H_2O)_{22}]^{60-}$, *Angew. Chem. Int. Ed.*, **48**, pp. 9088–9091.

168. Reinoso, S., Giménez-Marqués, M., Galán-Mascarós, J.R., Vitoria, P. and Gutiérrez-Zorrilla, J.M. (2010). Giant crown-shaped polytungstate formed by self-assembly of Ce^{III}-stabilized dilacunary Keggin fragments, *Angew. Chem. Int. Ed.*, **49**, pp. 8384–9091.

169. Artetxe, B., Reinoso, S., San Felices, L., Gutiérrez-Zorrilla, J.M., García, J.A., Haso, F., Liu, T. and Vicent, C. (2015). Crown-shaped tungstogermanates as solvent-controlled dual systems in the formation of vesicle-like assemblies, *Chem. Eur. J.*, **21**, pp. 7736–7745.

170. Wang, K.-Y., Bassil, B.S., Lin, Z., Römer, I., Vanhaecht, S., Parac-Vogt, T.N., Sáenz de Pipaón, C., Galán-Mascarós, J.R., Fan, L., Cao, J. and Kortz, U. (2015). Ln_{12}-containing 60-tungstogermanates: synthesis, structure, luminescence, and magnetic studies, *Chem. Eur. J.*, **21**, pp. 18168–18176.

171. Dolbecq, A., Dumas, E., Mayer, C.R. and Mialane, P. (2010). Hybrid organic–inorganic polyoxometalate compounds: from structural diversity to applications, *Chem. Rev.*, **110**, pp. 6009–6048.

172. Knoth, W.H. (1979). Derivatives of heteropolyanions. 1. Organic derivatives of $W_{12}SiO_{40}^{4-}$, $W_{12}PO_{40}^{3-}$, and $Mo_{12}SiO_{40}^{4-}$, *J. Am. Chem. Soc.*, **101**, pp. 759–760.

173. Izzet, G., Abécassis, B., Brouri, D., Piot, M., Matt, B., Serapin, S.A., Bo, C. and Proust, A. (2016). Hierarchical self-assembly of polyoxometalate-based hybrids driven by metal coordination and electrostatic interactions: from discrete supramolecular species to dense monodisperse nanoparticles, *J. Am. Chem. Soc.*, **138**, pp. 5093–5099.

174. Zhang, T., Mazaud, L., Chamoreau, L.-M., Paris, C., Proust, A. and Guillemot, G. (2018). Unveiling the active surface sites in heterogeneous titanium-based silicalite epoxidation catalysts: input of silanol-functionalized polyoxotungstates as soluble analogues, *ACS Catal.*, **8**, pp. 2330–2342.

175. Chen, W., Huang, L., Hu, J., Li, T., Jia, F. and Song, Y.-F. (2014). Connecting carbon nanotubes to polyoxometalate clusters for engineering high-performance anode materials, *Phys. Chem. Chem. Phys.*, **16**, pp. 19668–19673.

176. Lombana, A., Rinfray, C., Volatron, F., Izzet, G., Battaglini, N., Alves, S., Decorse, P., Lang, P. and Proust, A. (2016). Surface organization of polyoxometalate hybrids steered by a 2D supramolecular PTCDI/melamine network, *J. Phys. Chem. C*, **120**, pp. 2837–2845.

177. Villanneau, R., Marzouk, A., Wang, Y., Djamaa, A.B., Laugel, G., Proust, A. and Launay, F. (2013). Covalent grafting of organic–inorganic polyoxometalates hybrids onto mesoporous SBA-15: a key step for new anchored homogeneous catalysts, *Inorg. Chem.*, **52**, pp. 2958–2965.

178. Hampson, E., Cameron, J.M., Amin, S., Kyo, J., Watts, J.A., Oshio, H and Newton, G.N. (2019). Asymmetric hybrid polyoxometalates: a platform for multifunctional redox-active nanomaterials, *Angew. Chem. Int. Ed.*, **58**, pp. 18281–18285.

179. Carraro, M., Sartorel, A., Scorrano, G., Maccato, C., Dickman, M.H., Kortz, U. and Bonchio, M. (2008). Chiral Strandberg-type molybdates $[(RPO_3)_2Mo_5O_{15}]^{2-}$ as molecular gelators: self-assembled fibrillar nanostructures with enhanced optical activity, *Angew. Chem. Int. Ed.*, **47**, pp. 7275–7279.

180. Breen, J.M. and Schimitt, Dr. W. (2008). Hybrid organic–inorganic polyoxometalates: functionalization of V^{IV}/V^{V} nanosized clusters to produce molecular capsules, *Angew. Chem. Int. Ed.*, **47**, pp. 6904–6908.

181. Knoth, W.H. (1979). Derivatives of heteropolyanions. 2. Metal-metal-bonded derivatives, *J. Am. Chem. Soc.*, **101**, pp. 2211–2213.

182. Micoine, K., Hasenknopf, B., Thorimbert, S., Lacôte, E. and Malacria, M. (2009). Chiral recognition of hybrid metal oxide by peptides, *Angew. Chem. Int. Ed.*, **48**, p. 3466.

183. Ortiz, M., Debela, A.M., Svobodova, M., Thorimbert, S., Lesage, D., Cole, R.B., Hasenknopf, B. and O'Sullivan, C.K. (2017). PCR incorporation of polyoxometalate modified deoxynucleotide triphosphates and their application in molecular electrochemical sensing of *Yersinia pestis*, *Chem. Eur. J.*, **23**, pp. 10597–10603.

184. Hussain, F., Kortz, U. and Reicke, M. (2005). The ball-shaped heteropolytungstates $[\{Sn(CH_3)_2(H_2O)\}_{24}\{Sn(CH_3)_2\}_{12}(A\text{-}XW_9O_{34})_{12}]^{36-}$, *Angew. Chem. Int. Ed.*, **44**, pp. 3773–3777.

185. Ma, T., Yang, P., Dammann, I., Lin, Z., Mougharbel, A.S., Li, M.-X., Adăscăliţei, F., Mitea, R., Silvestru, C., Thorstenson, C., Ullrich, M.S., Cseh, K., Jakupec, M.A., Keppler, B.K., Donalisio, M., Caalli, R., Lembo, D. and Kortz, U. (2020). Tetra-(p-tolyl)antimony(III)-containing heteropolytungstates, $[\{(p\text{-tolyl})Sb^{III}\}_4(A\text{-}\alpha\text{-}XW_9O_{34})_2]^{n-}$ (X = P, As, or Ge): synthesis, structure, and study of antibacterial and antitumor activity, *Inorg. Chem.*, **59**, pp. 2978–2987.

186. Yvon, C., Surman, A.J., Hutin, M., Alex, J., Smith, B.O., Long, D.-L. and Cronin, L. (2014). Polyoxometalate clusters integrated into peptide chains and as inorganic amino acids: solution- and solid-phase approaches, *Angew. Chem., Int. Ed. Engl.*, **126**, pp. 3404–3409.

187. Vanhaecht, S., Jacobs, J., Van Meervelt, L. and Parac-Vogt, T.N. (2015). A versatile and highly efficient post-functionalization method for grafting organic molecules onto Anderson-type polyoxometalates, *Dalton Trans.*, **44**, pp. 19059–19062.

188. Xu, W., Pei, X., Diercks, C.S., Lyu, H., Ji, Z. and Yaghi, O.M. (2019). A metal–organic framework of organic vertices and polyoxometalate linkers as a solid-state electrolyte, *J. Am. Chem. Soc.*, **141**, pp. 17522–17526.

189. Lin, C.-G-, Fura, G.D., Long, Y., Xuan, W. and Song, Y.-F. (2017). Polyoxometalate-based supramolecular hydrogels constructed through host–guest interactions, *Inorg. Chem. Front.*, **4**, pp. 789–794.

190. Rosnes, M.H., Musumeci, C., Pradeep, C.P., Mathieson, J.S., Long, D.-L., Song, Y.-F., Pignataro, B., Cogdell, R. and Cronin, L. (2010). Assembly of modular asymmetric organic–inorganic polyoxometalate hybrids into anisotropic nanostructures, *J. Am. Chem. Soc.*, **132**, pp. 15490–15492.

191. Hasenknopf, B., Delmont, R., Herson, P. and Gouzerh, P. (2002). Anderson-type heteropolymolybdates containing tris(alkoxo)

ligands: synthesis and structural characterization, *Eur. J. Inorg. Chem.*, pp. 1081–1087.

192. Zhang, J., Li, Q., Huang, Y., Zhang, J., Hao, J. and Wei, Y. (2016). The proton-controlled synthesis of unprecedented diol functionalized Anderson-type POMs, *Chem. Commun.*, **52**, pp. 2378–2381.

193. Wang, Y., Li, B. Qian, H. and Wu, L. (2016). Controlled triol-derivative bonding and decoration transformation on Cu-centered Anderson–Evans polyoxometalates, *Inorg. Chem.*, **55**, pp. 4271–4277.

194. Zhang, J., Huang, Y., Hao, J. and Wei, Y. (2017). β-$\{Cr[RC(CH_2O)_3]_2 Mo_6O_{18}\}^{3-}$: the first organically-functionalized β isomer of Anderson-type polyoxometalates, *Inorg. Chem. Front.*, **4**, pp. 1215–1218.

195. Gumerova, N.I., Fraile, T.C., Roller, A., Giester, G., Pascual-Borràs, M., Ohlin, C.A. and Rompel, A. (2019). Direct single- and double-side triol-functionalization of the mixed type Anderson polyoxotungstate $[Cr(OH)_3W_6O_{21}]^{6-}$, *Inorg. Chem.*, **58**, pp. 106–113.

196. Pradeep, C.P., Long, D.-L., Newton, G.N., Song, Y.-F. and Cronin, L. (2008). Supramolecular metal oxides: programmed hierarchical assembly of a protein-sized 21 kDa $[(C_{16}H_{36}N)_{19}\{H_2NC(CH_2O)_3 P_2V_3W_{15}O_{59}\}_4]^{5-}$ polyoxometalate assembly, *Angew. Chem. Int. Ed.*, **47**, pp. 4388–4391.

197. Pradeep, C.P., Misdrahi, M.F., Li, F.-Y., Zhang, J., Xu, L., Long, D.-L., Liu, T. and Cronin, L. (2009). Synthesis of modular "inorganic–organic–inorganic" polyoxometalates and their assembly into vesicles, *Angew. Chem. Int. Ed.*, **48**, pp. 8309–8313.

198. Lachkar, D., Vilona, D., Dumont, E., Lelli, M. and Lacôte, E. (2016). Grafting of secondary diolamides onto $[P_2W_{15}V_3O_{62}]^{9-}$ generates hybrid heteropoly acids, *Angew. Chem. Int. Ed.*, **55**, pp. 5961–5965.

199. Zhu, Y., Huang, Y., Li, Q., Zang, D., Gu, J., Tang, Y. And Wei, Y. (2020). Polyoxometalate-based photoactive hybrid: uncover the first crystal structure of covalently linked hexavanadate-porphyrin molecule, *Inorg. Chem.*, **59**, pp. 2575–2583.

200. Müller, A., Meyer, J., Mögge, H., Stammer, A. and Botar, A.Z. (1995). *Cis-/trans*-isomerie bei bis-(trisalkoxy)-hexavanadaten: cis-$Na_2[V^{IV}_6O_7(OH)_6\{(OCH_2)_3CCH_2OH\}_2]\cdot 8H_2O$, cis-$(CN_3H_6)_3[V^{IV}V_5{}^{V}O_{13}\{(OCH_2)_3CCH_2OH\}_2]\cdot 4{,}5H_2O$ and trans-$(CN_3H_6)_2[V^{V}_6O_{13}\{(OCH_2)_3 CCH_2OH\}_2]\cdot H_2O$, *Anorg. Allg. Chem.*, **621**, pp. 1818–1831.

201. Du, Y., Rheingold, A.L. and Maatta, E.A. (1992). A polyoxometalate incorporating an organoimido ligand: preparation and structure of $[Mo_5O_{18}(MoNC_6H_4CH_3)]^{2-}$, *J. Am. Chem. Soc.*, **114**, pp. 345–346.

202. Zhang, J., Hao, J., Wei, Y., Xia, F., Yin, P. and Wang, L. (2010). Nanoscale chiral rod-like molecular triads assembled from achiral polyoxometalates, *J. Am. Chem. Soc.*, **132**, pp. 14–15.

203. Wei, Y., Xu, B., Barnes, C.L. and Peng, Z. (2001). An efficient and convenient reaction protocol to organoimido derivatives of polyoxometalates, *J. Am. Chem. Soc.*, **123**, pp. 4083–4084.

204. Zhang, J., Xiao, F., Hao, J. and Wei, Y. (2012). The chemistry of organoimido derivatives of polyoxometalates, *Dalton Trans.*, **41**, pp. 3599–3615.

205. Kwen, H., Young, V.G. and Maatta, E.A. (1999). A diazoalkane derivative of a polyoxometalate: preparation and structure of $[Mo_6O_{18}(NNC(C_6H_4OCH_3)CH_3)]^{2-}$, *Angew. Chem. Int. Ed.*, **38**, pp. 1145–1146.

206. Liu, H., Gómez-García, C.J., Peng, J., Sha, J., Li, Y. and Yan, Y. (2008). 3D transition metal mono-substituted Keggin polyoxotungstate with an antenna molecule: synthesis, structure and characterization, *Dalton Trans.*, pp. 6211–6218.

207. Artetxe, B., Reinoso, S., San Felices, L. Vitoria, P., Pache, A., Martín-Caballero, J. and Gutiérrez-Zorrilla, J. M. (2015). Functionalization of Krebs-Type polyoxometalates with N,O chelating ligands: a systematic study, *Inorg. Chem.*, **54**, pp. 241–252.

208. Fang, X., Anderson, T.M. and Hill, C.L. (2005). Enantiomerically pure polytungstates: chirality transfer through zirconium coordination centers to nanosized inorganic clusters, *Angew. Chem. Int. Ed.*, **44**, pp. 3540–3544.

209. Rousseau, G., Oms, O., Dolbecq, A., Marrot, J. and Mialane, P. (2011). Route for the elaboration of functionalized hybrid 3d substituted trivacant Keggin anions, *Inorg. Chem.*, **50**, pp. 7376–7378.

210. Zheng, S.-T., Zhang, J. and Yang, G.-Y. (2008). Designed Synthesis of POM–Organic Frameworks from $\{Ni_6PW_9\}$ Building blocks under hydrothermal conditions, *Angew. Chem. Int. Ed.*, **47**, pp. 3909–3913.

211. Ritchie, C., Moore, E.G., Speldrich, M., Kögerler, P. and Boskovic, C. (2010). Terbium polyoxometalate organic complexes: correlation of structure with luminescence properties. *Angew. Chem. Int. Ed.*, **49**, pp. 7702–7705.

212. Dolbecq, A., Mialane, P., Lisnard, L., Marrot, J. and Sécheresse, F. (2003). Hybrid organic-inorganic 1D and 2D frameworks with ε-Keggin polyoxomolybdates as building blocks, *Chem. Eur. J.*, **9**, pp. 2914–2920.

Chapter 2

Polyoxometalate Macroions in Solution

Jiahui Chen and Tianbo Liu

Department of Polymer Science,
The University of Akron, Akron, Ohio 44313, USA

tliu@uakron.edu

Polyoxometalate (POM) clusters, as examples of macroions, demonstrate special solution behaviors different from smaller simple ions or larger colloids, due to the size dispersity between POM and their counterions. The interactions of the POMs and the counterions are essential for their solution behavior. By manipulating the counterion–POM interactions, solution behaviors of POMs can be rationally controlled. The counterions of POMs can possess different degrees of association with the POMs, determined by the solvent polarity, ionic strength, counterion concentration, the charge density of the POMs, hydration level, and valence of the counterions, etc. In dilute solution, when counterions have modest degree of association, POMs can self-assembly into single-layered, hollow, spherical "blackberry"-type structures, mainly due to the counterion-mediated electrostatic attraction between the POM macroions. When other non-covalent interactions also contribute, or charge is not isotropically distributed on the POM surface, different self-assembly behaviors

Polyoxometalates: Advances, Properties, and Applications
Edited by Leire Ruiz Rubio, José Luis Vilas Vilela, Beñat Artetxe, and Juan Manuel Gutiérrez-Zorrilla

ISBN 978-981-4968-14-0 (Hardcover), 978-1-003-27744-6 (eBook)
www.jennystanford.com

can be observed. POMs and their blackberry structures show interesting properties usually believed to belong only to complex biomacromolecules, such as ion specific effects connecting to Hofmeister series, similar features with the virus capsid formation, selective cross-membrane ion transportation, the self-recognition and chiral selection phenomenon, making the POMs ideal models for understanding more complicated charged solution systems.

2.1 Introduction

Polyoxometalates (POMs) are a large group of metal oxide clusters consisting of transition metals (commonly form the groups 5 or 6) and heteroatoms (usually B, Al, Si, P, etc.) [1, 2]. The oxo-polyhedral of these elements forms POM by sharing their corners, edges, or faces. Versatile assembly of metal oxide polyhedrons result in rich structures diversities and abundant functionalities, which have drawn interests from different fields of applications including catalysis, photochemistry, electrochemistry, medicine, etc. [3]. Important POM structures and their formulas introduced in this chapter are listed in Fig. 2.1 and Table 2.1, respectively.

POMs show interesting solution behaviors. In solution, many POMs are negatively charged and fully hydrophilic, due to the excess of oxo ligands over metal ions. Some POMs are neutral in their crystalline form, but in solution their surface water ligands can partially deprotonate and make the cluster negatively charged as well [18]. Therefore, many types of POMs exist as macroanions in aqueous solution (or other polar solvents). With well-defined structures, no polydispersity, tunable charges, and no intra-molecular charge interactions [19], the POMs are valuable models shedding light on more complicated macroions including polyelectrolytes and biomacromolecules. The solution behaviors of macroions are different from smaller simple ions or larger colloids. Most importantly, the role of counterions is far beyond just charge balancing [20]. Compared with small, simple ions (e.g., Cl^-), as macroions are large enough to show the size disparity with their counterions, the interaction between macroions and counterions cannot be accurately described by point charges models used by Debye–Hückel theory [21].

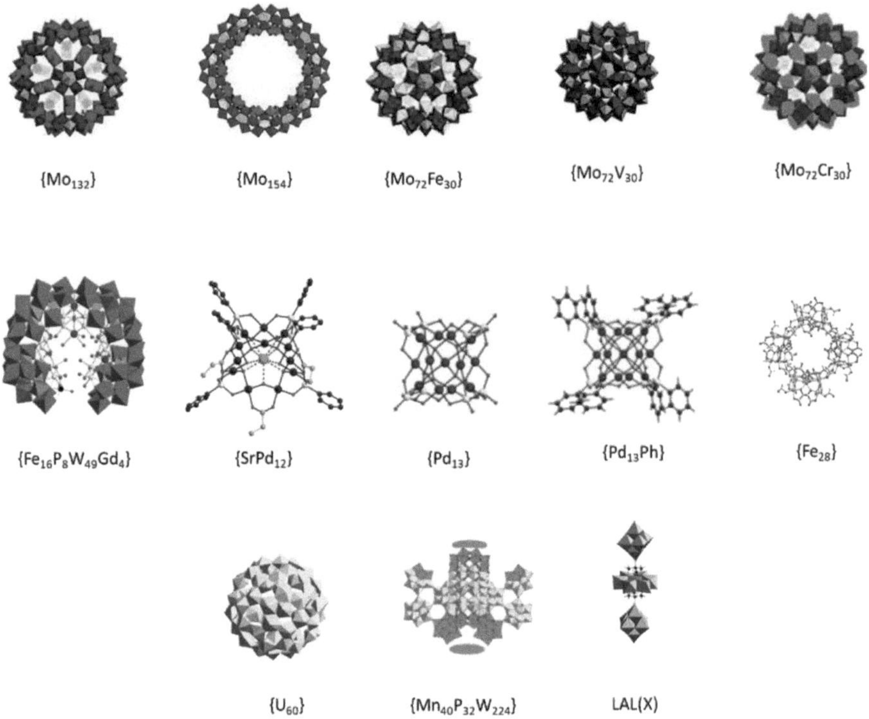

Figure 2.1 Structures of some typical POM clusters.

On the other hand, in Derjaguin–Landau–Verwey–Overbeek (DLVO) theory for colloids suspensions, the interactions among colloids are described as a combination of repulsive electrostatic interaction and attractive van der Waals forces [22]. Condensing around colloid surfaces, counterions form electric double layers along with co-ions and screen the electrostatic repulsion. However, macroions form "real" solutions, different from the thermodynamically unstable suspension of colloids, and the size disparity of counterions with macroions is not as significant as that with colloids. Therefore, counterions may not be suitable to apply point charges models and to omit the details of ions (hydration, size, electronegativity, etc.), as specific ion effects have been widely observed in macroionic systems [23, 24]. In addition to above common features of macroions, the formation of ion-pair between POMs and counterions has been observed to have great influence on the solution behavior of the POMs [20]. Overall in such dilute solutions we consider the interactions among the large POM anions, small countercations, co-ions (small anions), and the hydration layers of such ions. These unique features can lead to strong attraction among POM anions, resulting in a unique self-assembly process of POMs in solution, which will be covered in this chapter.

Table 2.1 Chemical formulae of some POM clusters

Abbreviation	Formula	Refs
$\{Mo_{154}\}$	$Na_{15}[Mo_{154}O_{462}H_{14}(H_2O)_{70}]_{0.5}[Mo_{152}O_{457}H_{14}(H_2O)_{68}]_{0.5} \cdot$ ca. 400 H_2O	[4]
$\{Mo_{132}\}$	$(NH_4)_{42}[\{(Mo^{VI})Mo_5^{VI}O_{21}(H_2O)_6\}_{12}\{Mo_2O_4(CH_3COOH)\}_{30}] \cdot$ ca. 300 $H_2O \cdot$ ca. 10 CH_3COONH_4	[5]
$\{Mo_{72}Fe_{30}\}$	$[Mo_{72}Fe_{30}O_{252}(CH_3COO)_{12}\{Mo_2O_7(H_2O)\}_2\{H_2Mo_2O_8(H_2O)\}(H_2O)_{91}] \cdot$ ca. 150 H_2O	[6–9]
$\{Mo_{72}Cr_{30}\}$	$[\{Na(H_2O)_{12}\} \subset \{Mo_{72}Cr_{30}O_{252}(CH_3COO)_{19}(H_2O)_{94}\}] \cdot$ ca. 120 H_2O	[8, 9]
$\{Mo_{72}V_{30}\}$	$Na_8K_{14}(VO)_2[\{(Mo^{VI})Mo_5^{VI}O_{21}(H_2O)_3\}_{10}\{(Mo^{VI})Mo_5^{VI}O_{21}(H_2O)_3(SO_4)\}_2\{V^{IV}O(H_2O)\}_{20}\{V^{IV}O\}_{10}(\{KSO_4\}_5)_2]$ ca. 150 H_2O	[10]
$\{Mn_{40}P_{32}W_{224}\}$	$K_{56}Li_{74}H_{22}[Mn_{40}{}^{III}P_{32}W_{224}{}^{VI}O_{888}]$ ca. 680 H_2O	[11]
$\{SrPd_{12}\}$	$Na_4[SrPd_{12}O_6(OH)_3(PhAsO_3)_6(OAc)_3]$	[12]
$\{Fe_{28}\}$	$Ba_{10}[Fe_{28}(u_3\text{-}O)_8(L/D\text{-}Tart)_{16}(HCOO)_{24}]$	[13]
LAL(X)	$((C_4H_9)_4N)_7[Mo_6O_{18}NC(OCH_2)_3XMo_6O_{18}(OCH_2)_3CNMo_6O_{18}]$, X=$Mn^{III}/Fe^{III}$	[14]
$\{Pd_{13}\}$	$[Pd_{13}^{II}As^{V}{}_8O_{34}(OH)_6]^{8-}$	[15]
$\{Pd_{13}Ph\}$	$[Pd_{13}^{II}(As^{V}Ph)_8O_{32}]^{6-}$	[15]
$\{Fe_{16}P_8W_{49}Gd_4\}$	$K_9LiNa[Fe_{16}O_2(OH)_{23}(H_2O)_9(P_8W_{49}O_{189})Gd_4(H_2O)_{20}]$	[11]
$\{U_{60}\}$	$Li_{48+m}K_{12}(OH)_m[UO_2(O_2)(OH)]^{60-}(H_2O)_n$ ($m \sim 20$ and $n \sim 310$)	[16, 17]

2.2 The Self-Assembly of POM Macroions into Blackberry Structures

2.2.1 Experimental Methods to Characterize the Self-Assembly of POM Macroions into Blackberry Structures

It has been observed that although some POMs are fully hydrophilic and give clear, stable solutions in polar solvents, they can unexpectedly self-assemble into vesicle-like supramolecular structures. The self-assembly process and the structure of the assembly can be characterized by laser light scattering (LLS) including static and dynamic light scattering (SLS and DLS), transmission electron microscopy (TEM), and atomic force microscopy (AFM) techniques. SLS study measures the scattered intensities from sample solution at different scattering angles, and SLS results analyzed by Zimm plot can provide information on weight average molecular weight (M_w) and radius of gyration (R_g) of the assembly in the solution. DLS study measures the intensity–intensity time correlation function. Analyzed by CONTIN method and transformed by Stokes–Einstein equation, DLS results can provide information about the average hydrodynamic radius (R_h) and particle size distribution of the assembly in solution. TEM and AFM allow imaging of assembly. Take a typical case [4] of POM self-assembling in dilute aqueous solution as an example: TEM study of 0.010 mg/mL $\{Mo_{154}\}$ aqueous solution at pH = 3.0 shows the existence of spherical objects with the size of ca. 90 nm (Fig. 2.2), which is consistent with the R_g and R_h values (45 nm) obtained from SLS and DLS studies, respectively. The R_g/R_h ratio is about 1, suggesting all the mass of these assemblies distributes on their surface, that is, they are hollow spheres. From the Zimm plot, the M_w of the assembly is about $(2.54 \pm 0.25) \times 10^7$ g/mol, equals to ~1150 $\{Mo_{154}\}$ in an assembly. Based on these results, a single-layered, hollow, spherical model to describe the assembly in the solutions was proposed, and a nickname of "blackberry" was given to such structures. The single-layered feature of blackberry was later confirmed by

AFM study on the blackberry structures formed by other POMs [25, 26]. Such self-assembly process has been well-known in POM communities, and similar self-assembly phenomenon has been observed in many macroionic systems, including POMs [19] (the major aspect of this chapter), POM-organic hybrid [27], metal–organic cages [28–31], actinides peroxide clusters [16, 32], functionalized C_{60} [33], functionalized cyclodextrin [34], and dendrimers [35], etc.

2.2.2 The Major Driving Forces of Blackberry Formation: Counterion-Mediated Attraction

As POM anions are usually highly negatively charged and are supposed to possess strong electrostatic repulsion between each other, such self-assembly is unexpected. Therefore, to identify the driving forces of the blackberry formation is the key to understand their solution behaviors. Although blackberry structures formed by hydrophilic POMs look similar to lipid vesicles driven by hydrophobic interaction, the driving forces are fundamentally different, as the hydrophilic POMs may not possess any hydrophobic domains.

Following studies have shown that, in most cases, the van der Waals forces (the major attractions between colloids, strongly dependent on the solute size) are not significant for macroions, and the major driving force of the self-assembly of macroions into blackberry is counterion-mediated attraction.

A typical experiment is the self-assembly of a Keplerate POM $\{Mo_{132}\}$ [36] clusters. $\{Mo_{132}\}$ has a hollow, spherical structure with a diameter of 2.9 nm. "Keplerate" refers to a series of spherical, hollow POMs structurally analogous to C_{60}, that is, constructed by 12 pentagons ($\{(Mo^{VI})Mo_5^{VI}\}$ or $\{(W^{VI})W_5^{VI}\}$ units) and 30 linkers (form 20 hexagons, for $\{Mo_{132}\}$ the linkers are $\{Mo_2^{V}O_4\}$). A thorough introduction of Keplerates can be found in literature [37]. Balanced by ammonium counterions, $\{Mo_{132}\}$ carries 42 negative charges in its crystalline form. In solution, $\{Mo_{132}\}$ are very soluble, and degree of the counterion association varies depending on solvent dielectric constant or the concentration of $\{Mo_{132}\}$, leading to $\{Mo_{132}\}$ macroanions of different charge densities [38]. In high dielectric constant

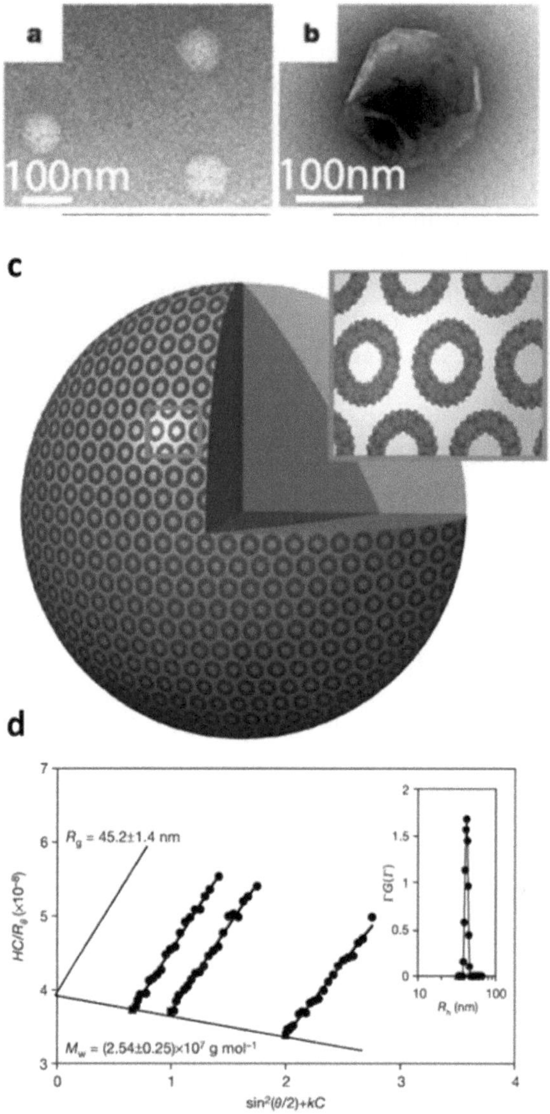

Figure 2.2 TEM images (a, b) of dilute aqueous solution of $\{Mo_{154}\}$ show the existence of spherical, ~45 nm radius assemblies. Zimm plot (c) based on the SLS study of the $\{Mo_{154}\}$ aqueous solutions; (inset) CONTIN analysis on the DLS result of the same solution. (d) Schematic present shows the supramolecular blackberry structure formed by $\{Mo_{154}\}$ macroions in aqueous solution. Adapted with permission from reference [4]. Copyright 2003 Springer.

In high dielectric constant solvents or low concentration of $\{Mo_{132}\}$, most of the counterions are free, result in $\{Mo_{132}\}$ anions of high charge densities. Meanwhile, in low dielectric constant solvents, for example, acetone, or $\{Mo_{132}\}$ concentration is high, the strong counterion condensation will result in $\{Mo_{132}\}$ with very low the effective charges. For 1.0 mg/mL $\{Mo_{132}\}$ acetone/water solutions, in pure water $\{Mo_{132}\}$ stays as discrete ion, in solvents containing 3–70 vol % acetone blackberry structures are observed, and with higher acetone vol % $\{Mo_{132}\}$ again stay as discrete ion (or neutral molecule) [5] (Fig. 2.2). If the van der Waals forces between $\{Mo_{132}\}$ anions are the major driving force for the self-assembly, due to the fact that in high vol % acetone solvents $\{Mo_{132}\}$ clusters are less charged and the electric repulsions between them are weaker, the overall inter-$\{Mo_{132}\}$ attraction should be stronger. However, this assumption is contradicted to the experimental observations, which exclude the possibility that van der Waals forces are the major driving forces for the self-assembly. The sizes of blackberries increase along with the increase of the ratio of acetone/water, that is, the decrease of charges carried by $\{Mo_{132}\}$, suggesting a charge regulated self-assembly process. Similar phenomena have been observed in the systems that the charge densities of macroions are also controlled by solvent dielectric constant, where the assembly sizes show linear relationships with inverse of the solvent dielectric constants (Figs. 2.3 and 2.4).

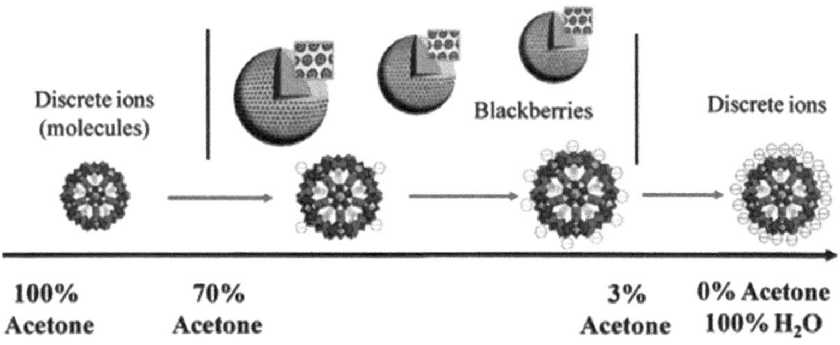

Figure 2.3 Self-assembly of $\{Mo_{132}\}$ in water/acetone mixture solvent. The presence of $\{Mo_{132}\}$ blackberries was found in 3 ~ 70% (volume percent) acetone. The average hydrodynamic radius of blackberries increased from 45 to 100 nm with increasing acetone content. Adapted with permission from reference [5]. Copyright 2007 ACS.

The correlation between the blackberry structure size and the pH-dependent charge densities of POM macroions is also observed [18]. $\{Mo_{72}Fe_{30}\}$ is another Keplerate POM (2.5 nm in diameter) with surface water ligands of weak acidity. It is charge neutral in its crystalline form, and in solution, the partially deprotonate of the water ligands make $\{Mo_{72}Fe_{30}\}$ negatively charged.

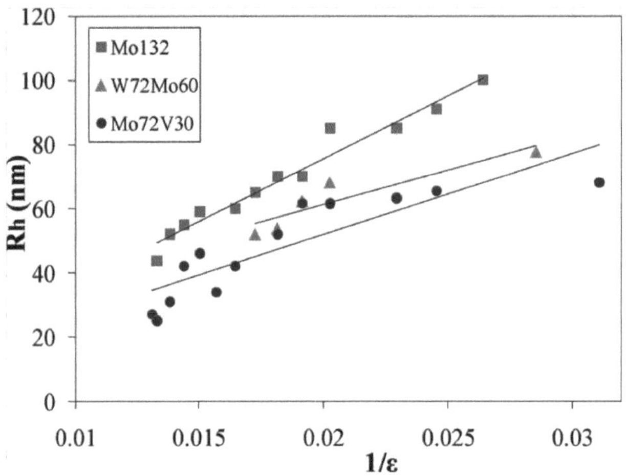

Figure 2.4 Linear relationship between the average blackberry radius (R_h) vs the inversed dielectric constant ($1/\varepsilon$) of the solvent for various POM macroions in water/acetone mixed solvents. Adapted with permission from reference [39]. Copyright 2009 ACS.

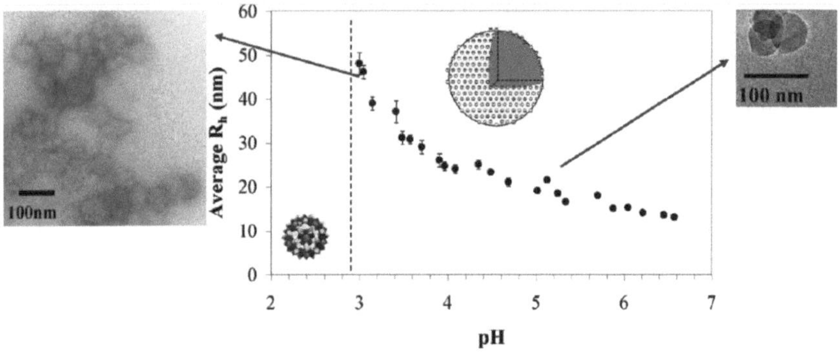

Figure 2.5 The average blackberry radius (R_h) vs the pH in 0.5 mg/mL aqueous solutions of $\{Mo_{72}Fe_{30}\}$ at different pH. TEM images of aggregates on carbon film formed at pH ~ 3.0 (left) and pH ~ 4.6 (right) Adapted with permission from reference [18]. Copyright 2006 ACS.

Therefore, the charge density of $\{Mo_{72}Fe_{30}\}$ depends on pH. There are some studies indicating hydrogen bonds and water bridges also contribute to intermolecular interactions of POMs in their crystalline form and in solutions [8, 40–43]. However, for 0.5 mg/mL of $\{Mo_{72}Fe_{30}\}$ aqueous solution, no blackberry was observed in solution of pH < 2.9 where $\{Mo_{72}Fe_{30}\}$ is almost charge neutral and yet still covered by water ligands, suggest both van der Waals forces and hydrogen bonds cannot be the major driving forces. The sizes of blackberries decrease with pH increasing, that is, $\{Mo_{72}Fe_{30}\}$ carrying more charges, which further confirms that the charge effects are important (Fig. 2.5).

Similar to polyelectrolytes and biomacromolecules [44, 45], counterion condensation around macroions and ion-pair formation between counterions and macroions have been observed [20]. More evidences from small-angle X-ray scattering (SAXS) measurements and molecular dynamics (MD) simulations show the importance of counterion–macroion interactions on the self-assembly and confirm that the major driving force of the blackberry formation is the counterion-mediated attractions.

Due to the high atomic numbers of the elements of the counterion and POMs, SAXS technique is able to resolve their spatial distributions and correlations with high resolutions. Increase of radius of gyration (R_g) obtained from Guinier plot and change of the shape of distance pair distribution functions ($p(r)$) by the partial Fourier transformation of the SAXS curves may indicate the existence of excess counterions around POMs. For example, various types of ion-pair formation between clusters of group V (Nb, Ta) have been confirmed by SAXS measurements [46, 47].

Figure 2.6 shows a SAXS experiment of a Keplerate POM, $\{Mo_{72}V_{30}\}$ (2.5 nm in diameter, carries 31 negative charges in its crystalline form), in water/acetone mixture solvents. For the dilute (0.1 mg/mL) aqueous solutions of $\{Mo_{72}V_{30}\}$, the $p(r)$ profiles match typical shapes of a hollow spherical object, indicate no significant counterion association can be observed. In the dilute aqueous solutions, $\{Mo_{72}V_{30}\}$ cannot self-assemble into blackberry as well. With the increase of acetone content in the solvents, additional peaks in $p(r)$ appear and become more significant in the range of 0.2–0.9 nm to the peak representing

$\{Mo_{72}V_{30}\}$, indicating more counterions are closely associated in that range. Under the same condition in the water/acetone mixed solvent, $\{Mo_{72}V_{30}\}$ was observed to self-assemble into blackberry, indicating the direct connections between the counterion association and the blackberry formation.

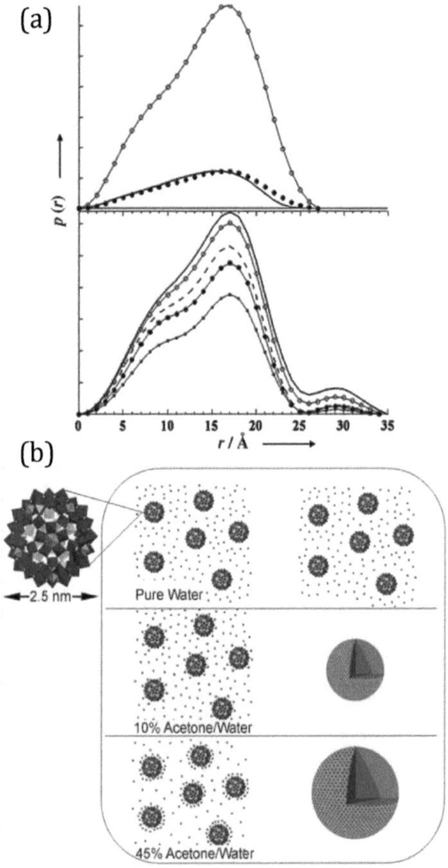

Figure 2.6 (a) Transition between discrete $\{Mo_{72}V_{30}\}$ and blackberries depends on the counterion condensation around the macroions, which can be controlled by decreasing the solvent polarity. (b) Distance distribution functions based on SAXS data for $\{Mo_{72}V_{30}\}$ solutions. Top: (○): 0.052 mM $\{Mo_{72}V_{30}\}$, (●): 0.013 mM $\{Mo_{72}V_{30}\}$, (—): $\{Mo_{72}V_{30}\}$ calculated. Bottom: 0.26 mM $\{Mo_{72}V_{30}\}$ in (—): 75% acetone/water, (○): 65% acetone/water, (- - -): 45% acetone/water, (···): 10% acetone/water, (□): in pure water. Adapted with permission from reference [10]. Copyright 2009 Wiley-VCH.

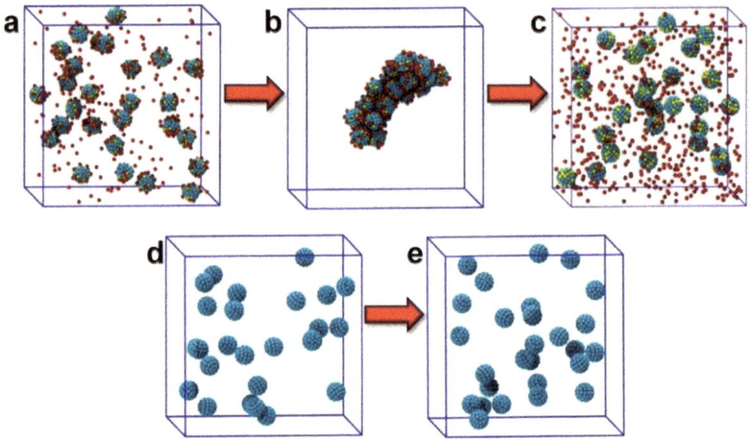

Figure 2.7 MD simulation snapshots for macroion in solution. (a) Initial status of charged macroions in solution. (b) Equilibrium status of charged macroions in solution. (c) Turning off the charges on the macroions and counterions after aggregate formation and running the simulation further results in a disassembled state. (d) Initial status of uncharged "macroions" in solution. (e) Equilibrium status of uncharged "macroions" in solution. Yellow: charged beads on macroions; cyan: uncharged beads on macroions; red: counterions. Reprint under Creative Commons Attribution 4.0 International License from reference [49].

Coarse-grained molecular dynamics (MD) simulations of macroionic systems also confirm some observations from experiments. There have been studies demonstrate solution behaviors of small POM ions by all-atom MD simulations [48]. However, due to large and complex structures, all-atom MD simulation of the POM macroions is difficult. Therefore, to demonstrate the nature of the attractive forces between macroions in solution, macroions were represented by spheres with small beads that are responsible for the electrostatic and van der Waals interactions, and systems containing macroions and counterions in solvent were simulated. After the equilibrium statuses were reached, a transit from the starting isolated macroions into aggregations was only observed in the system where macroions carried more than six charges. In the aggregations, counterions distributed among and around macroions and held the macroions together. Then by turning off all electrostatic interactions, the aggregation disassembled into isolated status (Fig. 2.7). These results confirm that counterion-mediated interactions, not van der Waals forces, are the major driving force for the self-assembly of macroions in solution. Further investigation by simulations show that purely isotropic charge distributions leads to three-dimensional (3D) aggregations, and a slightly anisotropy charge distribution preferring to the equator on the macroion may lead to the formation of two-dimensional (2D) monolayer structures, which are able to merge with each other and might be the precursor of a complete blackberry assembly (Fig. 2.8).

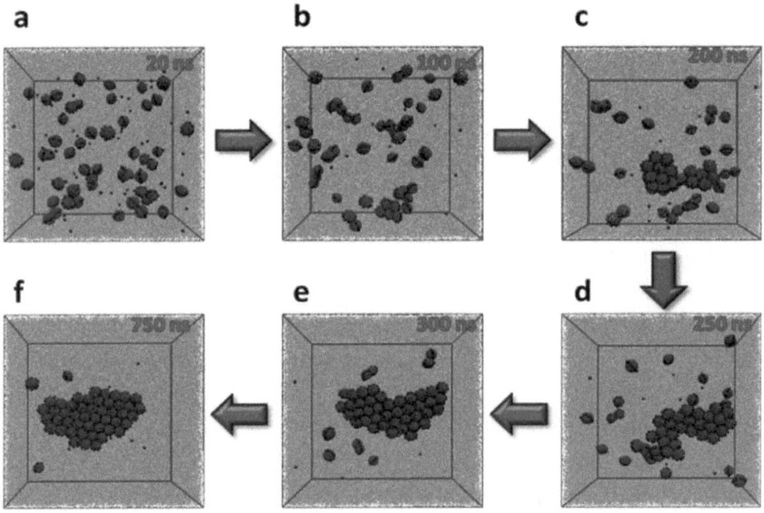

Figure 2.8 MD simulation snapshots for time evolution of the self-assembling process of 2D monolayers from macroion of anisotropic charge distribution. Reprint under Creative Commons Attribution 4.0 International License from reference [50].

The simulation and experimental evidences confirm the counterion-mediated attraction is the major driving force of the blackberry formation. The change of blackberry sizes can be reasoning as following [6]: the size of blackberry depends on the curvature of the blackberry "membrane." When the intermolecular distance between macroions on the membrane is short, the membrane become stiff and cannot tolerant high curvature, result in large blackberry, and the longer intermolecular distance result in small blackberry. The intermolecular distance is related to the intermolecular forces between macroion. When the solvent polarity decreases, stronger counterion association reduces the effective charges on the POM macroion, resulting in weaker electrostatic repulsion and stronger attraction between them. The weaker repulsion and stronger attraction leads to shorter intermolecular distance and consequently to larger blackberries [30, 31]. If there are other intermolecular interactions (besides counterion-mediated interactions) also involving in the self-assembly, the trends of blackberry may be different. Therefore, the size of blackberry can be rationally controlled, and POM macroions can be an effective model system to study the intermolecular interactions.

2.2.3 Counterion-Specific Effects in the Self-Assembly of POMs

In classical mean-field approach applied in DLVO theory and Debye–Hückel theory, counterions are usually treated as point ions, and solvents are implicitly described as continuous medium [51]. However, due to the obvious but not very significant size disparity between counterions and POM macroions, as well as the short intermolecular distance, such approach may be not valid in macroionic systems. Specific features of counterions, for example, hydration, size, and polarizability become important. The well-known phenomenon "Hofmeister Series" [52] (Fig. 2.9) ranks the impact of ions on solubilities of proteins, and later similar trends have been observed in many other systems in aqueous solutions [53]. The nature of the Hofmeister series is still in debate, but common believe is that such ranking is related to the hydration of counterions, co-ions, and macroions.

We also observed that different counterions can be recognized and selected by POM macroions, leading to specific effects on the self-assembly. Generally speaking, the ion association can be classified into three types of ion-pairs: (1) solvent-separated ion-pair, where the solvation shells of the ions remain intact; (2) solvent-shared ion-pair, where a solvent layer lays between the cation and anion; and (3) contact ion-pair, both the cation and anion lose their solvation shells upon association.

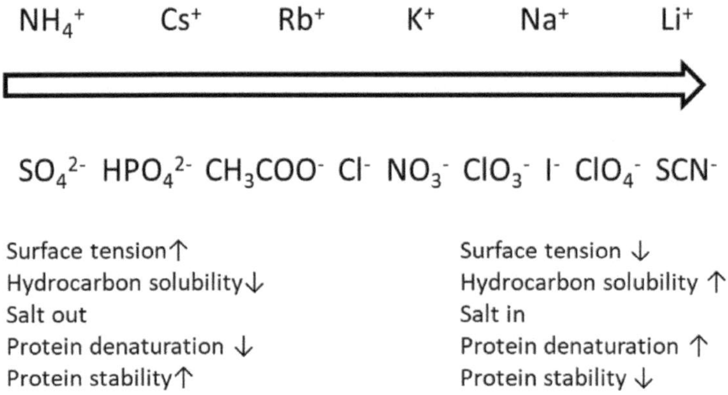

Figure 2.9 Scheme of Hofmeister series. Adapted with permission from reference [53]. Copyright 2014 Springer Nature.

Two major factors often control the interaction of counterions with a POM macroion: the valence and the hydration of the counterion. Higher valence of the counterion will result in stronger electrostatic interaction with the macroions. On the other hand, since the close association between counterions and POM macroions may break the hydration shell around them, counterions of the same valency but with thinner hydration shell are usually preferred by POM macroions. There are also some exceptions. For some POMs with nanoconfined binding sites, for example, the pores on Keplerate POMs surface [54] or lacunary sites on Keggin and Dawson POMs [55], they may prefer counterions of certain sizes that can fit into the pores. Recent study suggests that there can be some degrees of covalent features in the ion-pair between counterions and POM anions, leading to anomalous solubility trends of POMs with different cations [56]. The difference in the counterion–POM macroion interactions can also be reflected in the different self-assembly behaviors.

A study on the effects of multi-valent counterions on the solution behaviors of $\{Pd_{12}\}$ POM macroanions demonstrate the interplay between electrostatic interactions and the solvation of ions (Fig. 2.10) [57]. The LLS and isothermal titration calorimetry (ITC) measurements show that monovalent counterions (Na^+, K^+, Rb^+, and Cs^+) only weakly interact with $\{Pd_{12}\}$, resulting in solvent-separated ion-pairs or free ions; the interactions between some divalent counterions (Sr^{2+}, Ba^{2+}) with {Pd_{12}} are of moderate strength, making solvent-shared ion-pairs; trivalent counterions (Y^{3+}, Eu^{3+}, and Ce^{3+}) strongly interact with {Pd_{12}}, forming contact ion-pairs where the original water layer of {Pd_{12}} and counterions are destroyed. Only the divalent counterions (Sr^{2+}, Ba^{2+}) but not monovalent or trivalent counterions, can trigger the self-assembling of {Pd_{12}} into blackberry structures, suggesting that the formation of solvent-shared ion-pairs can provide stable counterion-mediated attraction for the self-assembly. {Pd_{12}} can only accommodate very few trivalent counterions, which is unlikely to fulfill multiple associated counterions required by the blackberry formation.

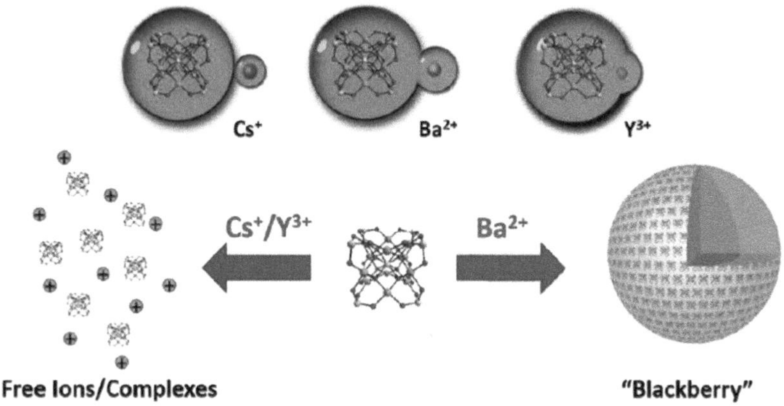

Figure 2.10 Schematic representation of connection between hydration, ion-pair, and blackberry formation. Top: three types of ion-pairs between counterions and {Pd_{12}}; bottom: solution behavior of {Pd_{12}} in presence of different counterions. Adapted with permission from reference [57]. Copyright 2018 Wiley-VCH.

Besides the dominant impacts of counterion valency, counterions with the same valency can also show distinct interactions with POM macroions and consequently affect the blackberry structure formation [58]. With small amount of extra salts, the sizes of {$Mo_{72}Fe_{30}$} blackberries showed different changes (Fig. 2.11). For the monovalent counterions, the sizes of blackberries in the solution with 1–20 mM added LiCl or NaCl were not significant different from that in salt-free solution.

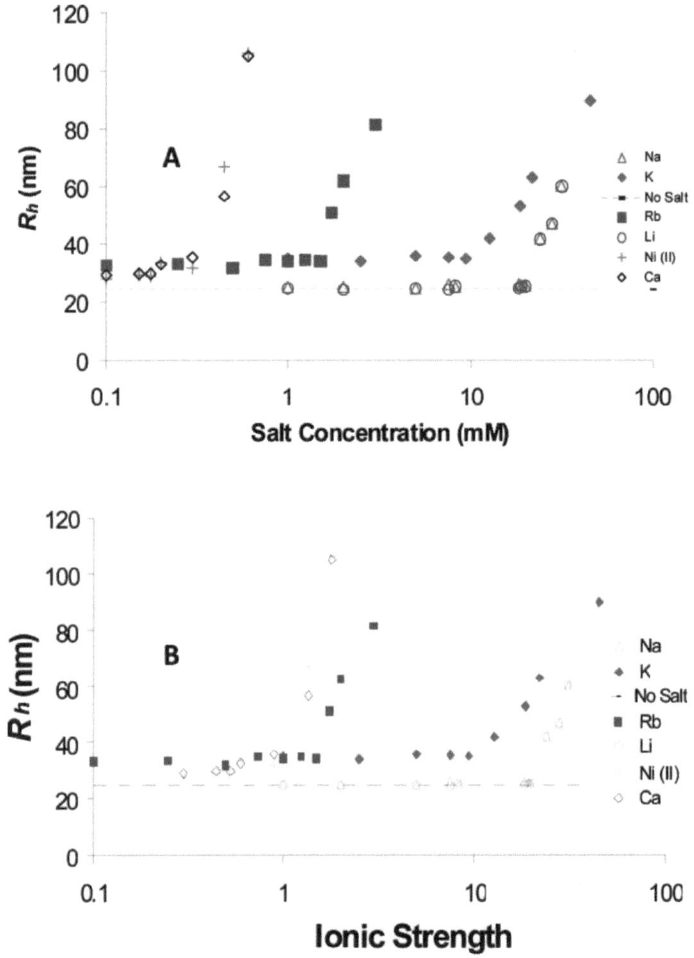

Figure 2.11 Change of blackberry size (R_h) with added chloride salt concentration (A) and total ionic strength (B) for 0.5 mg/mL $\{Mo_{72}Fe_{30}\}$ solutions. Adapted with permission from reference [58]. Copyright 2010 ACS.

However, the blackberry sizes became considerably larger with 0.1–10 mM KCl or RbCl than that with the same concentration of LiCl or NaCl, suggesting weak hydrated K^+ and Rb^+ can replace the original counterion of $\{Mo_{72}Fe_{30}\}$ (H^+), leading to stronger attractions between $\{Mo_{72}Fe_{30}\}$ and consequentially larger blackberries. Also, the critical point of blackberry size verse salt concentration is following the trend LiCl ~ NaCl > KCl > RbCl. ITC studies confirmed that the trend of binding affinities is $Li^+ \sim Na^+ < K^+ < Rb^+$. Titrating CsCl into $\{Mo_{72}Fe_{30}\}$ solution resulted immediate precipitation, suggesting the strongest binding affinity among above these cations. Titrating LiCl and NaCl into $\{Mo_{72}Fe_{30}\}$ solution resulted in no measurable heat, and the binding constants between K^+ and Rb^+ with $\{Mo_{72}Fe_{30}\}$ are 75 and 773, respectively. Similar trends of binding affinities with macroions among counterions of the same valency have been observed for other Keplerate POMs as well as uranyl clusters [24]. These results are consistent with the trend in the Hofmeister series originally found in protein solubilities. It is interesting that such unique features are shared by inorganic POM macroions and organic biomacromolecules, suggesting they might share some common features, which will be thoroughly discussed in section 2.2.5 as well.

2.2.4 Other Non-Covalent Interactions Contribute to Blackberry Structure Formation

Although the counterion-mediated attraction is the major driving force of blackberry formation, based on the functional groups on POMs, other different forces may contribute as well. The major driving force of POMs with functional groups is still counterion-mediated attractions in most case, as they self-assemble into blackberry structures. However, the additional forces usually lead to noticeable difference in assembly of morphologies or trends of the sizes, compared with that of the blackberries driving by only counterion-mediated attractions.

2.2.4.1 Hydrogen bond

Surface water ligands or small functional ligands on the POM macroions can make directional hydrogen bonding between them. The addition hydrogen bond may also contribute to regulating the self-assembly of POM macroions. With increasing of solvent polarity, that is, increasing water content or reducing organic solvent in the mixed solvent, the hydrogen bonds between functionalized POM macroions are stronger and may compensate the weaker counterion-mediated attractions, result in a larger or non-monotonically change in blackberry sizes (Figs. 2.12 and 2.13) [12, 59]. As a comparison, the blackberry formation driven by only counterion-mediated attractions will result in a smaller blackberry in the solvent of higher polarity, as demonstrated in section 2.2.2.

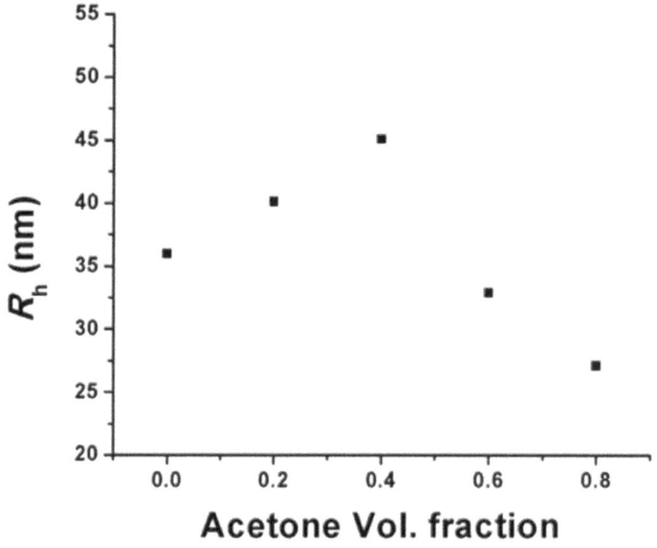

Figure 2.12 The average blackberry radius (R_h) in 0.2 mg/mL {$SrPd_{12}$} solution as a function of acetone content in different water/acetone mixtures at 40°C. "Volcano"-shaped size trend shows that hydrogen bonding along with counterion-mediated attraction can strongly affect the supramolecular self-assembly process. Adapted with permission from reference [12]. Copyright 2018 Wiley-VCH.

The effects of hydrogen bond can be further confirmed by adding strong hydrogen bond donor solvent, for example, hexafluoroisopropanol (HFIP), to cut off the intermolecular hydrogen bond between macroions [34, 59]. The reversed trends of blackberry sizes have been observed in the solvent with/without HFIP, confirming the hydrogen bond is the major reason responsible for the different trends of the POM macroions with ligands capable of forming hydrogen bonds (Fig. 2.13).

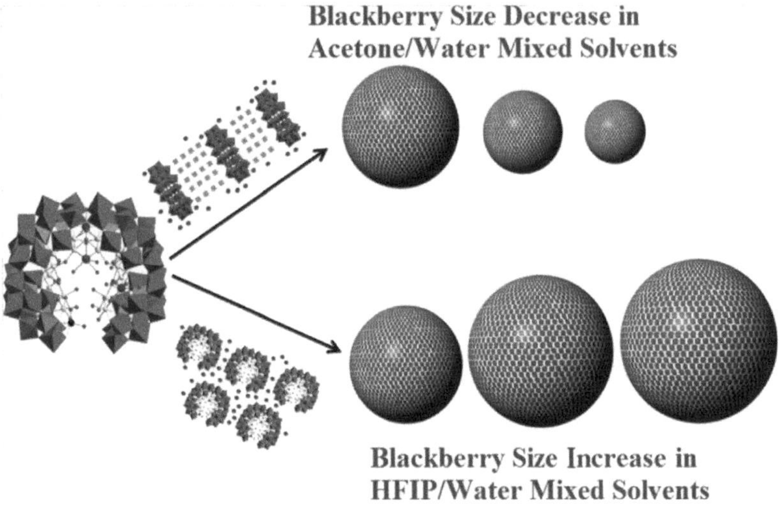

Figure 2.13 Graphic present of self-assembly behaviors of $\{Fe_{16}P_8W_{49}Gd_4\}$ in different solvents. Adapted with permission from reference [59]. Copyright 2016 Wiley-VCH.

Another method to study hydrogen bonding is to use deuterated solvents. In such solutions, most of active hydrogens will be replaced by deuterium, resulting in hydrogen bonds of different strength. $\{Mn_{40}P_{32}W_{224}\}$ POM macroion [11] self-assemble into rod-like structures in acetone/water mixture solvents, due to their anisotropic charge distributions. However, in methanol/water mixed solvents, $\{Mn_{40}P_{32}W_{224}\}$ behaves like POM macroions with isotropic charge distribution and self-assembles into spherical blackberry structures, due to the isotropic hydrogen bonding with solvents reduce the anisotropic charge effects (Fig. 2.14). Replacing regular methanol with deuterated methanol result in rod-like assemblies again, suggest the role of hydrogen bonding. It should be noted that for weak acid-type POM macroions, for example, $\{Mo_{72}Fe_{30}\}$, the isotope effects on the self-assembly of the POM macroions is usually complicated. In deuterated solvents, besides the differences in the strength of hydrogen bonds, the degree of deprotonation is usually lower, and hydration energy of macroion is higher than that in regular solvents (Fig. 2.15) [60].

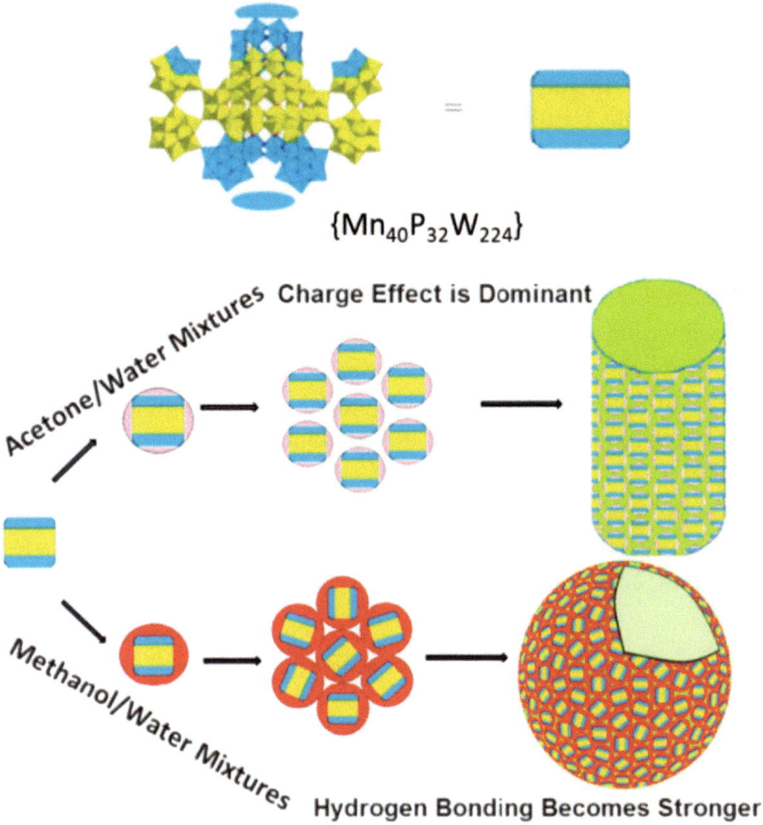

Figure 2.14 Top: crystalline structure of $\{Mn_{40}P_{32}W_{224}\}$ showing anisotropic surface charge density distribution. Some areas have high charge density (blue) and some others have lower charge density (yellow). Bottom: graphic present of the self-assembly behaviors of $\{Mn_{40}P_{32}W_{224}\}$ in different solvents combinations. Reproduced from reference [11] with permission from The Royal Society of Chemistry.

2.2.4.2 Hydrophobic interaction

Similar to hydrogen bond, intermolecular hydrophobic interaction introduced by small organic ligands may also contribute to the attraction and result in different trends of blackberry sizes upon changing solvent polarity (Fig. 2.16). However, longer or bulky organic ligands, for example, long hydrocarbon chains, may introduce strong hydrophobic (or solvophobic) interactions.

The dominant hydrophobic interactions between organic ligands, combined with the hydrophilic feature of POM building block, make such POM-organic hybrids behave like amphiphilic surfactants, for example, forming double layer vesicles in solutions. Depending on the relative strength of counterion-mediated attractions and hydrophobic interactions, the transit from macroion (forming blackberries) to surfactants (forming vesicles) in solutions have been observed (Fig. 2.17). Compared with common amphiphilic molecules, for example, surfactants with small charged, hydrophilic head groups, or block polymers consist of soft hydrophilic and hydrophobic chains, POMs as the head groups of POM-organic hybrids are rigid and large, which have shown interesting solution behaviors. Also, the abundant choices of large organic ligands and various molecular morphologies enable versatile functions of POM-organic hybrids. The topic of solution behaviors of inorganic-organic hybrids have been covered by a recent review [27].

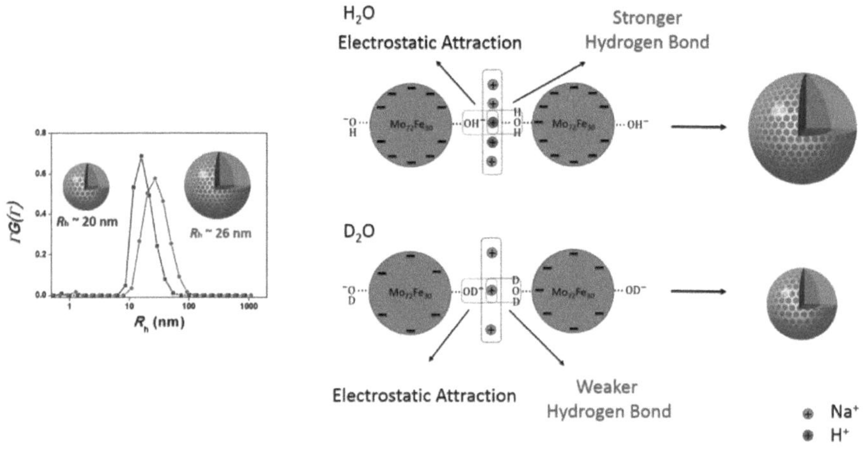

Figure 2.15 Left: size distributions of the assemblies in 0.5 mg/mL $\{Mo_{72}Fe_{30}\}$ solutions. Red: in H_2O, blue: in D_2O. Right: graphic representation of interactions during the $\{Mo_{72}Fe_{30}\}$ self-assembly process. In H_2O, stronger hydrogen bonds between protonated H_2O molecules lead to stronger attraction between $\{Mo_{72}Fe_{30}\}$, resulting in larger blackberries than that in D_2O. Adapted with permission from reference [60]. Copyright 2019 Wiley-VCH.

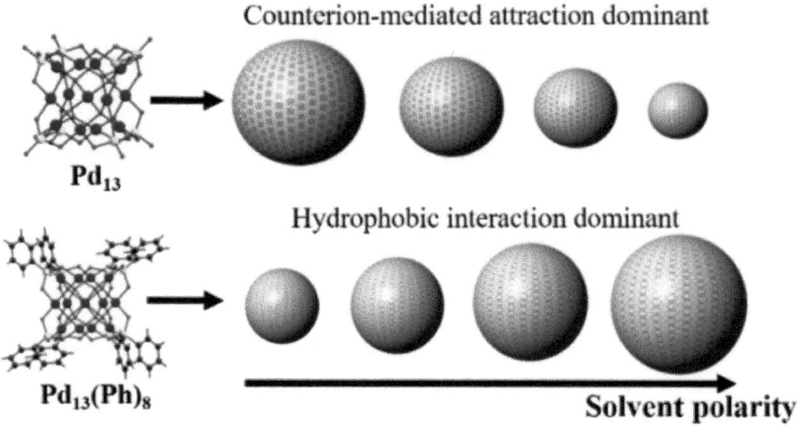

Figure 2.16 POM macroion with small hydrophobic ligands shows an increasing trend of blackberry sizes vs solvent polarity, opposite to the one regulated by only counterion-mediated attraction. Adapted with permission from reference [27]. Copyright 2019 ACS.

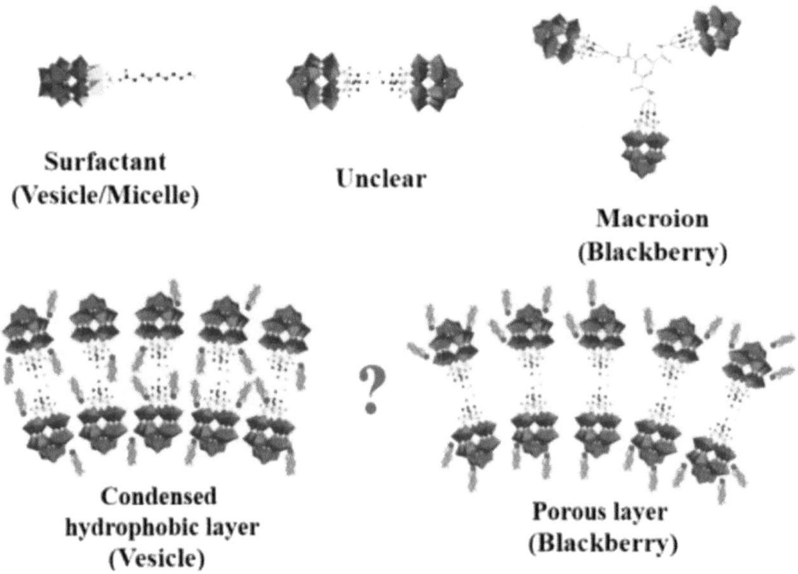

Figure 2.17 Based on the ratio between organic ligands and hydrophilic head groups, POM-organic hybrid can behave like macroions forming blackberries or surfactants forming vesicles. Adapted with permission from reference [27]. Copyright 2019 ACS.

2.2.5 The Connection Between the Self-Assembly of POM Macroions and Complex Biomacromolecule Assemblies

2.2.5.1 The self-assembly kinetics of POM macroions and connection to viral capsid formation

For the self-assembly of POM macroions driving by only counterion-mediated attractions, the rates of blackberry formation under certain circumstance are slow, which enables the detailed study on the kinetics and mechanism of the self-assembly. Figure 2.18 shows a typical study on kinetics of $\{Mo_{72}Fe_{30}\}$ blackberry formation [7]. Time-resolved SLS study can provide information about kinetics of the self-assembly. The continuous increase of scattered intensity recorded by SLS suggests the continuous formation of large structures in solution. Meanwhile, DLS studies indicate the existence of both discrete POM macroions ($R_h \sim 1.2$ nm) and blackberries ($R_h \sim 25$ nm) in the solution. The number of discrete macroions indicate by the area of the corresponding peak decreased while that of blackberries increased with the time, also confirm the continuous blackberry formation in the solution. The average blackberry size remained almost constant during the whole process. When large amount of NaCl existed in solution, the self-assembly of $\{Mo_{72}Fe_{30}\}$ was much slower, and a lag period become significant at the beginning of the self-assembly (Fig. 2.19) [6]. Sedimentation velocity experiments (Fig. 2.20) were performed on the $\{Mo_{72}Fe_{30}\}$ solution still in the lag period, and the coexistence of two species was found in the solution, corresponding to $\{Mo_{72}Fe_{30}\}$ monomers and oligomers. Therefore, the mechanism of the self-assembly of $\{Mo_{72}Fe_{30}\}$ can be described as an initial rate limiting step of oligomer formation, following by a rapid growth leading to the blackberry formation, resulting in a sigmoidal kinetic curve (Fig. 2.21). The assembling of oligomers into blackberries is fast, so no small blackberries can be observed, and the sizes of blackberry are constant during the self-assembly. By monitor the assembly rate verse temperate, the activation energies of blackberry formation can be determined by the Arrhenius Eq. [7]. The results (Fig. 2.22) show that the self-assembly of $\{Mo_{72}Fe_{30}\}$ with sodium salts of monovalent anions have similar activation energies, and the one with sodium sulfate has higher activation energy, because bivalent anions have a stronger capability to stabilize the $\{Mo_{72}Fe_{30}\}$ monomers by screening their intermolecular interactions.

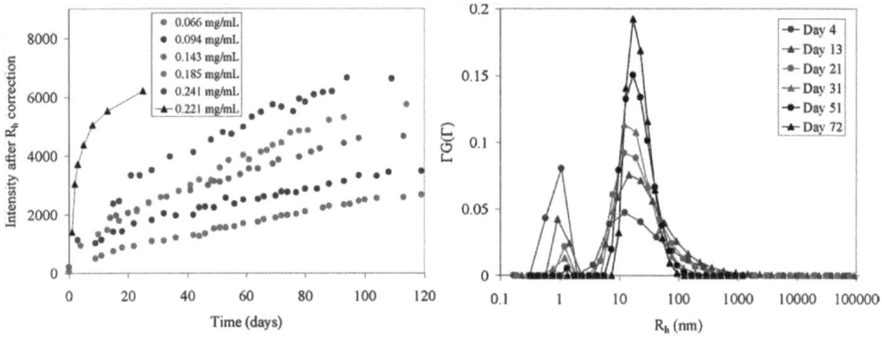

Figure 2.18 Left: time-resolved SLS result of $\{Mo_{72}Fe_{30}\}$ solutions at 25°C (except the data represent by triangles, the sample was kept at 45°C). Right: CONTIN analysis of DLS results of $\{Mo_{72}Fe_{30}\}$ aqueous solution at different times. Adapted with permission from reference [7]. Copyright 2005 ACS.

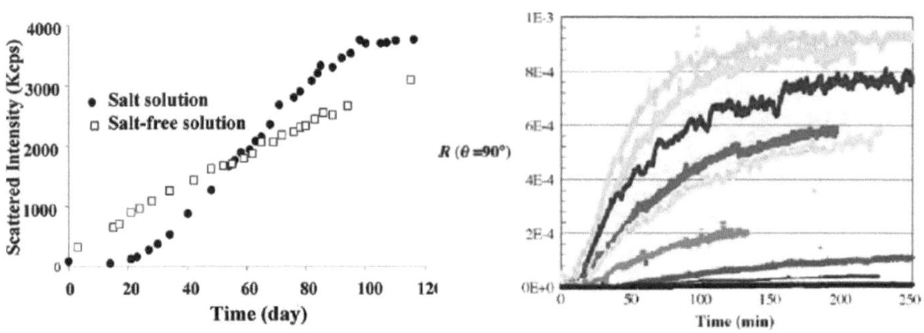

Figure 2.19 Left: $\{Mo_{72}Fe_{30}\}$ aqueous solution with 0.9 wt % NaCl shows much slower self-assembly rate and longer lag period than salt-free solution. Right: SLS studies of HPV capsid formation show sigmoidal kinetic curves similar to that of blackberry formation. Adapted with permission from reference [6]. Copyright 2009 ACS.

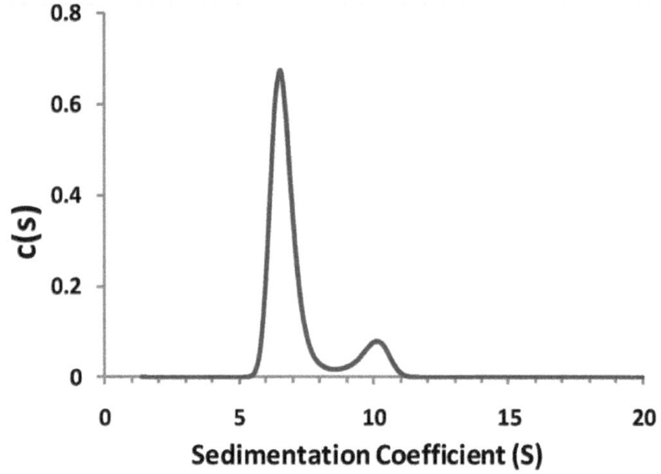

Figure 2.20 Continuous size distribution c(s) vs sedimentation coefficient, S. Experiments were performed at a $\{Mo_{72}Fe_{30}\}$ concentration of 10 mg/mL in 170 mM NaCl solution at 20°C. Adapted with permission from reference [6]. Copyright 2009 ACS.

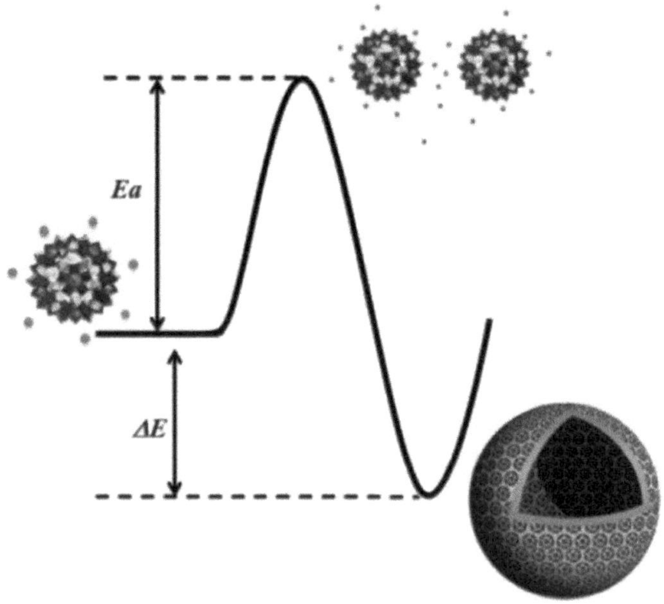

Figure 2.21 Graphic represent of the energy change during the blackberry formation. Adapted with permission from reference [6]. Copyright 2009 ACS.

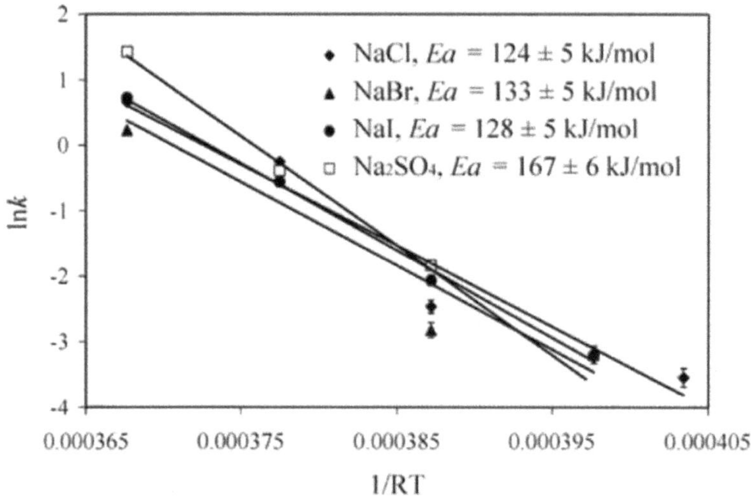

Figure 2.22 Calculation of the activation energy (E_a) for the blackberry formation of 0.5 mg/mL $\{Mo_{72}Fe_{30}\}/H_2O$ solutions containing various types of salts (NaCl, NaBr, NaI, and Na_2SO_4, respectively, with the concentrations of the cations all at 0.017 mol/L). Adapted with permission from reference [7]. Copyright 2005 ACS.

Sigmoidal kinetic curves have observed in many other systems sharing similar two-step kinetic processes: a lag period at the beginning due to the slow rates of the initial steps. Once the product of the initial steps has accumulated enough amount beyond a critical value, subsequent steps react quickly to get the final product. One of the important cases, *in vitro* self-assembly of viral proteins (e.g., human papillomavirus, HPV, Fig. 2.19 left) into capsid, displays sigmoidal kinetic curve and single-layered, hollow, spherical morphology on the final assembly, which are similar to the blackberry formed by POM macroions. In addition, the assembly of viral capsid is also sensitive to ionic strengths [61]. Considering the common features between viral capsid and POM macroions: they are both in the nanoscale, soluble, charged and able to self-assembly into single-layered, hollow, spherical assemblies, POM macroion can be good model system to understand the complex interactions in the self-assembly of proteins into viral capsid.

2.2.5.2 Cations transport across POM macroion and blackberry "membrane"

Cells are known for its selective transport small cations by their ion channels constructed by proteins on the cell membrane [62]. Studies on Keplerate POMs and their uranyl cluster resemblance-$\{U_{60}\}$ cluster have shown that the porous features on their surface can mimic the protein ion channels, achieving selective transportation across their shells. The encapsulation of counterions will consequently change the solution behavior of these clusters [16, 17, 54, 63]. Besides that, the surfaces of the blackberry structures are also porous, as neighboring POMs are not directly contact with each other, and the space between them may serve as ion channels like the ones on cell membranes. Figure 2.23 demonstrates a cross-membrane ion-transport experiment on $\{Mo_{72}Fe_{30}\}$ blackberries [64]. The ion-sensitive dyes were introduced into freshly prepared $\{Mo_{72}Fe_{30}\}$ solutions without any blackberry structures. Upon the blackberry formation, dye molecules were partially incorporated into the blackberries and showed wavelength shifts in fluorescent signals. After the completion of blackberry formation, additional salts were introduced into the solutions.

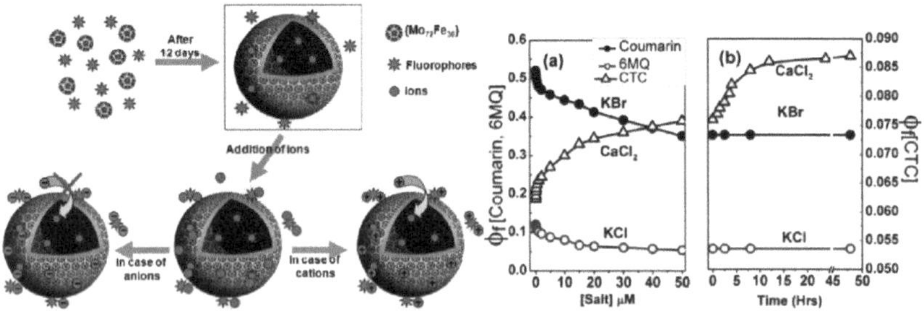

Figure 2.23 Left: graphic representation of the cross-membrane ion-transport experiments. The cations can enter blackberries and interaction with fluorophores inside, while the anions cannot cross the membrane. Right: change in fluorescence quantum yield of coumarin (sensitive to Br^-), 6-MQ (6-methoxyquinoline, sensitive to Cl^-), and CTC (chlorotetracycline, sensitive to Ca^{2+}) with addition of KBr, KCl, and $CaCl_2$, respectively; (a) instantaneous change of quantum yields with the addition of salts; (b) change in fluorescence quantum yield verse time, once the addition of salt was stopped. Adapted with permission from reference [64]. Copyright 2008 ACS.

A sudden jump followed by slow, continuous change in fluorescence signal from chlorotetracycline (CTC) (Ca^{2+}-sensitive dye) after the addition of $CaCl_2$ suggests process that the dye molecules in bulky solution were immediately saturated by Ca^{2+} ions, and then Ca^{2+} ions slowly transported across the blackberry membranes binding with the dye molecules inside the blackberries. Similar phenomena were observed using $MgCl_2$ and Mg-sensitive dyes. However, the experimental results show that small anions cannot pass through the blackberry membrane, likely due to the blackberry membrane is negatively charged and repels anions.

2.2.5.3 Self-recognition and chiral selection in the self-assembly of POM macroions, connecting to the origin of biological homochirality

Self-recognition is a unique behavior widely observed among biomacromolecules, and it is also a fascinating topic in fundamental supramolecular chemistry. Self-recognition is defined as narcissistic self-sorting behavior, that is, the high-fidelity affinity toward self, when two or more components coexist in solution [65, 66]. Another related topic is the homochirality in biological systems. If a molecule cannot be mapped to its mirror image by only rotational and translational transit, the molecule possesses chirality [67]. Asymmetric carbon atoms, or certain spatial conformation of molecules can cause chirality in molecules. Almost all chiral molecules in living organisms are found to be homochirality: L-amino acids, D-sugars, right-handed DNA helices, and so on. The puzzles around the homochirality lays down to two fundamental questions [68]: firstly, what serves as the "first" templates of chirality bias in the probably racemic environment at prebiotic times? Secondly, how was the bias sustained and amplified through the history leading the homochirality we observed today? As discussed earlier, POM macroions have shown their connections to biomacromolecules. They can serve as model systems to explore answers to these intriguing questions as well.

We recently observed a series of self-recognition behaviors in the self-assembly of structural similar POM macroions. $\{Mo_{72}Cr_{30}\}$

and $\{Mo_{72}Fe_{30}\}$ are both weak-acid type Keplerates of similar sizes and structures. Therefore, the two POM macroions are negatively charged, based on the deprotonation degrees. However, the mobility of their surface hydration as well as degree of surface water ligands deprotonation is different. The DLS study on their mixed aqueous solutions showed the presence of two types of large species with different sizes, corresponding to the sizes of blackberries in the individual solutions containing either $\{Mo_{72}Cr_{30}\}$ or $\{Mo_{72}Fe_{30}\}$ at the same pH value [8, 9]. The two assemblies in the mixed solution were separated by filtration and confirmed they were assemblies of the corresponding POM macroions by element analysis. Therefore, in this experiment the two POM macroion showed self-recognition behavior by forming their individual blackberries instead of a hybrid one. Changing temperate or solution pH, such self-recognition behavior could still be observed, indicating the self-recognition between the two POM macroion is a universal, independent from specific charge densities and free-energy controlled process (Fig. 2.24).

Similar self-recognition behaviors have been observed in other POM macroionic systems as well, including similar POM macroions but with different surface ligands chirality [13] (L/D-$\{Fe_{28}\}$, $Ba_{10}[Fe_{28}(u_3\text{-}O)_8(L/D\text{-}Tart)_{16}(HCOO)_{24}]$, Fig. 2.25), different functional groups introducing cation-pi interactions [69] (rod-shaped clusters with or without phenyl groups) or hydrophobic interactions($\{Pd_{13}\}$ and $\{Pd_{13}Ph\}$),[15] different central atom within rod-like POM macroions [14] (LAL(X), $((C_4H_9)_4N)_7[Mo_6O_{18}NC(OCH_2)_3\mathbf{X}Mo_6O_{18}(OCH_2)_3CNMo_6O_{18}]$, X = Mn^{III}/Fe^{III}, Fig. 2.26). The last case in the list demonstrates a very delicate self-recognition: two rod-like POM macroions of almost identical structures carries the same amount charges, and the only difference between the two POM macroions are just one atom in the center (Mn or Fe). The DFT calculation shows the difference center atoms lead to slight difference in the charge distribution, and such difference is amplified during the self-assembly process, resulting in the observed self-recognition behaviors.

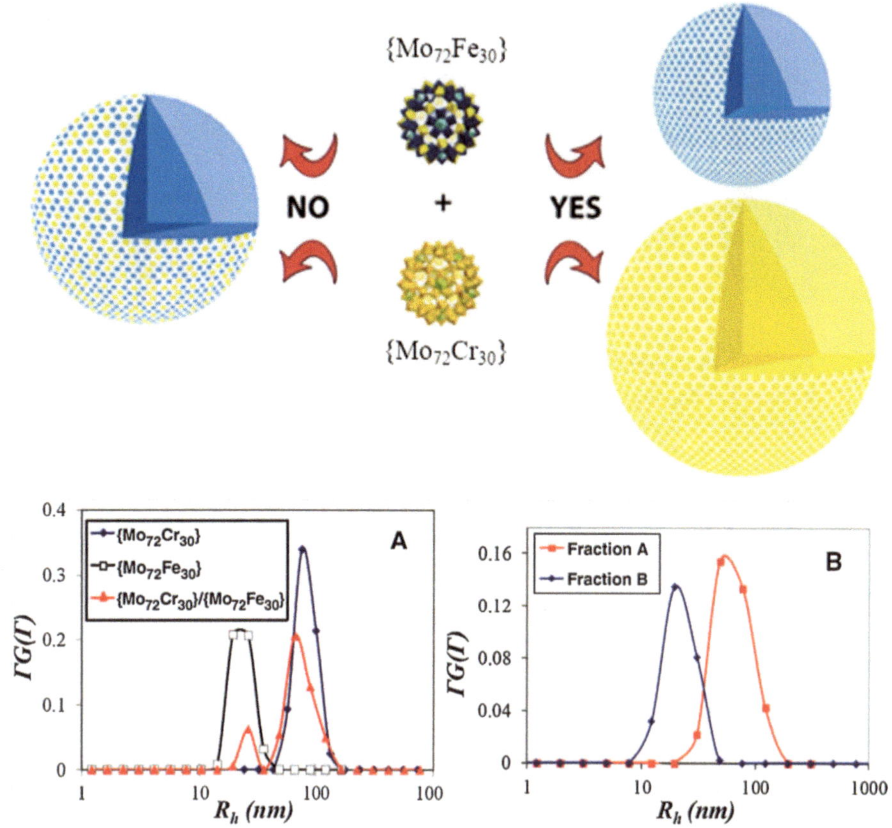

Figure 2.24 Top: graphic representation of the self-recognition in the self-assembly of $\{Mo_{72}Fe_{30}\}$ and $\{Mo_{72}Cr_{30}\}$. Bottom: (A) CONTIN analysis of the DLS studies on aqueous solutions containing $\{Mo_{72}Cr_{30}\}$ or $\{Mo_{72}Fe_{30}\}$ (0.1 mg/mL), as well as for a solution containing both species (0.1 mg/mL of each) (pH = 4.2). (B) CONTIN analysis of the DLS studies on the solution of the aqueous solution originally containing both $\{Mo_{72}Cr_{30}\}$ and $\{Mo_{72}Fe_{30}\}$ (0.1 mg/mL for each) is separated by a series of membranes with different pore sizes; fraction A (red): material retained after filtering by 100 000 Dalton-pore size membrane; fraction B (blue): material retained after filtering by 30 000 Dalton-pore size membrane. Adapted with permission from reference [8]. Copyright 2011 AAAS.

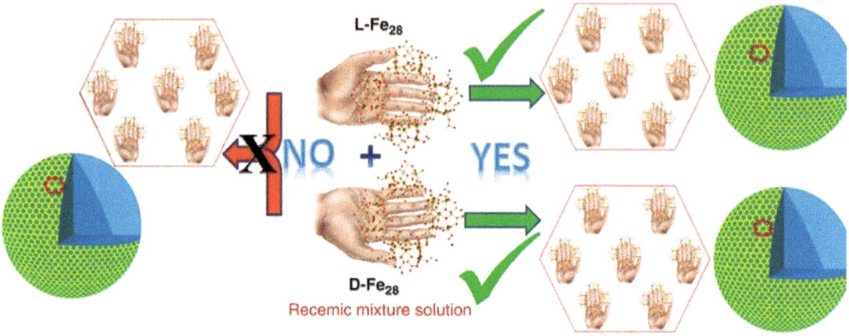

Figure 2.25 Graphical representation of the self-recognition behavior between $\{Fe_{28}\}$ enantiomers. Adapted with permission from reference [13].

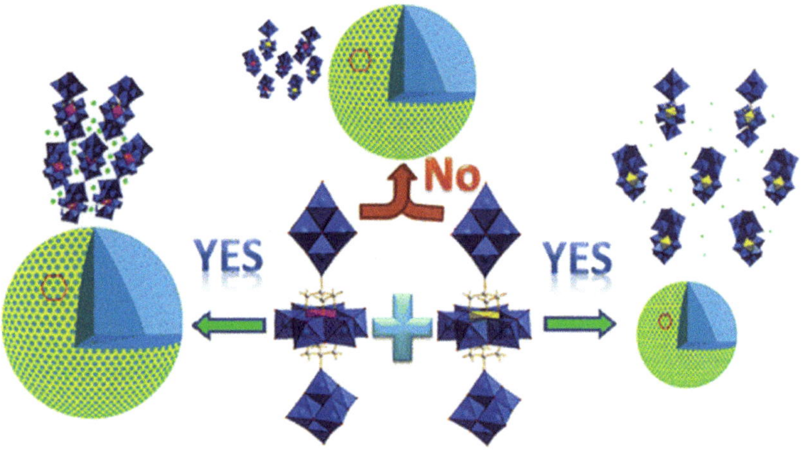

Figure 2.26 Graphical representation of the self-recognition behavior between rod-like POM macroions. Adapted with permission from reference [14]. Copyright 2013 ACS.

The detailed experiments suggest critical mechanism of the self-recognition should lay in the early stage of the blackberry formation. As introduced in the 2.2.5.1, the rate-determine step of blackberry formation is the initial step, oligomer formation. Time-resolved SLS studies at the early stage of self-assembly showed similar slopes of the scattered intensity increment of the solutions contains either $\{Mo_{72}Fe_{30}\}$ or $\{Mo_{72}Cr_{30}\}$, indication the oligomer formation of two POM macroions were kinetically similar (Fig. 2.27). However, in the mixed solution of the total concentration of $\{Mo_{72}Fe_{30}\}$ and $\{Mo_{72}Cr_{30}\}$ same to the individual solutions, the rate of oligomer formation was much slower than that in the individual solutions. Similar phenomena were also observed in the early stage of L/D-$\{Fe_{28}\}$ self-assembly, that two enantiomers self-assembled kinetically similar in their individual solutions but much slower in the mixed solution of the same total concentration. A possible explanation is that the POM macroions are less favorite by the other POM macroions in their mixed solutions, that is, the formation active energy for homo-oligomers might be slightly lower than the hybrid oligomers. Therefore, the collision between different macroions is ineffective (not able to form oligomers), leading to a slower assembling rate in the mixed solution. Later, the preference toward the homo-oligomers may be continuously amplified during the association and eventually leads to the dominance of homogeneous blackberries over heterogeneous ones. Small difference between two macroions may result in different directional, long-range intermolecular interactions, for example, electrostatic interactions, leading to the preferable oligomer formation. On the other hand, self-assembly driven by short, unidirectional intermolecular forces, for example, vesicles driven by hydrophobic interaction between surfactant tails, such self-recognition behavior usually cannot be observed.

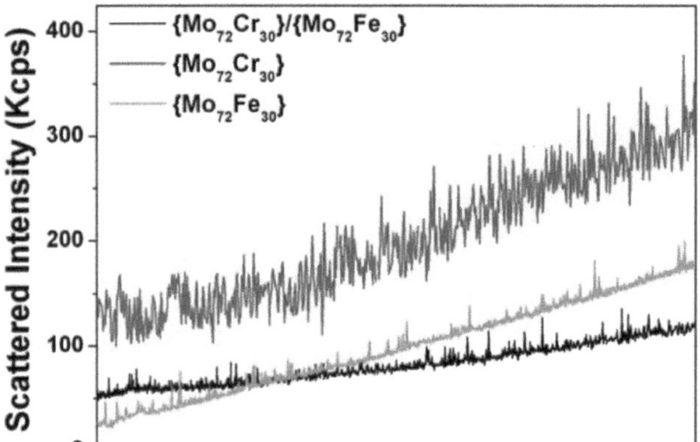

Figure 2.27 Time-resolved SLS results of 0.5 mg/mL $\{Mo_{72}Fe_{30}\}$, 0.5 mg/mL $\{Mo_{72}Cr_{30}\}$, and 0.25 mg/mL of each, respectively. Adapted with permission from reference [8]. Copyright 2011 AAAS.

Besides the chiral selection demonstrated between {Fe_{28}} enantiomers, the interactions between chiral POM macroions and small chiral molecules can also achieve chiral selections. Specifically speaking, the extra chiral counterions, co-ions, and solvents in solutions can trigger the chiral discrimination among the enantiomers of chiral POM macroion, resulting in the suppression of the self-assembly of one enantiomer.

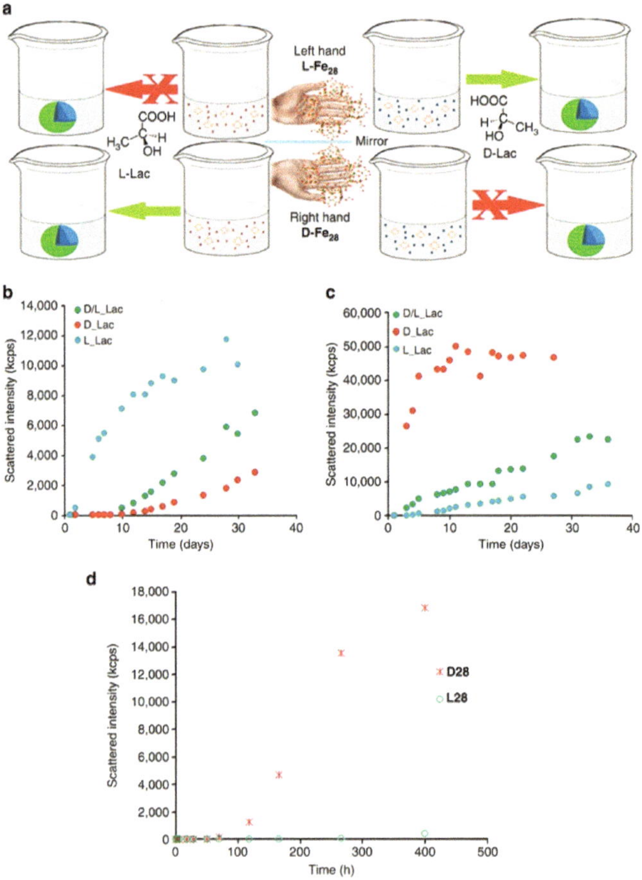

Figure 2.28 (a) Graphical representation of the chiral selection experiments by the introduction of lactic acid molecules. (b, c) SLS results of Ba-D-Fe_{28} (0.5 mg mL^{-1}) and Ba-L-Fe_{28} (0.5 mg mL^{-1}) in three different buffers, respectively. (d) SLS monitoring of Ba-D-Fe_{28} and Ba-L-Fe_{28} (0.5 mg mL^{-1}) in the buffer of L-lactic acid (33 μg mL^{-1}) over a period of 400 h. Adapted with permission from reference [13].

For $\{Fe_{28}\}$, the self-assembly of D-$\{Fe_{28}\}$ enantiomer was found to be significantly suppressed in the presence of D-lactic acids with a long lag period of ~10 days. In the presence of L-lactic acid, D-$\{Fe_{28}\}$ enantiomer self-assembled much faster and with a short lag period less than one day. The effects of racemic lactic acids stay between two pure enantiomers of lactic acids. Similar phenomena were observed for the combination of enantiomers of lactic acid with L-$\{Fe_{28}\}$ (see Fig. 2.28). The chiral selection during the self-assembly of $\{Fe_{28}\}$ could result from the compatibility between chiral co-ions and the chiral micro-environment confined by the macroions. LLS and TEM measurements indicate that the preferred pairs form larger blackberry structures than the nonpreferred pairs (e.g., D-lactic acid with D-$\{Fe_{28}\}$. As introduced earlier, of the same POM macroion, the larger blackberries indicate the stronger attractions between the macroions. Therefore, it can conclude that preferred partners interact more strongly with each other in the assembly structures. A recent investigation [70] on a different macroionic system, functionalized C_{60} with chiral ligands, demonstrate that the chiral counterions, co-ions and solvent molecules possess different strengths of chiral selection *via* the interplay of electrostatic interactions, broken of hydration shells, and molecular chirality. As the interactions between POM macroions with surrounding small ions or solvent molecules is also play important roles in regulating their solution behaviors, investigating the effects of small chiral molecules on the self-assembly of POM macroions may also lead to significant results.

2.3 Conclusion

Many POMs exist as macroanions in aqueous solutions and other polar solvents. Due to the size disparities between large POM macroions and their counterions, the unique interactions between POM macroions with counterions result in the self-assembly of POM macroions into single-layered, hollow, spherical blackberry structures. Experimental evidence as well as molecular dynamic simulations shows the driving force of the blackberry formation is the counterion-mediated attractions in the most cases. The sizes of blackberries are the calibrations of intermolecular interactions between macroions, which can be tunable with different solvent polarity, ionic strength, and pH for weak acid-type POM macroions. POM macroions can also distinguish counterions by the hydration sizes and valences, resulting in different types of ion-pairs between POM macroions and counterions. When other non-covalent interactions also contributing to intermolecular interactions between POM macroions, the self-assembly behaviors of the POM macroion can be different from that with only counterion-mediated attraction, reflected as different trends of blackberry sizes. POM macroions share many unique features with biomacromolecules, such as the self-assembly into viral capsid-like structure, selective ion transportation, self-recognition, and chiral selection. Therefore, POM macroions can be a good modeling system to understand complex biologic phenomena.

References

1. Long, D.-L., Burkholder, E. and Cronin, L. (2007). Polyoxometalate clusters, nanostructures and materials: from self assembly to designer materials and devices, *Chem. Soc. Rev.*, **36**(1), pp. 105–121.
2. Pope, M.T. and Müller, A. (1991). Polyoxometalate chemistry: an old field with new dimensions in several disciplines, *Angew. Chem. Int. Ed.*, **30**(1), pp. 34–48.
3. Proust, A., Thouvenot, R. and Gouzerh, P. (2008). Functionalization of polyoxometalates: towards advanced applications in catalysis and materials science, *Chem. Commun.*, (16), pp. 1837–1852.
4. Liu, T., Diemann, E., Li, H., Dress, A.W.M. and Muller, A. (2003). Self-assembly in aqueous solution of wheel-shaped Mo_{154} oxide clusters into vesicles, *Nature*, **426**(6962), pp. 59–62.
5. Kistler, M.L., Bhatt, A., Liu, G., Casa, D. and Liu, T. (2007). A complete macroion–"blackberry" assembly–macroion transition with continuously adjustable assembly sizes in $\{Mo_{132}\}$ water/acetone systems, *J. Am. Chem. Soc.*, **129**(20), pp. 6453–6460.
6. Zhang, J., Li, D., Liu, G., Glover, K.J. and Liu, T. (2009). Lag periods during the self-assembly of $\{Mo_{72}Fe_{30}\}$ macroions: connection to the virus capsid formation process, *J. Am. Chem. Soc.*, **131**(42), pp. 15152–15159.

7. Liu, G. and Liu, T. (2005). Thermodynamic properties of the unique self-assembly of $\{Mo_{72}Fe_{30}\}$ inorganic macro-ions in salt-free and salt-containing aqueous solutions, *Langmuir*, **21**(7), pp. 2713–2720.
8. Liu, T., Langston, M.L.K., Li, D., Pigga, J.M., Pichon, C., Todea, A.M. and Muller, A. (2011). Self-recognition among different polyprotic macroions during assembly processes in dilute solution, *Science*, **331**(6024), pp. 1590–1592.
9. Haso, F., Li, D., Garai, S., Pigga, J.M. and Liu, T. (2015). Self-recognition between two almost identical macroions during their assembly: the effects of pH and temperature, *Chem. Eur. J.*, **21**(38), pp. 13234–13239.
10. Pigga, J.M., Kistler, M.L., Shew, C.-Y., Antonio, M.R., and Liu, T. (2009). Counterion distribution around hydrophilic molecular macroanions: the source of the attractive force in self-assembly, *Angew. Chem. Int. Ed.*, **48**(35), pp. 6538–6542.
11. Haso, F., Fang, X., Yin, P., Li, D., Ross, J.L. and Liu, T. (2013). The self-assembly of a macroion with anisotropic surface charge density distribution, *Chem. Commun.*, **49**(6), pp. 609–611.
12. Yang, P., Li, H., Ma, T., Haso, F., Liu, T., Fan, L., Lin, Z., Hu, C. and Kortz, U. (2018). Rational design of organically functionalized polyoxo-palladates and their supramolecular properties, *Chem. Eur. J.*, **24**(10), pp. 2466–2473.
13. Yin, P., Zhang, Z.-M., Lv, H., Li, T., Haso, F., Hu, L., Zhang, B., Bacsa, J., Wei, Y., Gao, Y., Hou, Y., Li, Y.-G., Hill, C.L., Wang, E.-B. and Liu, T. (2015). Chiral recognition and selection during the self-assembly process of protein-mimic macroanions, *Nat. Commun.*, **6**, p. 6475.
14. Yin, P., Zhang, J., Li, T., Zuo, X., Hao, J., Warner, A.M., Chattopadhyay, S., Shibata, T., Wei, Y. and Liu, T. (2013). Self-recognition of structurally identical, rod-shaped macroions with different central metal atoms during their assembly process, *J. Am. Chem. Soc.*, **135**(11), pp. 4529–4536.
15. Haso, F., Yang, P., Gao, Y., Yin, P., Li, H., Li, T., Kortz, U. and Liu, T. (2015). Exploring the effect of surface functionality on the self-assembly of polyoxopalladate macroions, *Chem. Eur. J.*, **21**(25), pp. 9048–9052.
16. Gao, Y., Szymanowski, J.E.S., Sun, X., Burns, P.C. and Liu, T. (2016). Thermal responsive ion selectivity of uranyl peroxide nanocages: an inorganic mimic of K^+ ion channels, *Angew. Chem.*, **128**(24), pp. 7001–7005.

17. Gao, Y., Haso, F., Szymanowski, J.E.S., Zhou, J., Hu, L., Burns, P.C. and Liu, T. (2015). Selective permeability of uranyl peroxide nanocages to different alkali ions: influences from surface pores and hydration shells, *Chem. Eur. J.*, **21**(51), pp. 18785–18790.

18. Liu, T., Imber, B., Diemann, E., Liu, G., Cokleski, K., Li, H., Chen, Z. and Müller, A. (2006). Deprotonations and charges of well-defined $\{Mo_{72}Fe_{30}\}$ nanoacids simply stepwise tuned by pH allow control/variation of related self-assembly processes, *J. Am. Chem. Soc.*, **128**(49), pp. 15914–15920.

19. Yin, P., Li, D. and Liu, T. (2012). Solution behaviors and self-assembly of polyoxometalates as models of macroions and amphiphilic polyoxometalate–organic hybrids as novel surfactants, *Chem. Soc. Rev.*, **41**(22), pp. 7368–7383.

20. Misra, A., Kozma, K., Streb, C. and Nyman, M. (2020). Beyond charge balance: counter-cations in polyoxometalate chemistry, *Angew. Chem. Int. Ed.*, **59**, pp. 596–612.

21. Hückel, E. and Debye, P. (1923). The theory of electrolytes: I. lowering of freezing point and related phenomena, *Phys. Z*, **24** (185–206), p. 1.

22. Buschow, K.H.J., Cahn, R.W., Flemings, M.C., lschner, B., Kramer, E.J. and Mahajan, S. (2001) *Encyclopedia of Materials: Science and Technology*, Elsevier, The Netherlands.

23. Okur, H.I., Hladílková, J., Rembert, K.B., Cho, Y., Heyda, J., Dzubiella, J., Cremer, P.S. and Jungwirth, P. (2017). Beyond the Hofmeister series: ion-specific effects on proteins and their biological functions, *J. Phys. Chem. B*, **121**(9), pp. 1997–2014.

24. Chu, Y., Chen, J., Haso, F., Gao, Y., Szymanowski, J.E.S., Burns, P.C. and Liu, T. (2018). Expanding the Schulze–Hardy rule and the Hofmeister series to nanometer-scaled hydrophilic macroions, *Chem. Eur. J.*, **24**(21), pp. 5479–5483.

25. Liu, G., Cai, Y. and Liu, T. (2004). Automatic and subsequent dissolution and precipitation process in inorganic macroionic solutions, *J. Am. Chem. Soc.*, **126**(51), pp. 16690–16691.

26. Yin, P., Li, D. and Liu, T. (2011). Counterion interaction and association in metal-oxide cluster macroanionic solutions and the consequent self-assembly, *Isr. J. Chem.*, **51**(2), pp. 191–204.

27. Luo, J. and Liu, T. (2019). Competition and cooperation among different attractive forces in solutions of inorganic–organic hybrids containing macroionic clusters, *Langmuir*, **35**(24), pp. 7603–7616.

28. Li, D., Zhou, W., Landskron, K., Sato, S., Kiely, C.J., Fujita, M. and Liu, T. (2011). Viral-capsid-type vesicle-like structures assembled from $M_{12}L_{24}$ metal–organic hybrid nanocages, *Angew. Chem. Int. Ed.*, **50**(22), pp. 5182–5187.
29. Li, D., Zhang, J., Landskron, K. and Liu, T. (2008). Spontaneous self-assembly of metal–organic cationic nanocages to form monodisperse hollow vesicles in dilute solutions. *J. Am. Chem. Soc.*, **130**(13), pp. 4226–4227.
30. Li, H., Wang, R., Hong, Y.-l., Liang, Z., Shen, Y., Nishiyama, Y., Miyoshi, T. and Liu, T., (2019). Tuning the intercage distance in charge-regulated blackberry-type assemblies through host–guest chemistry, *Chem. Eur. J.*, **25**(22), pp. 5803–5808.
31. Li, H., Xie, T.-Z., Liang, Z., Shen, Y., Sun, X., Yang, Y. and Liu, T. (2019). Adjusting emission wavelength by tuning the intermolecular distance in charge-regulated supramolecular assemblies, *J. Phys. Chem. C*, **123**(37), pp. 23280–23286.
32. Soltis, J.A., Wallace, C.M., Penn, R.L. and Burns, P.C. (2016). Cation-dependent hierarchical assembly of U60 nanoclusters into macro-ion assemblies imaged *via* cryogenic transmission electron microscopy, *J. Am. Chem. Soc.*, **138**(1), pp. 191–198.
33. Yin, P., Lin, Z., Wu, J., Hsu, C.-H., Chen, X., Zhou, J., Lu, P., Eghtesadi, S.A., Yu, X., Cheng, S.Z.D. and Liu T., (2015). Charge-regulated spontaneous, reversible self-assembly of the carboxylic acid-functionalized hydrophilic fullerene macroanions in dilute solution, *Macromolecules*, **48**(3), pp. 725–731.
34. Zhou, J., Yin, P., Gao, Y., Hu, L. and Liu, T. (2015). Spontaneous self-assembly of γ-cyclodextrins in dilute solutions with tunable sizes and thermodynamic stability, *Chem. Eur. J.*, **21**(26), pp. 9563–9568.
35. Eghtesadi, S.A., Haso, F., Kashfipour, M.A., Lillard, R.S. and Liu, T. (2015). Supramolecular assembly of poly(propyleneimine) dendrimers driven by simple monovalent counterions, *Chem. Eur. J.*, **21**(51), pp. 18623–18630.
36. Müller, A., Krickemeyer, E., Bögge, H., Schmidtmann, M. and Peters, F. (1998). Organizational forms of matter: an inorganic super fullerene and keplerate based on molybdenum oxide, *Angew. Chem. Int. Ed.*, **37**(24), pp. 3359–3363.
37. Müller, A. and Gouzerh, P. (2012). From linking of metal-oxide building blocks in a dynamic library to giant clusters with unique properties and towards adaptive chemistry, *Chem. Soc. Rev.*, **41**(22), pp. 7431–7463.

38. Liu, G., Kistler, M.L., Li, T., Bhatt, A. and Liu, T. (2006). Counter-ion association effect in dilute giant polyoxometalate $[As^{III}_{12}Ce^{III}_{16}(H_2O)_{36}W_{148}O_{524}]^{76-}$($\{W_{148}\}$) and $[Mo_{132}O_{372}(CH_3COO)_{30}(H_2O)_{72}]^{42-}$ ($\{Mo_{132}\}$) macroanionic solutions, *J. Cluster Sci.*, **17**(2), pp. 427–443.

39. Liu, T. (2009). Hydrophilic macroionic solutions: what happens when soluble ions reach the size of nanometer scale?, *Langmuir*, **26**(12), pp. 9202–9213.

40. Chaumont, A. and Wipff, G. (2008). Ion aggregation in concentrated aqueous and methanol solutions of polyoxometallates Keggin anions: the effect of counterions investigated by molecular dynamics simulations, *Phys. Chem. Chem. Phys.*, **10**(46), pp. 6940–6953.

41. Anderson, R.E., Colorado, J.R., Crouse, C., Ogrin, D., Maruyama, B., Pender, M.J., Edwards, C.L., Whitsitt, E., Moore, V.C., Koveal, D., Sandu, C., Stewart, M.P., Smalley, R.E., Tour, J.M. and Barron, A.R. (2006). A study of the formation, purification and application as a SWNT growth catalyst of the nanocluster $[H_xPMo_{12}O_{40}\subset H_4Mo_{72}Fe_{30}(O_2CMe)_{15}O_{254}(H_2O)_{98}]$. *Dalton Trans.*,**25**, pp. 3097–3107.

42. Zhou, J., Hu, J., Li, M., Li, H., Wang, W., Liu, Y., Winans, R.E., Li, T., Liu, T. and Yin, P. (2018). Hydrogen bonding directed co-assembly of polyoxometalates and polymers to core–shell nanoparticles, *Mater. Chem. Front.*, **2**(11), pp. 2070–2075.

43. Wu, Y., Shi, R., Wu, Y.-L., Holcroft, J.M., Liu, Z., Frasconi, M., Wasielewski, M.R., Li, H. and Stoddart, J.F. (2015). Complexation of polyoxometalates with cyclodextrins, *J. Am. Chem. Soc.*, **137**(12), pp. 4111–4118.

44. Guilleaume, B., Ballauff, M., Goerigk, G., Wittemann, M. and Rehahn, M. (2001). Correlation of counterions with rodlike macroions as assessed by anomalous small-angle X-ray scattering, *Colloid. Polym. Sci.*, **279**(9), pp. 829–835.

45. Lipfert, J., Doniach, S., Das, R. and Herschlag, D. (2014). Understanding nucleic acid-ion interactions, *Annu. Rev. Biochem.*, **83**, pp. 813–841.

46. Antonio, M.R., Nyman, M. and Anderson, T.M. (2009). Direct observation of contact ion-pair formation in aqueous solution, *Angew. Chem. Int. Ed.*, **48**(33), pp. 6136–6140.

47. Fullmer, L.B., Molina, P.I., Antonio, M.R. and Nyman, M. (2014). Contrasting ion-association behavior of Ta and Nb polyoxometalates, *Dalton Trans.*, **43**(41), pp. 15295–15299.

48. Poblet, J.M., López, X. and Bo, C. (2003). Ab initio and DFT modeling of complex materials: towards the understanding of electronic and

magnetic properties of polyoxometalates, *Chem. Soc. Rev.*, **32**(5), pp. 297–308.

49. Liu, Z., Liu, T. and Tsige, M. (2016). Elucidating the origin of the attractive force among hydrophilic macroions, *Sci. Rep.*, **6**, p. 26595.
50. Liu, Z., Liu, T. and Tsige, M. (2018). Unique symmetry-breaking phenomenon during the self-assembly of macroions elucidated by simulation, *Sci. Rep.*, **8**(1), p. 13076.
51. Trefalt, G., Szilagyi, I. and Borkovec, M. (2013). Poisson–Boltzmann description of interaction forces and aggregation rates involving charged colloidal particles in asymmetric electrolytes, *J. Colloid Interface Sci.*, **406**, pp. 111–120.
52. Hofmeister, F., (1888). About the science of the effect of salts, *Arch. Exp. Pathol. Pharmakol.*, **24**, pp. 247–260.
53. Jungwirth, P. and Cremer, P.S. (2014). Beyond Hofmeister, *Nat. Chem.*, **6**, p. 261.
54. Rehder, D., Haupt, E.T.K. and Müller, A. (2008). Cellular cation transport studied by 6, 7L i and 23Na NMR in a porous Mo132 Keplerate type nano-capsule as model system, *Magn. Reson. Chem.*, **46**(S1), pp. S24–S29.
55. Luo, J., Ye, S., Li, T., Sarnello, E., Li, H. and Liu, T. (2019). Distinctive trend of metal binding affinity *via* hydration shell breakage in nanoconfined cavity, *J. Phys. Chem. C*, **123**(23), pp. 14825–14833.
56. Segado, M., Nyman, M. and Bo, C. (2019). Aggregation patterns in low- and high-charge anions define opposite solubility trends, *J. Phys. Chem. B*, **123**(49), pp. 10505–10513.
57. He, J., Li, H., Yang, P., Haso, F., Wu, J., Li, T., Kortz, U. and Liu, T. (2018). Tuning of polyoxopalladate macroanionic hydration shell *via* countercation interaction, *Chem. Eur. J.*, **24**(12), pp. 3052–3057.
58. Pigga, J.M., Teprovich Jr., J.A., , Flowers, R.A. Antonio, M.R. and Liu, T. (2010). Selective monovalent cation association and exchange around keplerate polyoxometalate macroanions in dilute aqueous solutions, *Langmuir*, **26**(12), pp. 9449–9456.
59. Haso, F., Luo, J., Bassil, B.S., Artetxe, B., Zhou, J., Yin, P., Reinoso, S., Gutiérrez-Zorrilla, J.M., Kortz, U. and Liu, T. (2016). Effect of directional hydrogen bonding on the self-assembly of anisotropically-shaped macroions, *ChemistrySelect*, **1**(14), pp. 4345–4349.
60. Li, H., Shen, Y., Yang, P., Szymanowski, J.E.S., Chen, J., Gao, Y., Burns, P.C., Kortz, U. and Liu, T. (2019). Isotope and hydrogen-bond effects

on the self-assembly of macroions in dilute solution, *Chem. Eur. J.*, **25**(71), pp. 16288–16293.

61. Sun, X., Li, D., Wang, Z., Yin, P., Hu, R., Li, H., Liu, Q., Gao, Y., Ren, B., Zheng, J., Wei, Y. and Liu, T. (2018). Role of protein charge density on hepatitis B virus capsid formation, *ACS Omega*, **3**(4), pp. 4384–4391.
62. Noskov, S.Y., Bernèche, S. and Roux, B. (2004). Control of ion selectivity in potassium channels by electrostatic and dynamic properties of carbonyl ligands, *Nature*, **431**(7010), pp. 830–834.
63. Müller, A., Das, S.K., Talismanov, S., Roy, S., Beckmann, E., Bögge, H., Schmidtmann, M. Merca, A., Berkle, A., Allouche, L., Zhou, Y. and Zhang, L. (2003). Trapping cations in specific positions in tuneable "artificial cell" channels: new nanochemistry perspectives, *Angew. Chem. Int. Ed.*, **42**(41), pp. 5039–5044.
64. Mishra, P.P., Pigga, J. and Liu, T. (2008). Membranes based on "keplerate"-type polyoxometalates: slow, passive cation transportation and creation of water microenvironment, *J. Am. Chem. Soc.*, **130**(5), pp. 1548–1549.
65. Rebek, J., (2009). Introduction to the molecular recognition and self-assembly special feature, *Proc. Natl. Acad. Sci. U.S.A*, **106**(26), p. 10423.
66. Safont-Sempere, M.M., Fernández, G. and Würthner, F. (2011). Self-sorting phenomena in complex supramolecular systems, *Chem. Rev.*, **111**(9), pp. 5784–5814.
67. Wagnière, G.H., (2007). *On Chirality and the Universal Asymmetry: Reflections on Image and Mirror Image*. John Wiley & Sons, USA.
68. Blackmond, D.G., (2010). The origin of biological homochirality. *Cold Spring Harb Perspect Biol*, **2**(5), pp. a002147–a002147.
69. Luo, J., Chen, K., Yin, P., Li, T., Wan, G., Zhang, J., Ye, S., Bi, X., Pang, Y., Wei, Y. and Liu, T. (2018). Effect of cation-π iinteraction on macroionic self-assembly, *Angew. Chem. Int. Ed.*, **57**(15), pp. 4067–4072.
70. Luo, J., Ye, S., Ustriyana, P., Wei, B., Chen, J., Raee, E., Hu, Y., Yang, Y., Zhou, Y., Wesdemiotis, C., Sahai, N. and Liu, T. (2020). Unraveling chiral selection in the self-assembly of chiral fullerene macroions: effects of small chiral components including counterions, co-ions, or neutral molecules, *Langmuir*, **36**(17), pp. 4702–4710.

Chapter 3

Rational Design and Self-Assembly of Polyoxometalate-Peptide Hybrid Materials

Elena Atrián-Blasco,[a,b,*] **Héctor Soria-Carrera,**[a,b,*] **Rafael Martín-Rapún,**[a,b,c] **and Scott G. Mitchell**[a,b]

[a]*Instituto de Nanociencia y Materiales de Aragón (INMA), Consejo Superior de Investigaciones Científicas-Universidad de Zaragoza, 50009 Zaragoza, Spain*
[b]*CIBER de Bioingeniería, Biomateriales y Nanomedicina, Instituto de Salud Carlos III, 28029 Madrid, Spain*
[c]*Departamento de Química Orgánica, Universidad de Zaragoza, 50009 Zaragoza, Spain*

elenaab@unizar.es, scott@unizar.es

The synergistic combination between biomolecules and polyoxometalates (POMs) has recently been considered as an effective approach to construct nano-biomaterials with diverse structures and morphologies toward biological applications. Different synthetic routes for the preparation and characterization of covalent POM-peptide hybrids are discussed within the first part of this chapter, together with a comprehensive description of some of their unique properties such as stereochemistry and self-assembly ability. The second part is devoted to ionic compounds that have been used as catalysts and adhesive

*These authors contributed equally.

Polyoxometalates: Advances, Properties, and Applications
Edited by Leire Ruiz Rubio, José Luis Vilas Vilela, Beñat Artetxe, and Juan Manuel Gutiérrez-Zorrilla

ISBN 978-981-4968-14-0 (Hardcover), 978-1-003-27744-6 (eBook)
www.jennystanford.com

materials and have shown interesting biomedical and anti-microbial applications.

3.1 Introduction

POMs are typically self-assembled from one-pot aqueous acidic media through acid-condensation reactions of simple metal-oxide ions. One inherent advantage of the syntheses of these robust and stable molecular metal-oxide species is that the architectures that are generated generally possess high solution and thermal stability along with unparalleled redox properties from simple and quick high-yielding and scalable reactions. The combination of their unique properties combined with ease of synthesis has made them interesting for a variety of catalytic and biomedical applications. Furthermore, POMs can be derivatized through the direct conjugation of organic or bio-molecules to further enhance their physicochemical properties and unique structural and morphological characteristics as well as to diversify the chemistry that can be carried out with the POM post-functionalization.

The synthetic approaches, fundamental properties, and emerging applications of hybrid POM materials have recently been reviewed in depth by Parac-Vogt [1] and Proust and co-workers [2]. In broad terms, POM hybrids can be divided into two classes depending on the nature of interaction with the (bio)organic counterpart, where Class I hybrid POMs represent those where predominantly ionic interactions exist between the POM and the corresponding (bio)organic component(s), while Class II hybrid POMs possess a covalent or coordination interaction with their (bio)organic counterpart(s) [3]. It follows that each class of hybrid POM material possesses its own unique set of advantages and disadvantages. For example, the Class I hybrid POMs based on electrostatic interactions offer the potential to self-assemble functional materials in high yield through relatively simple building blocks, which therefore, effortlessly lend this approach to applied chemistry. Meanwhile for the Class II hybrids, the covalent or coordination bond provides a stronger union of the inorganic and (bio)organic components, which—in theory at least—allow for a more precise modification of the electronic and spectroscopic properties, not to mention a more controlled means of carrying out any relevant post-functionalization of other surfaces.

The self-assembly of hybrid POM materials offers chemists a unique opportunity to generate functional nanoscale materials for a variety of applications. Although organic molecules, polymers, and biomolecules such as DNA possess the ability to self-assemble into functional architectures, peptides have the intrinsic ability to self-assemble into highly versatile nanoscale biological constructs with pre-programmed functionality. Many different kinds of peptides are known and can be synthesized on a large scale to exert biological functions that include, but are not limited to, antibacterial, antifungal, cytotoxic, adhesive, anti-cancer, anti-amyloid, anti-inflammatory peptides, and antioxidant [4–8].

The majority of the hybrid POM-peptide materials found in the literature are obtained *via* direct reaction of POMs with (bio)organic components, where the four "classical" POM molecules- Lindqvist $[M_6O_{19}]^{q-}$, Anderson–Evans $[XM_6O_{24}]^{q-}$, Keggin $[XM_{12}O_{40}]^{q-}$, and Wells–Dawson $[X_2M_{18}O_{62}]^{q-}$–are of course the most studied. The most significant advances in POM-peptide chemistry are barely a decade old and the physicochemical properties and versatile chemistry of POM-peptide hybrids have been attracting more attention in the five-year period from 2015 to 2020, where the number of publications has doubled from 9 to 28 (Fig. 3.1).

The ionic self-assembly of Class I POM-peptide materials represents a facile scalable route to functional nanomaterials from simple building blocks. Importantly, both the POM and peptide components represent highly modular families and by simply tuning the stoichiometry of these modular components, chemists can access a wide variety of highly tailorable materials for a number of applications. The few known examples concerning the self-assembly of POM-peptides and their bio-applications were summarized by Wu and co-workers in 2016 [9]. On the other hand, the synthesis of Class II covalent hybrids represents the challenging multistep chemistry, where only a limited number of examples have been reported and until now have demonstrated limited applications. In this respect, the development of controllable synthesis for preparing functional Class II POM-peptide materials represents one of the grand challenges.

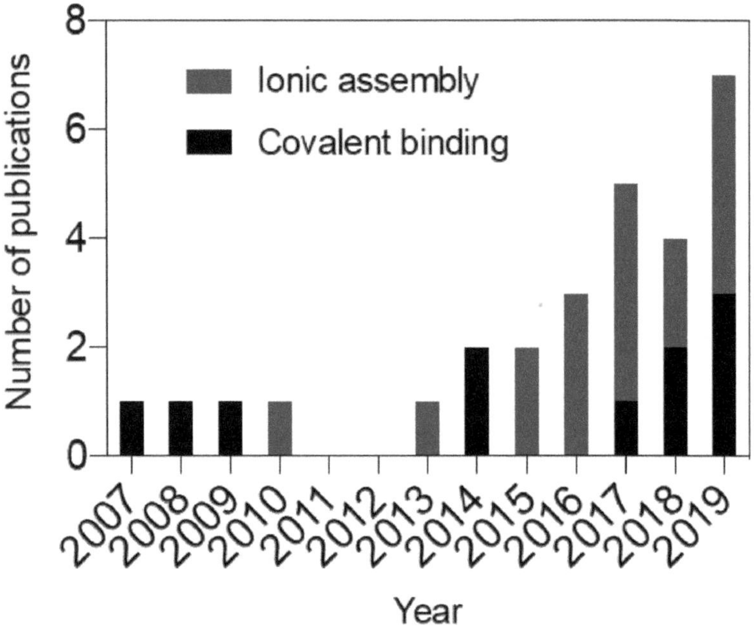

Figure 3.1 Number of articles covering POM-peptide hybrids published by year.

We have divided this book chapter according to the Class I and Class II hybrid categorization. Class II POM-peptide hybrids will be covered first since they allow us to explain some fundamental chemistry concepts regarding the combination of POM and peptides that can be used later for discussing the Class I hybrids. For instance, in a solution-based self-assembly, Class II hybrids only display the stoichiometry provided by the synthesis, but the POM-to-peptide ratio in Class I hybrids can be tuned. Until now, Class I research has been focused on applied materials research, whilst Class II hybrids have generally been viewed as more of a synthetic chemistry challenge.

3.2 Covalent POM-Peptide Hybrids

3.2.1 Synthesis and Characterization of Covalent Hybrids

To date, two main types of POM-peptide hybrids formed by covalent conjugation (Class II POM hybrids) have been described in the literature. The first involves the use of the tris-(hydroxymethyl)aminomethane (TRIS-NH_2) moiety as a bridging linker (see section 3.2.1.1) and the second uses an organotin group grafted to the POM to link amino acids and peptides (see section 3.2.1.2). All of the currently available literature on covalently grafted POM-peptide hybrids has been summarized in Table 3.1 and the four principal types of POM structures used in these examples are shown in Fig. 3.2.

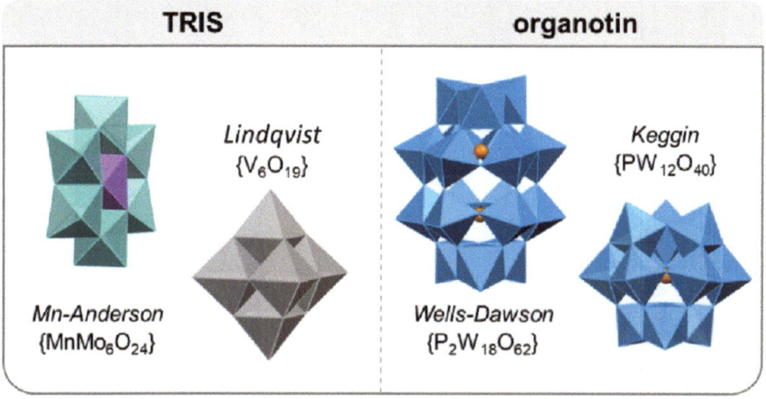

Figure 3.2 Polyoxometalate precursors used for the preparation of covalent POM-peptide hybrids.

3.2.1.1 Synthesis *via* TRIS functionalization

TRIS-NH_2 is a small organic molecule that has acted as a physical and metaphorical bridge to connect the inorganic POM component with the versatile—and arguably more controllable—world of organic synthesis. POMs have been functionalized with TRIS-based ligands with the aim of creating more complex

Table 3.1 Summary of published literature on POM-peptide hybrids by covalent conjugation†

Entry	Reference	Functionalization	POM type	Number of amino acid residues	Amino acid(s)	Properties (Application)
1	[10]	TRIS	Mn-Anderson	1 to 15	Diverse	Secondary structure self-assembly
2	[11]	TRIS	Mn-Anderson	8	Demobensin-1 peptide	Self-assembly (Anticancer)
3	[12]	TRIS	Mn-Anderson	1	L/D-**H**, L/D-**D**, L/D-**L**, L/D-**F**	Chirality (Antiamyloid)
4	[13]	TRIS	Mn-Anderson	1, 2, and 3	**G**, **F**, **A**	Self-assembly
5	[14,15]	TRIS	Lindqvist	1	β-**A**-OEt, **A**-OMe **G**-OEt	(Anticancer)
6	[16]	TRIS	Lindqvist	1	**L**-O^{t}Bu	Chirality
7	[17]	Organotin	α_1/α_2-Wells–Dawson	1	**Y**-O^{t}Bu	—
8	[18]	Organotin	α_1/α_2-Wells–Dawson	1	**F**, **Y**-O^{t}Bu, **Y**	Diastereoisomers resolution
9	[19]	Organotin (click-chemistry)	α_2-Wells–Dawson	1 and 3	**F**-OMe, **Y**, **V**	—
10	[20]	Organotin (oxoacylation)	α_2-Wells–Dawson	1 and 3	**Y**-O^{t}Bu, **V**	Chirality
11	[21]	Organotin (oxoacylation)	α_1-Wells–Dawson	3	Tripeptides-OMe of **W**, **A**, **L**, **M**, **S**, **V**, **R**	Chirality
12	[22]	Organotin (oxoacylation)	α_1/α_2-Wells–Dawson	3, 4, 5, and 6	**G**	Folding
13	[23]	Organotin (oxoacylation)	α_1/α_2-Wells–Dawson	3	Tripeptides-OMe of **V**, **W**, **A**, **L**, **F**	(SOX transcription factors inhibitors)
14	[24]	Organotin (oxoacylation)	Keggin	1 and 2	**F**-OMe & **FF**-OMe	Self-assembly

†Configuration of the amino acids at their α-carbon is only indicated in the cases where this is studied. If not specified, amino acids are in their L-configuration. The same applies for *N*-term or *C*-term modifications.

architectures around the metallic cluster. Once functionalized with TRIS, the POM can be used as a conventional organic molecule, where it is possible to exploit the reactivity of terminal TRIS amine to endow new functionalities to the cluster. The literature describing the synthesis of TRIS-functionalized POMs during the last decade has been prolific. The first example of the modification of POM with a TRIS derivative was reported by J. Zubieta and Q. Chen in 1990, where they successfully achieved the modification of a Wells–Dawson polyoxovanadate with tris(hydroxymethyl)nitromethane (TRIS-NO_2) [25]. However, it was the pioneering work of Hasenknopf *et al.* [26] that truly developed the field by reporting the now well-known Anderson–Evans TRIS-functionalization and making this platform one of the firm favorites for preparing novel POM materials with properties akin to metal-organic frameworks (MOFs) [27] or supramolecular gels [28].

Recently, strategies to introduce organic linkers in POMs have been extensively reviewed [1, 2, 29, 30]. Briefly, the most common POMs functionalized with TRIS-based ligands are Anderson–Evans, Wells–Dawson, Lindqvist, and Keggin-type clusters (see Fig. 3.2). In all of them, the oxygen in TRIS forms a stable covalent bond with the metal-oxide center. Furthermore, depending on the reaction conditions, it is possible to conjugate more than one type of TRIS-based ligands to the inorganic cluster.

The Anderson–Evans POM hybrids have been postulated as the ultimate "Bio-POM" in a recent review by A. Blazevic and A. Rompel, due to the unique possibility of chemically modulating both the metallic cluster and the functional organic component to display, for example, desired redox, magnetic, or photocatalytic activity, among other properties [29]. Although these POM archetypes have been extensively used in protein crystallography, the advances produced from the syntheses of relevant bio-relevant hybrids remain low [29]. Consequently, to further develop the field of hybrid POM-peptide materials, true synergistic properties between the biological moiety attached to the cluster and the POM should be designed and evaluated. The Rompel group has vast experience in studying the interactions between biomolecules and POMs and has pointed out the need to gain access to robust methodologies for preparing different types of POM regioisomers to determine the influence of the geometry in biological systems.

With respect to POM regioisomerism, the Anderson–Evans POM doped with Mn is one of the most explored in the literature. As mentioned before, Hansenknopf *et al.* [26] reported the first synthesis of a symmetrically modified TRIS-based Anderson POM in 2002 and, thereafter, the seminal work of Song *et al.* [31] in 2008 showed the potential that these hybrid structures had to offer to the field. The next level of complexity pursued the synthesis of asymmetrically modified Mn-Anderson [32]. In broad terms, two different approaches to prepare Mn-Anderson TRIS asymmetric hybrids can be identified and from now on, we may refer to them as type **I** (from the symmetric hybrid) and type **II** (from the inorganic precursors) (see Fig. 3.3).

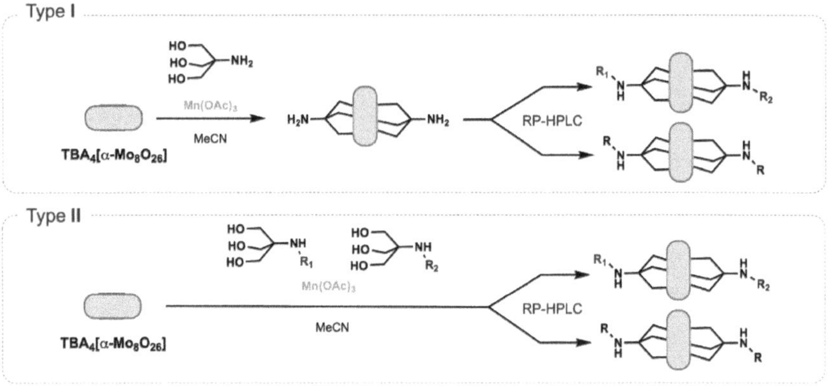

Figure 3.3 Preparation of asymmetric TRIS based POMs following Type I and Type II strategies.

Asymmetrically modified Mn-Anderson single crystals are difficult to obtain in one-pot reactions, since the crystallization energies are similar for the asymmetric and symmetric products, which are obtained in statistical ratio. The first example of asymmetrically functionalized POM relied on the separation of isomers attending to differences in solubility [31]. Following the type **II** synthesis, they submitted to reaction TRIS-NH_2 and TRIS-NO_2 with the Mo source to form a statistical mixture of both ligands. The authors were able to isolate the asymmetric hybrid, $(TBA)_3[MnMo_6O_{18}(TRIS\text{-}NH_2)(TRIS\text{-}NO_2)]$, by monitoring the crystallization process through mass spectrometry (MS) of the crystallizing solutions (ESI-MS). Although this process provided the asymmetric Mn-Anderson, the throughput was limited by the crystallization times.

Later on, Yvon *et al.* [33] developed a chromatographic separation of the symmetrical and unsymmetrical TRIS-POM compounds. Using normal-phase chromatographic procedures, tetrabutylammonium (TBA) cations were substituted by protons, due to the acidic properties of silica, yielding a hydrophilic hybrid with compromised solubility [34], but the authors managed to avoid that problem by using reverse-phase high-performance liquid chromatography (RP-HPLC). As the authors chose TRIS-ligands with different polarities, the separation was favored—RP-HPLC separates compounds depending on their affinity to the hydrophobic stationary phase—the more hydrophobic, the slower the elution. RP-HPLC is therefore a robust methodology to access asymmetrically modified POMs in moderate yield. What is more, both functionalization strategies (type I and type II) were compatible with this separation. The authors pointed out that type I strategy is more convenient when the modification of the amino group may compromise the stability of the cluster or can be readily prepared by standard organic synthesis. In turn, type II is generally preferred when reaction conditions are compatible with the POM, as it offers a faster and more convenient synthesis.

This synthetic approach permitted the authors to introduce a base-labile protecting group on one side, which is a recurrent tool in organic synthesis. After grafting onto the inorganic cluster, a cleavage deprotection step was used to release the hidden functionality. This strategy expands the scope of these building blocks for creating new complex structures.

Despite these relatively recent successes in the organic derivatization of TRIS-functionalized POMs, examples in which a peptide is covalently attached to the inorganic core are rare and underexplored compared with the conjugation of small molecules to the inorganic scaffold. In 2014, Yvon *et al.* [10] studied the grafting of peptides to a Mn-Anderson POM following two main strategies (see Fig. 3.4). The first strategy was based on an asymmetric Mn-Anderson TRIS hybrid (**a-Hyb**) equipped with a Fmoc (9-fluorenylmethyloxycarbonyl)-protected amino group on one side and a carboxylate *N*-hydroxysuccinimidyl ester (NHS) on the other. Thus, the authors obtained an amino acid-like moiety, which could be incorporated into a solid phase peptide synthesis (SPPS) workflow. In turn, the second approach used a symmetric Mn-Anderson TRIS hybrid (**s-Hyb**), whose carboxylic acids were derivatized symmetrically as NHS esters, thus the peptide was directly grafted through its *N*-terminus.

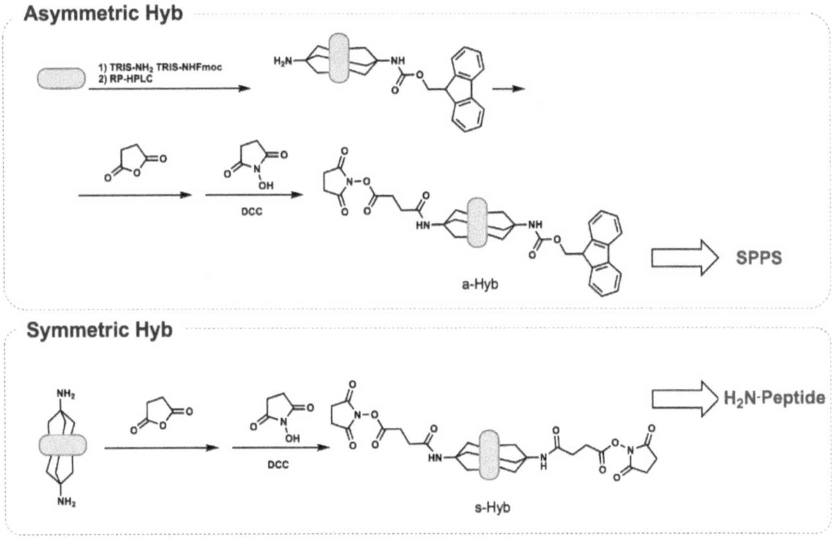

Figure 3.4 General synthesis of POM-peptide hybrids *via* TRIS-functionalization [10].

The **a-Hyb** approach was a stepwise process where an activated carboxylic acid was added, in large excess, to an amine-based solid resin. The advantage of the heterogeneous reaction permitted simple purification through filtration and washing of the resin. In a second step, the Fmoc protected-amine group was deprotected, releasing a free amine that initiated the cycle again. The final peptide was obtained after the sequential addition of the corresponding amino acids and ended up cleaving the peptide from the resin. Such cleavage typically occurs under harsh conditions that may decompose the inorganic clusters. Given this concern, the authors verified the stability of the POM toward deprotection conditions (20% hexafluoroisopropanol, HFIP, in dichloromethane, DCM) and the results according to ESI-MS indicated that the POMs remained intact.

Even if SPPS workflow employs vast excess of reactants, the unique nature of the **a-Hyb** might hamper the synthesis workflow making the coupling less efficient because of steric hindrance. If the amino acid was not successfully incorporated after **a-Hyb**, the final sequence may change from the forecasted one. The authors, thus, analyzed the sequence of the peptide by MS after the coupling of **a-Hyb**. It was found that, indeed, the steric hindrance of the POM avoided the coupling of further amino acids. However, increasing the amount of activated amino acid by up to six times could circumvent such a problem.

This example was one of the few publications at that date in which peptides are incorporated onto POMs. The working principles behind this work have settled the bases for the preparation of new hybrid materials based on the combination of both worlds. The **a-Hyb** is a highly versatile approach that potentially gives access to materials with elevated complexity (defined sequence and molecular weight) from readily available reagents. Although the number of articles containing covalent POM-peptide hybrids is continuously increasing (see Fig. 3.1 and Table 3.1), none have thus far exploited such a platform. In general, the peptide is symmetrically linked to the cluster limiting the scope of the materials prepared.

Thereafter, more examples have been published based on the same inorganic scaffold of the Mn-Anderson structure. Despite the emergence of new examples, all of the reports simplify the model to symmetric hybrid structures, avoiding the chromatographic separation proposed by Cronin's group.

In 2018, Ventura *et al.* [11] adapted the previous strategy to covalently graft an active peptide, i.e., demobensin-1. The Bombesin peptides are a family of neuropeptides known for their affinity toward some types of the Gastrin Releasing Peptide Receptor (GRP-R). These peptides have been used in cancer therapy because such receptors are overexpressed in some types of cancer such as colon, lung, or prostate. To address the linkage of TRIS Mn-Anderson and the peptide, they first derivatized the amine group by reaction with succinic anhydride [35]. After this simple step, the carboxylic acid that was generated could be further activated to react with the *N*-terminus of the peptide. In this way, the peptide was symmetrically grafted to both sides of the metallic cluster (for a general synthesis overview see **s-Hyb** approach in Fig. 3.4).

Demobensin-1 is rich in hydrophobic amino acids (FQWAVGHL-NHEt). For that reason, these systems behave like surfactants where the POM is the polar head group, whereas the peptides act as hydrophobic tails. When the hybrid was dissolved in water, it showed a tendency to segregate at the nanoscale forming nanoparticles of *ca.* 30 nm. The core-shell particle morphology obtained from the self-assembly process was consistent with already reported in the literature [36].

Using a similar approach, Luo *et al.* [13] attached hydrophobic peptidic sequences to the POM (symmetric fashion). In this case, the Mn-Anderson was linked to peptides based on the hydrophobic amino acid phenylalanine (F). Dipeptides based on this amino acid represent a common feature in self-assembly and are known to induce the formation of superstructures like ribbons or fibers [37]. Therefore, adding these chains to the inorganic cluster may drive the self-assembly of the hybrids. A detailed description of the self-assembly study can be found in section 3.2.2.

In 2019, N. Gao *et al.* [12] published another example of symmetrical grafting of amino acids to the Mn-Anderson scaffold *via* covalent bonding. As in the other examples, the TRIS-POM was derivatized to convert the amine into a carboxylic acid, which was then coupled with the α-amino group of diverse amino acids (e.g., H, D, L, and F) and their enantiomers (for a general synthesis overview see **s-Hyb** approach in Fig. 3.4). In this work, which exceeds the scope of the book chapter, the authors explored the impact of the chirality of the amino acid grafted to the POM on the aggregation of the amyloid-β peptide. It has been postulated that POMs are excellent candidates to prevent the folding of amyloidogenic peptides since they tend to inhibit the formation of fibrils, which are precursors to the amyloid plaques found in brain tissue of Alzheimer's patients [38]. In doing so, the authors highlighted the importance of paying attention to chirality in biological processes, since each POM-hybrid enantiomer has a different affinity toward amyloid aggregates (see section 3.2.3 for discussion on chirality).

While most literature examples concern the grafting of Mn-Anderson POMs, there are a few examples with other POM architectures. Among all structures, Lindqvist and Wells–Dawson ones are the most reported in bibliography and are widely explored using other organic substituents rather than peptides [1]. Regarding the synthesis of peptide-POM hybrids, however, the publication record is very limited and restricted to just three examples.

Although the systems developed by Xiao and collaborators relied on the same working principle as in the aforementioned work of Ventura *et al.* [11], the combination of the metallic core and the organic moiety aimed to explore synergistic effects between the Lindqvist hexavanadate $\{V_6O_{13}\}$ POM and amino acids grafted to the metal-oxide [14–16]. The authors have focused on studying such synergy in cancer cells comparing with well-known antitumoral drugs like 5-fluorouracil and obtained modest *in vitro* results.

3.2.1.2 Synthesis *via* organotin functionalization

Different approaches other than functionalization of POMs with TRIS-derivatives have been explored for producing covalently modified POM hybrids [1]. The Wells–Dawson and Keggin polyanion structures are known to form stable lacunary species, where the exposed terminal oxo atoms are highly nucleophilic. Therefore, p-block based organo groups can be grafted to these oxo atoms to produce organo-functionalized POMs. Such organo groups can be {Sn-R}, {R-X-O-X-R}, and $\{R\text{-}P(O)O_2\}$ (X = P, Si, Ge). To the best of our knowledge, only the organotin derivatives have been used to produce peptide-POMs hybrids [17–21, 23, 24, 39], where the inorganic part of these hybrids is composed of mono-lacunary polyoxotungstate (POT) species of the Wells–Dawson or Keggin structure. To obtain all of these tin-functionalized hybrids, direct reaction of $SnRCl_3$ with the mono-lacunary POT species was used to graft organic groups (R), as first reported by Pope and co-workers [40, 41].

In 2003 and 2005, Bareyt *et al.* first proposed the use of the trichlorotin moiety to graft $-CH_2R$ groups onto POMs to be used as linkers for further functionalization with bioactive molecules such as peptides (see Fig. 3.5A) [17, 18]. In these works, two Wells–Dawson anions were synthesized, α_1- and α_2-$[P_2W_{17}O_{61}\{Sn(CH_2)R\}]^{7-}$ and obtained as their TBA salts. The difference between these two products is the placement of the tin atom, and therefore, the original placement of the lacuna, and the impact it has in the chirality of the molecule. While for the α_2 isomer, the Sn atom is at the cap and has C_S symmetry, for the α_1 isomer, the Sn atom is at the belt making the POM chiral. However, $[\alpha_1\text{-}P_2W_{17}O_{61}\{Sn(CH_2)R\}]^{7-}$ enantiomers interconvert in an aqueous solution, which has been the key for subsequent work by these authors as discussed in section 3.2.3. This characteristic influences the reaction conditions for the grafting of the organotin moiety, and therefore, two different strategies were used for the $-SnCH_2R$ grafting.

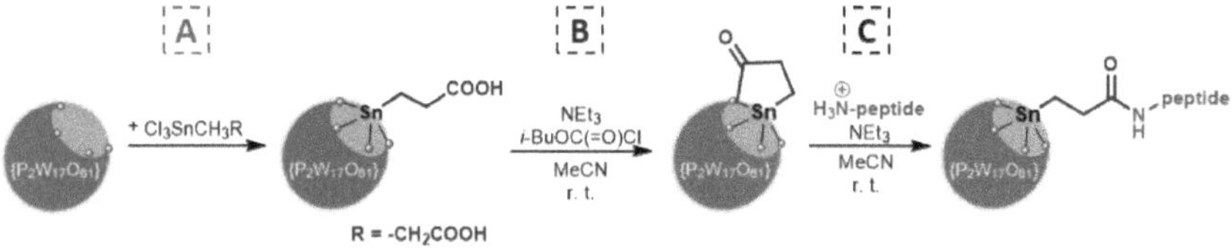

Figure 3.5 Reaction pathway toward peptide-POT hybrids *via* cyclization.

Strategy A, firstly reported by Pope and co-workers [40, 41], consists of the treatment of a buffered-aqueous solution of $K_{10}[\alpha_2\text{-}\{P_2W_{17}O_{61}\}]$ with $Cl_3Sn(CH_2)R$. The addition of tetrabutylammonium bromide (TBA·Br) allowed the isolation of the $(TBA)_7[\alpha_2\text{-}(P_2W_{17}O_{61}\{Sn(CH_2)R\})]$ hybrids with R = $-CH_2COOH$ and $-CH{=}CH_2$. The reaction by strategy A with the $\{\alpha_1\text{-}P_2W_{17}O_{61}\}$ precursor led to thermal isomerization to the α_2-isomer, and a second strategy had to be developed.

Strategy B consists of a solid-to-organic phase transfer: $Cl_3Sn(CH_2)R$ is added to a solution of TBA·Br in acetonitrile and the product was obtained by addition of the lacunary POMs $K_{10}[\alpha_2\text{-}P_2W_{17}O_{61}]$ or $K_9[\alpha_1\text{-}LiP_2W_{17}O_{61}]$. Strategy B turned out to be a more versatile approach than strategy A, not only for the milder conditions, but also for its compatibility with both α_1 and α_2 isomers. The authors further explored the functionalization of the new α_1 and α_2 hybrids by reaction of primary and secondary amines with both $(TBA)_7[\alpha_{1/2}$-$\{P_2W_{17}O_{61}(Sn(CH_2)_2COOH)\}]$, where the grafted $Sn(CH_2)R$ arm acted as a linker [18]. Again, different strategies—activating systems and conditions—were explored. Two methods proved to be successful for the formation of amides: method C used EEDQ and method D used chloroformate as coupling agents.

From these two works, Lacôte and co-workers have explored more sophisticated strategies for the functionalization of POTs with biomolecules: (i) the use of copper-catalyzed azide/alkyne cycloaddition (click chemistry) was presented in 2007 for ligation of organic and biological molecules to Dawson and Keggin POTs [19], and (ii) the regioselective mono-oxoacylation of Wells–Dawson POTs by conversion of the O_4-$Sn(CH_2)_2COOH$ group into an inorganic lactone (see Fig. 3.5B and section 3.2.3 for discussion on regioselectivity) in 2008 [20]. Peptide-POT hybrids have been obtained using both strategies. Through click chemistry, a trivaline (VVV) peptide was grafted to the azido derivative of α_2-$[P_2W_{17}O_{61}SnR]$ at 68% yield [19]. Production of peptide-POT hybrids *via* oxoacylation has been explored in more detail producing peptidic chains with three [20, 21, 39] and up to six amino acid residues [39]. In this case, hybrid amides were obtained through nucleophilic attack of the carbonyl lactone by amines in presence of triethylamine (NEt_3) as a base. Since the POM framework acts as an acyl-activating agent, no nucleophilic catalysts were needed [20]. Furthermore, hybrid amides were obtained under mild reaction conditions, primarily using acetonitrile as a solvent, at room temperature and with relatively short reaction times from 4 to 24 hours. As with the TRIS-based approach, the peptide was grafted by its N-terminus. In 2014, tripeptide-POM hybrids $(\alpha_1/\alpha_2$-$[P_2W_{17}O_{62}Sn(CH_2)_2$ CO-NH tripeptide$])^{7-}$ were studied, together with their precursors and other POMs, as inhibitors of SOX transcription factors in the context of cancer and regenerative biomedicine [23]. The negative charge of the POMs could offer high affinity to the DNA binding domains of the transcription factors, providing a great activity as inhibitors. Indeed, the organic-modified Wells–Dawson POMs had the greatest inhibitory activity of all the tested compounds, but not important differences were found between the peptide-hybrids and the other types of hybrids.

L-phenylalanine (F) and di-L-phenylalanine (FF) hybrids with Keggin POTs were obtained by Nikoloudakis *et al.* [24] *via* amide bond formation between the precursor $TBA_4[PW_{11}O_{39}\{Sn(C_6H_4)C{\equiv}C(C_6H_4)COOH\}]$ [42] and the carboxyl-protected (methyl-ester) di-L-phenylalanine or L-phenylalanine (H-FF-OMe or H-F-OMe) derivatives, using isobutyl chloroformate as a coupling reagent (see Fig. 3.6). In contrast to the propionyl group, this linking group presents a more rigid structure, which could influence the supramolecular chemistry of these hybrids.

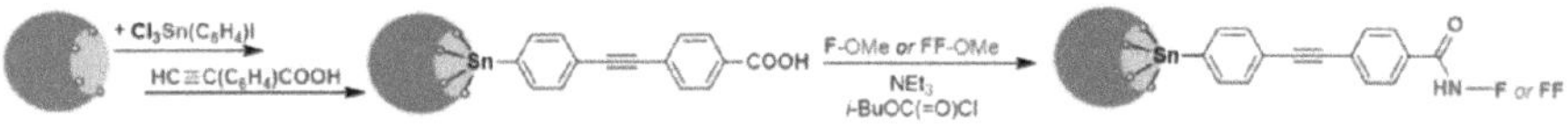

Figure 3.6 General synthetic approach to obtain phenylalanine-Keggin hybrids.

3.2.1.3 Characterization of POM-peptide hybrids

As we have seen in the previous sections, the synthesis of the POM-peptide hybrids employs techniques similar to those used in traditional organic and inorganic chemistry. One key advantage of dealing with the ionic inorganic metal-oxide clusters is that, in most of the cases, crystals can be obtained which, through single-crystal X-ray diffraction, offer a direct method of confirming the chemical structure of the molecule that has been synthesized. The vast majority of examples use TBA counter cations, thus the crystallization is mainly driven by slow-ether diffusion into a DMF or MeCN solution, an approach that enables the isolation of high-quality crystals. In addition to single-crystal X-ray diffraction, thermogravimetric analysis (TGA) can be used as a tool to determine the amount of organic linker attached to the inorganic core. This technique, however, is not fully quantitative and the bare POM must be submitted to the same analysis, since metal-oxide may sublimate yielding an additional mass loss. For instance, Yvon *et al.* [10] employed TGA analysis to calculate the formula of the molecule.

On the other hand, traditional organic chemistry techniques permit further exploration and understanding of the interaction between both entities. For instance, nuclear magnetic resonance (NMR) spectroscopy is an interesting tool to check not only the structure of the hybrid, but also the dynamics. When dealing with Mn-Anderson derived hybrids, an interesting phenomenon is produced due to the paramagnetic behavior of Mn(III). Due to this property, signals of the methylene protons of TRIS, which are close to the Mn core, are downfield shifted up to 65 ppm. This signal (12H as broad singlet) may shift due to the conjugation of the peptides following the aforementioned strategies. NMR is a powerful tool to elucidate the dynamics of organic molecules in a solution, from small molecules up to proteins. For this reason, non-routine techniques in NMR have been applied in POM-peptide science like Diffusion Ordered SpectroscopY (DOSY) or Nuclear Overhauser Effect SpectroscopY (NOESY). The latter is especially relevant in protein science since it can be used to estimate the distance of protons within the molecule up to a few Å and the structure that the peptide displays in a solution. Peptides fold in the solution due to different driving forces mainly non-covalent interactions like hydrogen bonding, electrostatic interactions, and hydrophobic effect.

For instance, Ventura *et al.* [11] took advantage of this spectroscopic technique to demonstrate the interaction of the peptidic chain with the succinate linker, which is directly attached to the TRIS moiety. This proximity is related to the secondary structure that the peptide displays once attached to the POM. In addition to peptide conformation, Luo *et al.* [13] used NOESY to demonstrate that their assumption of the release of TBA cations to the solution is true, since no cross peak was observed between the peptide chains and the TBA signals. NMR has also been used in the studies of the stereochemistry of Sn-based POM hybrids, and will be described in section 3.2.3.

MS is often employed in the field of peptide chemistry, where high-resolution MS techniques determine not only the purity of the peptide, but also the amino acid sequence. In a similar way, MS has been a useful tool in POM chemistry to characterize the metal-oxide cluster formula as well as its self-assembly.

Other conventional but infrequently employed chemical characterization techniques are also being used to tease chemical and structural information on how these hybrids behave in a solution. As mentioned previously, peptides can fold in the solution with established structures and traditional biochemical techniques like circular dichroism (CD) spectroscopy can be used to monitor the formation of those secondary structures. An interesting finding was obtained by two distinct groups. In the works of Ventura *et al.* [11] and Yvon *et al.* [10], the conjugation of small peptides to the Mn-Anderson core led to the formation of α-helix-based structural assemblies in contrast to the same POM-free peptides. Although the explanation is not clear yet, it points to the key role of these inorganic clusters in the assembly of peptides.

Lastly, non-covalent interactions such as hydrogen bonding, electrostatic interactions, and hydrophobic interactions can drive the assembly of the hybrids where nanoparticles can be formed. Therefore, the effects of modifying experimental variables (e.g., temperature and salt concentration) can be monitored through dynamic light scattering (DLS), which uses light scattering to calculate the diffusivity of a given particle, thus permitting to calculate the hydrodynamic diameter by the Einstein–Stokes equation.

As a consequence of the assembly process, it is possible to investigate the morphology of the assembly by means of electron microscopy techniques either transmission electron microscopy (TEM) or scanning electron microscopy (SEM) depending on the size. Traditional microscopy of peptide materials requires staining with heavy metal ions, but in the case of these hybrids, the built-in electronic contrast is provided by the POM clusters. This feature facilitates hugely the characterization by electronic microscopy because the sample will be largely free from staining artifacts.

3.2.2 Self-Assembly, Folding, and Supramolecular Chemistry

3.2.2.1 Charge and morphology of the POM

Proteins and peptides adopt complex structures once in a solution, driven by non-covalent interactions. Among all possible contributions, hydrogen bonding, electrostatic interactions, and hydrophobic interactions play a pivotal role in protein folding. POMs can also self-assemble in the solution giving rise to more sophisticated structures through non-covalent interactions (see Fig. 3.7).

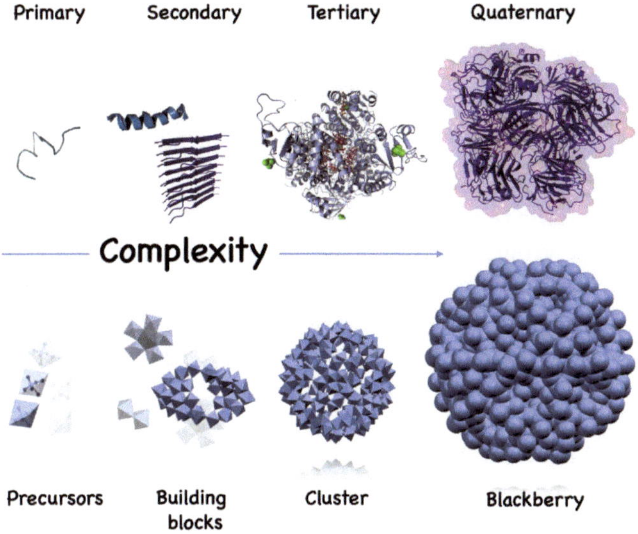

Figure 3.7 General illustration comparing the assembly of biological systems and the assembly of polyoxometalates. The degree of complexity increases in each step where both covalent and non-covalent interactions drive the formation of more complex structures. Protein Images were made online using 3D Protein Imaging [43].

Due to the high anionic charge of POMs, electrostatic interactions are the main driving force that triggers self-aggregation, and the pioneering work of the Tianbo Liu group has provided a solid base for studies on structure–assembly relationships [44–46]. POMs, like proteins, increase in complexity, due to self-assembly processes triggered by both covalent and non-covalent interactions. For instance, in the case of the highly charged wheel-type POM as $\{Mo_{132}\}^{42-}$, these clusters appear as single ions in pure water but upon addition of acetone, hollow particles of *ca.* 100 nm were formed. The counter-cations played a crucial role during the self-assembly process, because in concentrated solutions they were in close proximity surrounding the surface of the metal-oxo cluster. This phenomenon was favored by the addition of a co-solvent in which the POM was less soluble, since in pure water the highly charged POM was unable to aggregate due to electrostatic repulsion. However, once a miscible less polar solvent was added, the negative charge of the POM anion was gradually screened by the counterions, increasing the "effective concentration" of POMs able to self-assemble. If hydrophobic forces drove the self-assembly, by increasing the acetone content, larger structures would be obtained, due to the less charged nature of the POM. However, this phenomenon was not observed in the system. According to the authors, only electrostatic interactions steered the assembly of these types of POMs.

The self-assembly of hybrid POMs derivatized with TRIS or organotin groups has been widely explored, since the organic moiety endows new functionality to the system. In this case, not only the organic functionality, but also the geometry and charge of the POM as well as the associated counterions all play a pivotal role in the self-assembly process.

For the self-assembly of POM-peptide hybrids, it is worth citing the research from the groups of Cronin on TRIS-based hybrids, and Coutsolelos and Lacôte on Sn-derived hybrids. Each of these groups has explored and reported the assembly of a number of different POM-peptide systems. In each case, the inorganic center (Mn-Anderson, Keggin, and Wells–Dawson) and the peptide are different, so we will use those examples to deduce rules and conclusions on the assembly of these complexes. All POMs, as mentioned before, are highly negatively charged, so the assembly is expected to be driven by electrostatic interactions. In this case, though, the added peptidic fragment may influence the association. Furthermore, it can be studied as the other way around, the charge and the morphology of the cluster can impact the peptide folding.

As an example of how the metal-oxide cluster can influence the structure of the peptide, Lâcote and co-workers studied the hybrids $\alpha_{1/2}$-[$P_2W_{17}O_{61}${Sn(CH_2)CO($NHCH_2CO$)$_n$-NH_2}] by high field NMR and showed how a simple poly-glycine (poly-G) chain folded around the tin-substituted Wells–Dawson POT [39]. They observed how poly-G oligopeptides fold toward the metal-oxide surface was established by a "zipper hydrogen bond network". The rigidity of these structures was directed mainly by the polarity of the solvent, obtaining more relaxed structures by increasing the polarity. In the case of longer chain peptides, where there was not enough space around the metal-oxide surface, the rest of the peptidic chain wrapped around the initial loop forming β-turns. This folding was also topology-dependent, since two different regioisomers of the same POM resulted in different peptide arrangements. Consequently, it was possible to establish some rules to bear in mind for future designs, such as how the interaction of the peptide with the POM depends on the morphology of the hybrid and also the length of the sequence. Interestingly, this is the only case of all the studies reported here (see Table 3.1) in which the peptidic chain folds toward the metallic cluster.

The organic linker is not a mere bridge between the POM and the peptide. It is another valuable tool to program the self-assembly of hybrids, since it confers key properties to the forming structures, such as flexibility and rigidity. For instance, the hybrids obtained by Coutsolelos and co-workers contained a highly rigid linker $\{Sn(C_6H_4)C{\equiv}C(C_6H_4)CO\}$ (see Fig. 3.6) that separates the peptide motif from the POM [24]; while in the Sn-substituted cluster design by Lacôte and co-workers, the peptide was grafted to the metallic cluster through a shorter $\{Sn(CH_2)_2CO\}$ spacer. Although this aspect was not discussed in their work, the rigid linker could prevent folding of the dipeptide toward the metallic cluster.

Also, in the case of the Keggin hybrids reported by Nikoloudakis *et al.* [24], their self-assembly in a solution was especially relevant. By SEM, supramolecular nanospheres were observed in different solvent mixtures for all compounds. The main difference between the peptide-hybrids ($PW_{11}O_{39}$-F and $PW_{11}O_{39}$-FF) and the precursor hybrid ($PW_{11}O_{39}$-COOH) was the diameter of the nanospheres obtained, bigger for the peptide-hybrids. The nanospheres presented micellar structures, in which the organic arms were placed at the most internal part of the aggregates and the POM heads got more exposed to the solvent due to their higher polarity. The TBA countercations acted, in this case, as mediators by enabling the communication between polyoxoanions located at the hydrophobic interface. Molecular dynamic simulations confirmed the hypothesis of hydrophobic forces being the key for the assembly to take place.

Interestingly, these nanospheres resembled those obtained by Yan *et al.* [47] by electrostatic interaction of FF and POM anions.

Luo *et al.* [13] induced the self-assembly process by means of the addition of a non-solvent to an organic solution of the hybrid. This process was slow, though, and required up to two weeks to reach the thermodynamic equilibrium. As a result, the authors found nanospheres that were readily observable by means of TEM.

The authors studied how the modification of the preparation condition influences the nanoparticle size, measured by DLS. When the authors gradually modified the amount of water added to the system, they found that there was no linear relationship with the changes in the reciprocal of the dielectric constant. A self-assembly system that exhibits such a linear relationship could not be only ionically driven, if not, the mechanism would involve multiple contributions.

On the other hand, the authors proved the dissociation of TBA cations from the inorganic cluster by DOSY. It is an NMR technique that allows measuring the diffusion coefficient of a given spin system. Substances of different molecular weight have different diffusion coefficient, so they will appear in different lines at the bidimensional spectrum. What they found was that TBA was partially dissociated from the POM in pure acetonitrile solutions—this depends on the strength of the ion pair—but when water was added the dissociation of TBA from the cluster was total. The authors suggested that TBA cations were free in the bulk of the solution, yielding a highly negative POM. Similar experimental findings were observed by Nikoloudakis *et al.* [24].

The authors aimed to ascertain whether either hydrophobic interaction or hydrogen bonding was minor contribution to the electrostatic interactions. In order to demonstrate that hydrophobic interactions were not predominant in the assembly, the authors selected three hybrids where the number of F residues varies, from one to three (bearing in mind that these numbers should be multiplied by two due to the symmetry of the hybrid). At 50% water content, all three hybrids had similar hydrodynamic diameter, while the scattered intensity was different and decreased with the number of F units. Provided the scattered intensity increases with the concentration and the molecular weight/hydrodynamic size of the aggregates, the result showed that the tendency of the hybrids to aggregate diminished upon increasing the number of F in the peptides. As the hybrids with more F amino acids are more hydrophobic, hydrophobic interactions were discarded as a potential major driving force for the self-assembly. Finally, to exclude the role of hydrogen bonding, the authors registered the hydrodynamic diameter of the aggregates in both H_2O and D_2O, while adding NaCl to favor the assembly. D_2O forms hydrogen bonds with different strength than H_2O thus it is very convenient to study processes driven by this interaction. What the authors extracted from this experiment was that hydrogen bonding played only a minor role in the hybrid assembly and only arose at high salt content. This paper settled the bases for the elucidation of the self-assembly of peptide-POM hybrids, suggesting that the main driving force was based on electrostatic interactions of the inorganic clusters. This finding was supported since the counterion (TBA) was not playing any role while assembling. It is worth remarking that Ventura *et al.* [11] obtained a similar behavior for $\{MnMo_6O_{18}\}$-(TRIS-demobensin-1)$_2$ hybrid.

3.2.2.2 Peptide properties

In all the literature examples on covalent POM-peptide hybrids, it was found that the peptides, always hydrophobic, played a secondary role in the self-assembly of the system. However, in works from Luo *et al.* [13] and Nikoloudakis *et al.* [24], the authors observed that the length of the peptide influenced the size of the particles obtained.

CD can be used to monitor the formation of secondary structures and an interesting phenomenon was revealed when the hybrids were studied by this technique. Ventura *et al.* [11] and Yvon *et al.* [10] saw that conjugation of small peptides to the Mn-Anderson core led to the formation of structure assemblies based on an α-helix. Although the explanation is not clear yet, it evidences the role of these inorganic clusters on the secondary structure of peptides. According to Lacôte and co-workers, the presence of an inorganic cluster might make the system more rigid favoring the folding of the peptide.

3.2.3 Stereochemistry in POM-Peptide Hybrids: Study and Application

Several groups have focused on studying the chirality of POMs chemistry and its impact on their applications [48]. This is one of the main interests of the Lacôte and Thorimbert groups who routinely use $\{\alpha_{1/2}\text{-}P_2W_{17}O_{61}\}$ POT precursors to access various hybrids. As mentioned in section 3.2.1.2, the functionalization of the Wells–Dawson POM will have different effects on the chirality of the hybrid, mainly depending on which lacuna the Sn atom is placed. While for the α_2 isomer, the Sn atom is at the cap and has C_S symmetry; for the α_1 isomer, the Sn atom is at the belt and is chiral, isomerizing in aqueous solution. Their main interests are the resolution of the enantiomers of $\{\alpha_1\text{-}P_2W_{17}O_{61}\}$ and its hybrids (see Fig. 3.8) and the regioselectivity of the oxoacyl derivatives $[\alpha_{1/2}\text{-}P_2W_{17}O_{61}\{SnCH_2CH_2C(=O)\}]^{6-}$ (see Fig. 3.5B).

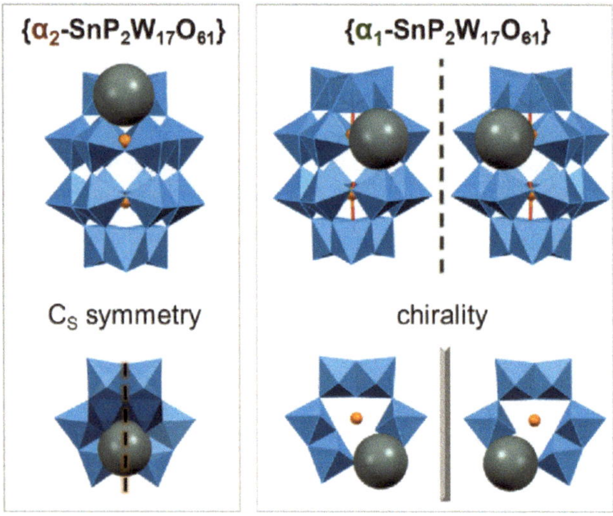

Figure 3.8 Stereochemistry of α_1 and α_2 isomers of $\{SnP_2W_{17}O_{61}\}$ moieties, based on reference [21].

In 2005, they accomplished the first direct observation of the two pure α_1-diastereoisomers, without averaging signals from different compounds, through the splitting of ^{31}P-NMR signals of hybrids formed with chiral amino acids [18]. This splitting takes place when the spacer folds facilitating a hydrogen bond between the free proton of the amide and the negatively charged oxygen atoms of the surface of the POMs. This interaction approaches the two stereogenic elements and also allows possible secondary interactions between the side chain of the amino acid and the POM surface. The splitting of the ^{31}P-NMR was due to the presence of the two diastereoisomers, formed from the racemic synthesis of the lacunary POM and not a rotameric equilibrium in the amide bond.

This hypothesis was further explored in the work of Micoine *et al.* [21]: multiple hydrogen bonds between organic molecules and the negatively charged inorganic cluster would be the key for the chiral recognition and kinetic resolution of $[\alpha_1\text{-}P_2W_{17}O_{61}\{SnCH_2CH_2C(=O)\}]^{6-}$ enantiomers. A series of tripeptides was screened in the amide bond formation reaction with the racemic mixture. After optimization of the kinetic resolution conditions, 20% of the starting material was recovered as unreacted enantiopure (+)-$[\alpha_1\text{-}P_2W_{17}O_{61}\{SnCH_2CH_2C(=O)\}]^{6-}$, while the POT-peptide hybrid was obtained in 79% yield and 65/35 diastereomeric ratio. The reaction of the isolated compound with any chiral nucleophile yielded a single diastereomer, which was the final proof for the > 99% enantiopurity of the product of the kinetic resolution. Pure (-) enantiomer was obtained from the reaction of the same racemic mixture with the D-tripeptide.

As previously mentioned, in the work of Boglio *et al.* [20] in 2008, the authors explored the regioselectivity of the intramolecular oxoacylation reaction (see Fig. 3.5B) to form the hybrids $(TBA)_6[\alpha_{1/2}\text{-}P_2W_{17}O_{61}\{SnCH_2CH_2C(=O)\}]$, as well as the chirality of the new products. Since the acylated POM-hybrids could not be crystallized, determination of the regioselectivity was made from spectroscopic data (NMR and ESI-MS) and theoretical studies (DFT calculations). In the case of the α_2 hybrid, whose precursor is achiral, the privileged regioisomer arises from the reaction of the activated carboxyl group with a cap/cap bridging Sn-O-W oxygen (see Fig. 3.9, isomer B). For the α_1 hybrid, whose starting complex is chiral, all oxygen atoms are non-equivalent and there are four different positions for the reaction. Theoretical calculations demonstrated that only two of them are possible, being the one with the organic ligand within an edge-sharing Sn-O-W group the most plausible as for the α_2 hybrid (see Fig. 3.9, isomer D).

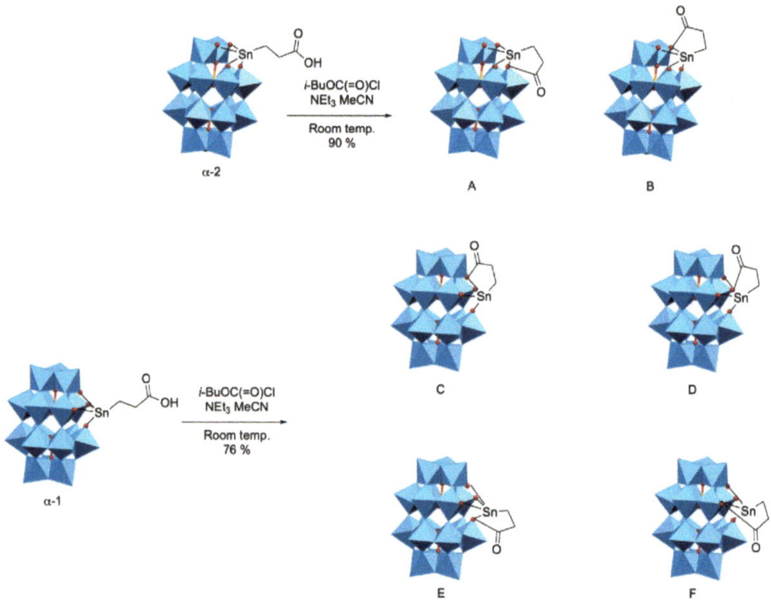

Figure 3.9 Acylation of $\alpha_{1/2}$-$P_2W_{17}O_{61}\{SnCH_2CH_2C(O)\}$ and possible regioisomers, based on reference [20].

Chirality broadens the already interesting potential applications of POM-hybrids: from selective catalysis, materials science to biomedicine. In the latter case, there are numerous examples of chiral biomolecules and stereoselectivity has an important role in biochemical reactions, e.g., proteins are formed only by L-amino acids. As described above, new POM hybrids with L/D-amino acids, (L/D-aa){TRIS-$MnMo_6O_{18}$-TRIS}(L/D-aa), were recently prepared [12]. A wide variety of POMs and POM-hybrids have been used as *in vitro* inhibitors of amyloid-β (Aβ) aggregation [38, 49]. The authors aimed at studying if chirality of the hybrids would have a positive impact on their activity and selectivity, since amyloid-β fibrils prepared *in vitro* present left-handed helical structures [50]. First, the binding affinity of the Mn-Anderson hybrids to the Aβ peptide was measured by isothermal titration calorimetry (ITC) and fluorescence titration. The (F)-modified POMs showed the highest affinity and the first proof of enantioselectivity: (D-F)-POM bonded 8-fold stronger to Aβ than (L-F)-POM. Results from ^{1}H-NMR spectroscopy and molecular dynamic simulations confirmed the binding site of the hybrids to the peptide, found in the segment $A\beta_{13-20}$ (HHQKLVFF). Although all hybrids could prevent the aggregation of the Aβ peptide to fibrils, the (F)-modified POMs, and more specifically the (D-F)-POM, showed stronger activity.

3.2.4 Future Perspectives

A thorough review of the published literature of the covalent Class II POM-peptide hybrids reveals the robust methodology developed for the preparation of asymmetric Mn-Anderson POMs, as yet no more reports have been published according to this approach. The scope for this type of asymmetric hybrid, specifically in the field of medicine—where POM-peptide hybrids remain practically unexplored—is significant and we anticipate that the development of this area will provide a number of relevant and useful materials.

In addition to the often-employed Mn-Anderson core, there are numerous possible POM architectures available, all of them with different topology and redox properties. Thus, the use of other POM architectures is highly desirable to expand the chemical space in the design of new POM-peptide hybrids. Furthermore, increasing the peptide chain length would allow for more complex hybrid assemblies.

As a general observation, the covalent conjugation of peptides to the POM cluster in search for synergistic effects between both structures, combined with the attachment of biologically relevant peptides like those reported by Ventura *et al.* [11], paves the way for using these hybrids in real-world application. Such systems would be ideal for linking enzymes to tune and/or adapt their catalytic activity. In many enzymes, metallic center cofactors are found to play a pivotal role in catalysis [51, 52] and indeed, nature has evolved to use molybdenum storage proteins in bacteria [53, 54].

Another possible direction would consist of the covalent assembly of distinct POMs with different properties. In this way, it would be possible to explore synergistic effects between the metallic clusters. Further, the modularity of a peptide spacer may permit the system to fold, creating interesting structures or functional materials.

What is clear is that future steps in this field will involve passing from fundamental synthetic chemistry to functional materials so that structure–property relationships feed fluidly into innovative design principles. In this regard, new chemistries like supramolecular polymerization reactions or dynamic covalent bonds would offer an appealing boost to the preparation of new covalent POM-peptide materials.

3.3 Ionic POM-Peptide Hybrids

Throughout the previous section, we discussed how POMs can influence the assembly of peptides covalently attached to the cluster in Class II-type POM-peptide hybrids. In spite of the limited number of publications on this topic until now, we can draw some general conclusions: (i) thermodynamics and electrostatic interactions constitute the main driving forces in the self-assembly of these supramolecular architectures; (ii) the size and charge of the POM cluster anion play a major role on the assembly process; and (iii) the counter cations are partially dissociated from the anionic scaffold. Here we will review the literature examples of the Class I-type POM-peptide hybrids, which are based on the purely ionic assembly between amino acids or peptides and POMs. Please note that several examples discussed throughout this section of the chapter have been revised in a review from 2016 [9].

One key feature of ionic assembly of POM-peptide hybrids compared to the covalent molecular POM-peptides is the POM peptide stoichiometry, which is fixed in the latter material, whereas it can be easily tuned in the ionic hybrid materials without any extra synthesis, i.e., an excess of peptide/positive charges can be added in the material. We will explore this feature as well as how the components of the hybrids affect the self-assembly and therefore the key physicochemical properties giving rise to a variety of relevant applications (see Fig. 3.10).

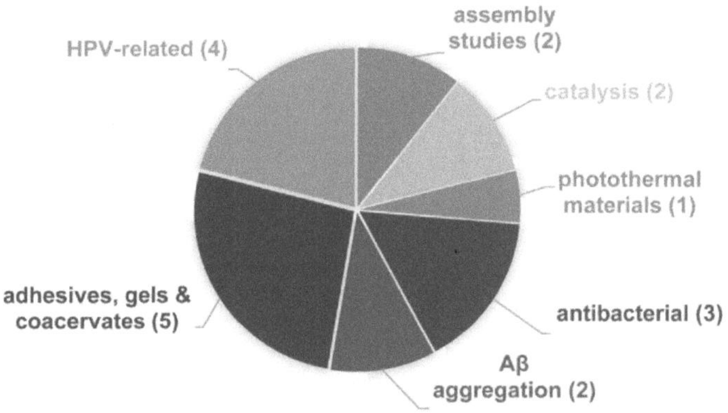

Figure 3.10 Class I POM-peptide hybrids formed by ionic self-assembly present interesting properties for a range of different applications, as described in the 18 articles covered in this chapter (see Table 3.2). (HPV = Human Papilloma Virus, Aß = amyloid-ß peptide).

Both POM and peptide systems can self-assemble from small building blocks to larger supramolecular structures—from 1D to 3D structures—where both are characterized by novel physicochemical properties associated with their increasing structural complexity and new dynamic structural arrangements, as we have discussed before (see Fig. 3.7). Therefore, the combination of both entities may open a vast number of possibilities of assembly or synergistic properties. Nature is expert at such combinations gathering the rigidity and redox properties of metallic clusters with the elevated flexibility and self-assembly properties of proteins; for example, the iron-sulfur clusters found in enzymes that assess vital roles like energy production [55].

Recently Streb & Nyman and co-workers reviewed the importance of counter-ions in the solubility of POMs and therefore self-assembly of these clusters in a solution [56]. Normally, alkali cations and bulky organic cations have driven the research in POM assembly; however, more exotic cationic species like lysine-/arginine-rich peptides have gained the attention lately as readily available building blocks, which provide access to complex structures with highly tunable physicochemical properties.

3.3.1 Building Blocks: POMs & Peptides

Table 3.2 Overview of the POM clusters and peptides used in formation of Class I POM-peptide hybrids

Entry	Reference	Year	POMs			Peptide	Application
			POM type	Formula	Abbreviation		
1	[47]	2010	Keggin heteropolyacid	$H_3PW_{12}O_{40}$	PTA	CDP (FF- NH_2)	encapsulation
2	[57]	2013	Co-substituted Wells–Dawson	$K_8[P_2CoW_{17}O_{61}]$	P_2CoW_{17}	$A\beta_{15\text{-}20}$ (Ac-QKLVFF- NH_2)	inhibition Aβ aggregation
3	[58]	2015	Weakley/bi-Lindqvist	$Na_9[EuW_{10}O_{36}]$	EuW_{10}	HPV18Ctb (SSKPAKRVRVRARK) HPV16Ctb (SSTSTTAKRKKRKL)	HPV related
4	[59]	2015	Weakley-type Keggin	$K_{13}[Eu(SiW_{10}MoO_{39})_2]$	EuSiWMo	HPV16Ctb (SSTSTTAKRKKRKL)	HPV related
5	[60]	2015	Keggin heteropolyacid	$H_3PW_{12}O_{40}$	PTA	(FF-OMe)	catalysis
6	[61]	2016	Keggin heteropolyacid	$H_4SiW_{12}O_{40}$	HSiW	KAzoKAzoKAzoK-NH_2[†]	antibacterial
7	[62]	2016	Decavanadate tri-lacunary Keggin Keggin heteropolyacid Wells–Dawson Preyssler Wheel Keplerate	$Na_6[V_{10}O_{28}]$ $Na_9[PW_9O_{34}]$ $H_4SiW_{12}O_{40}$ $K_6[P_2W_{18}O_{62}]$ $K_{12.5}Na_{1.5}[NaP_5W_{30}O_{110}]$ $K_{28}Li_5[H_7(P_8W_{48}O_{184})]$ $(NH_4)_{42}[Mo_{132}O_{372}(CH_3COO)_{30}(H_2O)_{72}]$	$V_{10}O_{28}$ PW_9 HSiW P_2W_{18} NaPW P_8W_{48} Mo_{132}	KAzoKAzoKAzoK-NH_2 KAzoKAzoKAzoKAzoK-NH_2 KAzoKAKAzoK-NH_2 KFKFKFK-NH_2 SAzoKAKAzoS-NH_2 SFKFKFS-NH_2	1D assembly
8	[63]	2016	Keggin heteropolyacids	$H_4SiW_{12}O_{40}$ $H_3PW_{12}O_{40}$ $H_5PMo_{10}V_2O_{40}$	HSiW PTA HPMoV	KFKFKFK-NH_2 KLKLKLK-NH_2 KVKVKVK-NH_2 KAKAKAK-NH_2 KFKFK-NH_2 KFKFK-NH_2 KVKVK-NH_2 KAKAK-NH_2 KFK-NH_2 KVK-NH_2	hydrogels

9	[64]	2017	Keggin heteropolyacid	$H_4SiW_{12}O_{40}$	HSiW	H	adhesives
10	[65]	2017	Keggin heteropolyacids Keggin	$H_4SiW_{12}O_{40}$ $H_5PMo_{10}V_2O_{40}$ $K_6[CoW_{12}O_{40}]$	HSiW HPMoV CoW_{12}	Ac-EE**MQ**RRAD- NH_2 Ac-EE**MQ**READ- NH_2 Ac-EE**MQ**EEAD- NH_2 Ac-EE**MQ**RR- NH_2 **RR**DV**Y** AH	adhesives
11	[66]	2017	Decavanadate Keggin Wells–Dawson	$Na_6[V_{10}O_{28}]$ $K_6[CoW_{12}O_{40}]$ $K_6[P_2W_{18}O_{62}]$	$V_{10}O_{28}$ CoW_{12} P_2W_{18}	L-**S**FKFKF**S**-NHOAc D-**S**FKFKF**S**-NHOAc L-**S**FKFKL**S**-NHOAc L-**S**FKFKL**S**-NHOAc L-**S**VKVKV**S**-NHOAc L-**S**WKWKW**S**-NHOAc	1D assembly
12	[67]	2017	Weakley / bi-Lindqvist	$Na_9[EuW_{10}O_{36}]$	EuW_{10}	KKKKKKKK	photocatalysis
13	[68]	2018	Weakley-type Keggin Weakley / bi-Lindqvist Lacunary Keggin sandwich	$K_{13}[Eu(SiW_{10}MoO_{39})_2]$ $Na_9[EuW_{10}O_{36}]$ $K_{11}[Eu(PW_{11}O_{39})_2]$	EuSiWMo EuW_{10} $EuPW_{11}$	HPV16L1Ctb (**SSTSTTA**KRKKRK**L**) HPV16L2Nt (**M**RHKR**SA**KR**T**KR**A**)	HPV related
14	[69]	2018	Weakley / bi-Lindqvist	$Na_9[EuW_{10}O_{36}]$	EuW_{10}	HPV16 Virus-like particles (VLPs) 72 x L1-pentamers	HPV related
15	[70]	2019	Mono-lacunary Keggin Tri-lacunary Keggin Keggin	$K_8[\alpha\text{-}SiW_{11}O_{39}]$ $Na_9[PW_9O_{34}]$ $K_6[CoW_{12}O_{40}]$	SiW_{11} PW_9 CoW_{12}	**G**HK **G**FK **G**VK	adhesives
16	[71]	2019	V-capped polymolybdate	$(H_2O)_3(NH_4)_{18}$ $[Mo_{57}V_6(NO_3)_6O_{183}$ $(H_2O)_{183}]\cdot{}_{56}H_2O$	Mo-POM	$A\beta_{15\text{-}20}$ (Ac-**Q**K**LV**FF-NH_2)	inhibition Aβ aggregation
17	[72]	2019	Keggin heteropolyacid	$H_3PMo_{12}O_{40}$	HPMo	Soybean pentapeptide (**S**H**CMN**)	Photothermal materials Antibacterial
18	[73]	2019	Mono-lacunary Keggin	$K_8[\alpha\text{-}SiW_{11}O_{39}]$	SiW_{11}	R	anticorrosive coating

Notes: (a) Key amino acid residues have been highlighted, the color coding is red, hydrophobic; green, basic; blue, acidic; (b) only *N*-terminal and *C*-terminal modifications are labeled. †azo = azobenzene as side chain.

3.3.1.1 POM clusters

In most of the works here reviewed, small and simple POM structures have been used (see Table 3.2), namely the Keggin anion and some of its different varieties, such as heteropolyacids ($H_3PW_{12}O_{40}$ and $H_4SiW_{12}O_{40}$), lacunary ($Na_9[PW_9O_{34}]$ and $K_8[\alpha\text{-}SiW_{11}O_{39}]$), and Weakley-type Keggin ($K_{13}[Eu(SiW_{10}MoO_{39})_2]$). POMs are nano-sized polyanions. Each unit has a nanometrical size, e.g., the Keggin anion $[SiW_{12}O_{40}]^{6-}$ measures 1 nm on each of its sides [74]. Furthermore, POMs are highly negatively charged. Being dense polyanions render them with characteristics such as their ability to interact with positively charged molecules. This is going to be the key for their interaction with peptides and the formation of ionic self-assembled hybrids.

As mentioned before, one of the most used types of POMs is the Keggin-type heteropoly acids (HPAs), which are the acid forms of the heteroPOMs. HPAs not only present all the interesting properties of POMs, but also have other advantages. They are strong Brønsted acids and, contrary to their salts, HPAs can be dissolved in polar organic solvents [75]. HPAs have been used in ionic self-assembly with peptides for different applications, from catalysis [60, 72] to adhesives [64, 65]. POMs are more frequently used, in a broad approach, in their conjugated salt forms for all types of structures. The Europium-containing POMs, which add the fluorescence to their properties, deserve a special mention in this section. Modulation of the luminescence intensity in co-assembly with peptides provides new applications for these hybrids.

3.3.1.2 Peptides

Given the negatively charged nature of the most common POMs, it follows that cationic peptides, or positively charged amino acids, have been chosen for the self-assembly formation of Class I POM-peptide hybrids. Aromatic amino acids, such as phenylalanine, have also been widely used in the works reviewed here. These residues provide hydrophobicity to the system and also present self-assembling and packing characteristics, as in the case of the previously seen di-L-phenylalanine peptide [24, 47].

An interesting type of peptide widely used by Wen Li and co-workers is the facial-like peptide that is characterized by its amphipathic properties; that is, when the peptide folds, hydrophobic and hydrophilic side chains locate in different sides of the peptide giving the name of facial-type peptide. Furthermore, this motif is especially relevant in natural antimicrobial peptides that display a combination of hydrophobic and positively charged (e.g., K or R) residues and it is broadly accepted that such particular face segregation is responsible for their antimicrobial activity [76, 77]. Many naturally occurring peptides have been used for the assembly of ionic POM-peptide hybrids; for example, arginine-/lysine-rich peptides, also known as nuclear localization signals (NLSs), from the capsid proteins of HPV-16 and HPV-18 (human papilloma virus); and the antioxidant soybean pentapeptide. Importantly, naturally occurring peptides can also confer biocompatibility to the hybrids: a key parameter for the development of materials with prospective biomedical applications.

3.3.2 Mechanisms of Assembly: A Case Study

In 2010, X. Yan *et al.* [47] reported—to the best of our knowledge—the first assembly compromising both entities, POMs and peptides (Fig. 3.11). Herein, they selected a nearly spherical phosphotungstic acid (PTA) and a dipeptide of phenylalanine (FF), we have previously mentioned some examples with this motif and its impact on protein assembly. The authors added an aqueous PTA solution into a peptide HFIP (1,1,1,3,3,3-hexafluoro-2-propanol) solution, at 1:5 charge ratio. The electrostatic interactions between the positive *N*-terminus of FF and the anionic PTA trigger the encapsulation of these anionic clusters inside peptide capsules. These structures of *ca.* 1.4 nm, that compromise one PTA surrounded by FF, can further self-assemble producing larger structures of *ca.* 150 nm. When the authors varied the PTA-to-FF ratio, they did not observe significant variation in the nanoparticle size. Moreover, upon a certain threshold from the system was no longer stable, leading to precipitation. This occurs due to the excess of peptide that may crosslink nanoparticles. The ζ-potential, in turn, varied according to the amount of peptide added shifting toward more positive values. The simple self-assembling system that Li and co-workers developed herein displayed structure–property responsiveness to external stimuli like pH and temperature. For instance, increasing the pH led, again, to an unstable system as a consequence of the loss of positive charge of the amino groups. Regarding temperature, upon heating to 70 °C turbidity of the colloidal suspension disappeared and was ascribed to the disassembly the ionic interactions holding together the nanospheres. Beyond stimuli responsiveness, these nanoparticles were used as scaffold to encapsulate different molecules like fluorescent dyes, which were incorporated no matter their charge, natural polymers, and dextran, which is a neutral polysaccharide. In summary, the stepwise assembly process first of all consists of charge recognition between the anionic POM and positive peptide and then a subsequent association and assembly process *via* non-covalent interactions. A similar phenomenon in the covalent hybrids was described by Luo *et al.* [13], where the assembly was also described as mainly ionically driven through POM association. In general terms, this type of assembly is common for all the examples we are going to focus on during this section.

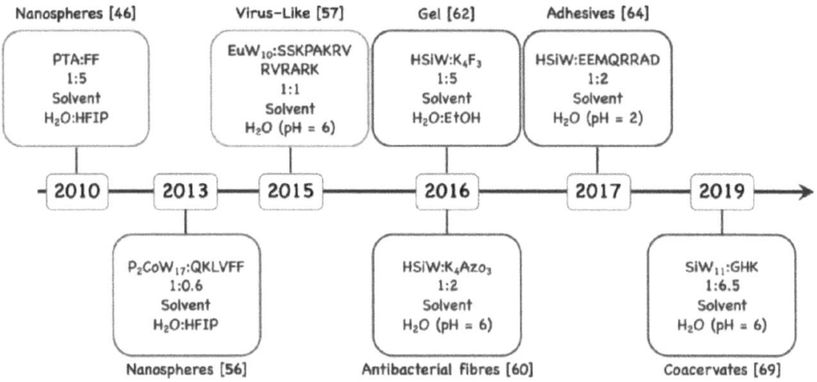

Figure 3.11 Timeline of different Class I (ionic) POM-peptide hybrids summarizing the experimental conditions and morphology of the materials obtained. The color code depicts the applications as in Fig. 3.10. The references of each work are indicated between brackets.

Since then, the Li group has made a number of significant contributions to the field of active POM-peptide materials. This unique combination allowed the authors to obtain a library of different types of structures by changing the preparation conditions, e.g., adding a co-solvent, varying the peptide length, or the inorganic cluster.

One notable example is the hybrids formed from facial-like peptides. The combination of hydrophobic and hydrophilic residues in these peptides adopt a β-sheet structure, however a certain number of amino acids is required for the peptide to arrange into a secondary structure. Thus, when the peptide cannot fold into either an α-helix or a β-sheet, it adopts a random coil conformation. The addition of POMs to these particular peptide arrangements might favor the assembly, according to the ionically-driven processes we have discussed previously. Therefore, Li *et al.* [61] pushed the idea into a further level by combining facial-like peptides, consisting of hydrophilic (lysine, K) and hydrophobic (azobenzene, Azo) residues, and the hetero-polyacid $H_4SiW_{12}O_{40}$ (HSiW).

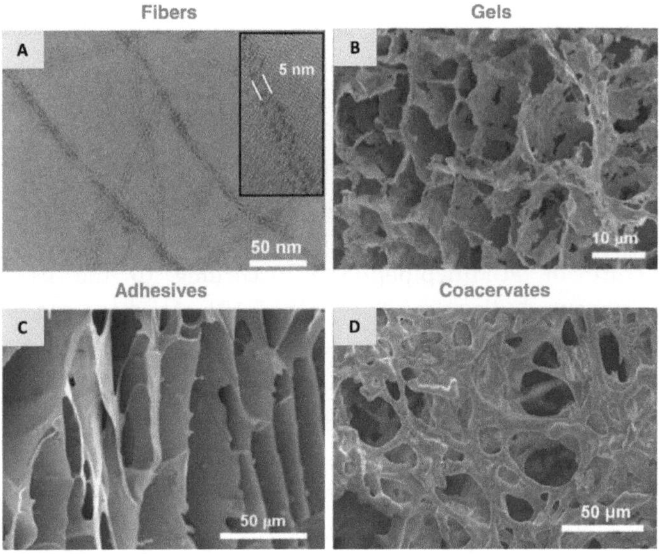

Figure 3.12 Micrographs of different POM-peptide assemblies. The sequence of the peptide, POMs characteristics, and preparation conditions decide the type of material. Reproduced with permission from references [61, 63, 65, 70].

The peptide depicted in Table 3.2 (entry 6) did not display any preferential secondary structure when free in a solution, but a random coil. However, as soon as HSiW was added the electrostatic interactions between both entities triggered the formation of β-sheets, as monitored by CD. Indeed, the β-sheet content rose upon increasing the equivalents of POM, up to a maximum POM-to-peptide ratio of 1:2. With a further increase in the amount of POM, the system started to precipitate. At this point, it is clear that the anionic cluster was triggering some sort of aggregation because conformational changes may microphase separate the hydrophilic amino acids—that may be exposed to the solvent from the hydrophobic residues—buried inside the structure. The POM, here, will be acting as a bridge connecting different peptides. Electron microscopy allowed the authors to observe the formation of long fibers with a uniform diameter of *ca.* 13 nm. In these fibers, the POM anions could be clearly seen due to the electron contrast in TEM, and by HR-TEM, the authors were able to measure the layer distance, which was *ca.* 5 nm (see Fig. 3.12A). ^{183}W-NMR and EDX showed that the inorganic cluster remained stable after the assembly process and the high association constant obtained by ITC demonstrated that the process is mainly driven by electrostatic interactions.

According to the experimental results, the authors suggested that the fibers might be composed of a core-shell structure of POM-peptide. The core of the structure was defined as a layered structure of POM|pep|pep|POM, according to the calculated and measured distances (see Fig. 3.13). As discussed before, the hydrophobic moieties are located at the core of the fibers—interacting by hydrophobic interactions and π-π stacking—while positively charged lysine residues interact with the POM anions. It should be noted that the POM-to-peptide ratio plays an important role in the stabilization of the structure. Each POM anion can interact with four positively charged residues of the peptide, setting a 2:1 peptide-to-POM molar ratio, and therefore, the excess of peptide was accumulated at the surface of the fiber, generating a positively charged structure.

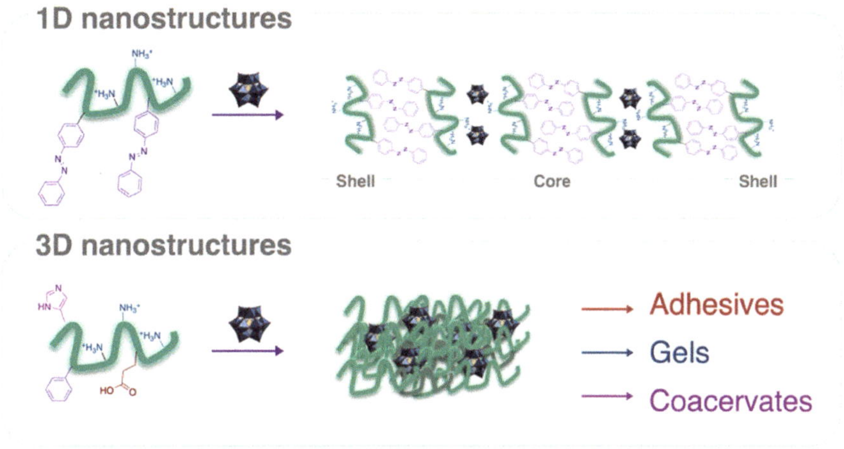

Figure 3.13 General overview of the materials formed by combining facial-like peptides with anionic heteropolyacids.

The hydrophobic residues of the peptide were non-canonical amino acids based on an azobenzene, which undergoes *trans*-to-*cis* isomerization upon irradiation with UV light. The authors observed that when the fibers were submitted to UV irradiation, the fibrillar structures broke down, as a consequence of disrupting the packing at the hydrophobic core. In a recent follow-up publication, the same authors managed to prepare transient nanostructures that undergo morphological change from spheres to sheets using this same stimulus [78].

Other peptides prepared with either phenylalanine or valine—as hydrophobic residues—did not exhibit enough hydrophobic packing to trigger the formation of fibrils, instead, irregular and unstable spheres were observed. With these experiments, one can anticipate that both the peptide design and the nature of the POM will be crucial to obtain stable structures.

Provided these few guidelines, the Li group systematically explored the influence of POM and peptide characteristics to generate new hybrid nanostructured systems. The experimental conditions (e.g., pH or cosolvents) and the POM-to-peptide ratio steer the formation of either 1D or 3D structures (see Fig. 3.12).

For example, in their continuation work, seven different types of POMs and six variations of the facial-like peptides were combined and their assemblies studied [62]. Briefly, the POM size plays a pivotal role in the type of assembly, since it will decide which non-covalent interactions drive the self-assembly process. For a given peptide, when the cluster size was increased, the steric hindrance inhibited peptide folding into fibrils and instead nanospheres were observed. In these structures, for a given type of POM, the length and hydrophobic content of the peptide drive the type of assembly. For example, when the authors kept the same lysine content (four residues), but the hydrophobic content was reduced; the electrostatic repulsions did not allow the peptide to cover and stabilize the nanoparticle surface. Whereas, peptides with less lysine content experimented less repulsion, leading to an optimum coverage and stabilization of nanosphere.

In line with the importance of the hydrophobicity, the same authors described in a recent paper its role in the assembly of the fibrils [66]. Using the decavanadate $V_{10}O_{28}$ as an inorganic cluster, the authors observed that peptides with different chirality—either L- or D-amino acids were used—induce twisted fibers where the helical sense depends on the peptide chirality. Compared with the previous report, the fibers twisted as a consequence of an increased steric hindrance of a larger cluster, causing the peptide layer to bend.

The helical pitch, the distance between two consecutive twists, was found to be dependent on the hydrophobicity of the hydrophobic residue. The authors noticed that a decrease in the hydrophobicity induced larger pitch distances, indeed spherical structures were observed when valine was employed. Tryptophan, in turn, was not sufficiently hydrophobic, and the more effective π-π stacking overcomes the steric hindrance of the POM, resulting in straight tapes.

We have briefly mentioned that the peptide sequence and its length can be precisely modified to obtain different nanostructures. Further, the POM size and its charge might influence the assembly as well, since it will drive how the peptides are arranged in the core of the nanostructure. In addition to the hydrophobicity of the POMs, the addition of another POM (CoW_{12} or P_2W_{18}) finely tuned the pitch of the fibrils. These secondary POMs may disrupt the stacking of the decavanadate inside the fibrils, so the pitch varied accordingly.

Besides nanoparticle or fiber formation, the system developed by the Li group is versatile enough to produce multiple outcomes only varying a few experimental conditions like pH, type of POM, or the characteristics of the peptide. For instance, peptides can form gels, which have been widely described in literature. In order to form a gel, the peptides need to polymerize forming a 3D network, a process that can be achieved either by covalent cross-linking or by supramolecular polymerization.

As we have already seen, HPAs can be used as supramolecular cross-linkers to peptides bearing basic residues to form a 3D structure [61]. The Li group also explored this possibility, by mixing HPAs—Keggin archetype—with facial-like peptides in 2016 (see Table 3.2, entry 8) [63]. The expertise gathered in previous works let them predict that supramolecular polymerization, which forms either fibrils or spheres, can be scaled up to a 3D structure (see Fig. 3.13). We have previously described that increasing the POM-to-peptide ratio led to precipitation due to the cross-linking of multiple nanostructures, e.g., fibers.

Therefore, a trace of an organic solvent modifies the dielectric constant of the medium, thus the ionization of the POM may prevent the system to collapse. It was found, indeed, that EtOH successfully triggered gel formation; other organic solvents did not favor gel formation. The gelation process was carried out by mixing the POM with the peptide (1:1 molar ratio), at relatively high concentrations of both (100 times more concentrated), in presence of EtOH (EtOH:H_2O 1:5).

An interesting observation was that for the system to form a gel, POMs had to be in their protonated forms. In addition, alkali salts of HPA did not induce gelation, due to their scarce solubility in EtOH. When the authors characterized the materials, they observed that the structure formed upon complexation depicted a similar behavior as in their previous reports. HPA induced the transition of the peptides from random coils to β-sheets. This process rendered the formation of irregular nanospheres—observed using diluted solutions—and stack forming sheets and finally the 3D hydrogel structure—observed by SEM (see Fig. 3.12B).

Encouraged by the fact that supramolecular polymerization of POMs and peptides can form 3D network gels, the Li group explored the formation of adhesive ionic POM-peptide hybrids in 2016 [65]. In this work, the peptide design was slightly different from the facial-like peptides, due to the introduction of glutamic acid residues (E), as the researchers hypothesized that carboxylic groups could favor adhesion properties meaning that acidic pH was required. The supramolecular polymerization, as well as in the aforementioned examples, was driven by electrostatic interactions between the positive amino acids (R or K) and the POM. The addition of the inorganic cluster modifies the folding of the peptide, again from coil to β-sheet. If we pay attention to the microscopic network, we can extract some similarities with the structure observed in gels, such as sheet-like structures with 3D pores (see Fig. 3.12C). The strength of adhesive was found to be dependent on the sequence of the peptide. A balance between negatively and positively charged residues was required, since peptides with a high number of anionic amino acids avoided POM complexation [65].

The examples we have described depicted a common motif in the peptide sequence—basic amino acids in an amphipathic peptide—that were responsible for the POM complexation and therefore the self-assembly of the system. However, the introduction of other amino acids like histidine (H), which is well known for its metal binding capacities, could provide different properties to the hybrid material. In recent work, X. Li *et al.* [70] inserted H in small peptides to prepare coacervates—colloidal dispersions with liquid-liquid phase separation. A simple peptide design GHK, permitted to prepare coacervates with a Keggin POM in a 6.5:1 peptide-to-POM molar ratio. The formation was carried out at pH 6.5 in which histidine was not protonated; however, when the pH decreased, the adhesive strength of the coacervates increased. This phenomenon was explained, due to the imidazole degree of protonation that improved the ionic cross-linking and created stiffer structures. Examination of the gel formed at pH 4.5 under SEM showed again a rich 3D network with fibrillar morphology (see Fig. 3.12D). A similar trend was observed when adding metal ions, e.g., Co^{2+} and Ni^{2+}. These metals ions, once coordinated to H, favored the entanglement creating more rigid structures.

Regarding the characterization of the Class I POM-peptide hybrids, various techniques have been repeatedly used in these works, both for the study of the interactions of the precursors in the hybrid and the supramolecular structures formed. For the latter, as previously seen in section 3.2.1.3, electron microscopy, both TEM and SEM, is the most effective technique and the primary source of structural information. The use of high resolution-TEM allows even to observe the small clusters formed by one sole POM unit and the peptide shell, which form part of larger structures such as fibers (see Fig. 3.12A) [61] and nanospheres [47].

Contrary to the examples with covalent hybrids, these self-assembled materials do not crystallize in a solution. Therefore, energy dispersive X-ray spectroscopy (EDX) coupled to EM together with X-ray diffraction has been used instead of X-ray crystallography to gain insight on the chemical composition and structure of the materials. Common and readily available techniques such as FTIR can provide meaningful information of the chemical composition and structure, from slight changes in the bonding energy of POMs upon binding to the peptides to the secondary structures of the peptides based on the vibrations of the amide bonds. ITC is a routine technique in biochemistry that permits to calculate the affinity constant of a given system and can provide valuable information on how the POM-peptide precursors bind to form the resulting hybrid material.

Altogether the ionic POM-peptide hybrids represent an elegant approach for the design of soft materials, and we can draw some general guidelines to prepare such assemblies. Regarding the peptide design, the only condition required to interact with the POM anions is the presence of positively charged residues, which can be either natural or tailor made. This electrostatic recognition is the base of the assembly, and then hydrophobic interactions, hydrogen bonding, and the POM archetype will define the type of nanostructure, thus the application.

3.3.3 Applications

3.3.3.1 Biomedical applications

Most of the published literature describing ionic POM-peptide hybrids include applications in the biomedical field: from antibacterial activity to inhibition of the aggregation of the amyloid-β peptide, including an important group of publications related to the use of peptides from the human papilloma virus (HPV). Their main properties and applications are described below; taking into consideration their supramolecular structures can have an impact on them.

HPV related

The groups of Yuqing Wu and Lixin Wu have performed extensive studies on hybrids formed from Europium-substituted-POMs with peptides derived from the capsid proteins of two pathogenic strains of HPV. In their works published in 2015, the authors described a two-step assembly between HPV peptides and EuW_{10} or EuSiWMo POMs (see Table 3.2, entries 3 and 4) [58, 59]. The peptides used in the studies, also called NLS, are fragments of the HPV-16 and HPV-18 capsid proteins rich in arginine and lysine and therefore, positively charged. As seen in the previous section, the addition of the peptides to a POM solution triggers first the formation of clusters by electrostatic interaction, and further assembly of the clusters leads to the formation of nanospheres. The most exploitable property of these hybrids is the enhanced fluorescence of the Eu-POMs upon binding to the positively charged amino acids of the peptides, which remove the solvating water around the POMs. With this idea in mind, in 2018, the authors developed a protocol for the detection of a type of cell receptors, glycosaminoglycans (GAGs), which allow the HPV to enter cells [68]. In a competitive recognition of the peptides and GAGs, they studied the fluorescence emission of the ionic POM-peptide hybrids in the presence of different types of GAGs: heparin, chondroitin sulfate, and hyaluronic acid. Whenever the peptides showed a high affinity for a specific GAG, the peptide would "leave" the assembly with POMs to bind the GAG, leading to a decrease in the fluorescence of the Eu-POM.

The method was further tested by using the recombinant protein HPV 16L1, confirming its effectiveness to identify the cell receptors, which allow the virus to enter the cells: heparin and chondroid sulfate. This protocol constitutes an example in the use of POM-peptide hybrids for virus screening. In the same year, Wu and co-workers described a different type of assembly again using Eu-substituted-POMs with VLPs of HPV [69]. This type of VLP is formed by 72 pentamers of the peptide HPV16Ctb and has been used for the development of HPV vaccines [79]. With the formation of VLPs-POMs hybrids, the authors aim at improving the thermal and storage stabilities of the VLPs as well as their biological activities. The main characteristic of this work is the difference between the mechanisms of assembly: encapsulation *vs.* adhesion. Encapsulation of EuW_{10} into HPV 16 VLPs (EuW_{10}@VLPs) takes place when the POM solution is added to an aqueous solution of pentamers of the peptide, at 10:1 POM-to-peptide molar ratio. In this case, the assembly of the pentamers into VLPs happens at the same time as the assembly with the POM leading to a co-assembly of the precursors and the encapsulation of EuW_{10} into the VLPs. On the other hand, when the POM is added to the already formed VLPs, its encapsulation is not possible. Instead, the EuW_{10} anions adhere to the surface of the VLPs to form the VLPs@EuW_{10} hybrids. The different mechanisms of assembly and the hybrids formed will impact their properties. For example, the thermal stabilities of VLPs were studied by UV cloud point measurement, which traces the transition temperature. HPV 16 VLPs start to decompose at 68.83 °C, and this critical temperature is increased by *ca.* 2 °C, in the presence of both hybrids, EuW_{10}@VLPs and VLPs@EuW_{10}. The mechanism involved the ionic binding of the negatively charged POMs and the positively charged Arg and Lys amino acid residues of the VLPs peptides would be enough to promote thermal stability. A so-called "charge neutralization" assembly of POMs, both in the cavity and at the surface of VLPs, increased the thermal and storage stabilities of HPV 16 VLPs by *ca.* 2 °C and at least 15 days. The EuW_{10} POM acts as a "charge neutralizer", avoiding the aggregation and decomposition of HPV 16 VLPs through the ionic interaction with the positively charged arginine and lysine amino acid residues of the peptides forming the VLPs.

Antimicrobial

There are several reports on the antimicrobial activity of Class I POM-peptide hybrids in the literature. All of them have used the Gram-negative and potentially pathogenic bacteria *Escherichia coli*. The first example of an antimicrobial ionic POM-peptide hybrid was described in 2016 by Li *et al.* [61]. The precursors, HSiW and KAzoKAzoKAzoK, were not active by themselves against *Escherichia coli*, but the nanofibers HSiW@(L1)$_3$ (see Fig. 3.12A) showed antibacterial activity, with a minimum inhibitory concentration (MIC) of 60 μM. Confocal microscopy studies showed that the positively charged surfaces of the nanofibers bind and accumulate on the bacterial cell membrane causing lysis and cell death. This behavior is commensurate with other cationic antimicrobial peptides [76] and to the antimicrobial activity proposed for the amyloid-β peptide [80], which also forms β-sheet rich fibers.

A second example was the different hybrids described in the previous paragraph, formed by Eu-POMs and VLPs. The HPV 16 VLPs were not active against *E. coli* growth; the EuW_{10} POM had a concentration-dependent activity, and its activity was increased when in the EuW_{10}@VLPs hybrid form (co-assembly) but not for the VLPs@EuW_{10} hybrids (EuW_{10} adhered to the surface of VLPs). Therefore, co-assembly and encapsulation of POMs in VLPs favor the growth inhibition of *E. coli*, while POMs are easily released when on the surface of VLPs@EuW_{10} hampering the antibacterial activity. Growth inhibition activity: VLPs (0) < VLPs@EuW_{10} < EuW_{10} < EuW_{10}@VLPs. SEM imaging showed that morphological changes of the bacteria in presence of EuW_{10}@VLPs were indicative of cell death. Another example of inhibition of growth of *E. coli* is found for the reduced supramolecular material (rSP) formed by HSiW and the antioxidant soybean peptide [72]. This rSP worked as a photothermal agent against tumoral cells. As in the previous example, while the precursors did not show any effect on the growth of *E. coli*, no bacterial growth could be observed in presence of the hybrids for 8 hours. However, the authors do not specify the conditions, such as concentration of compounds and initial bacteria concentration, used for the assay.

Inhibition of amyloid-β peptide aggregation

There have been several literature reports on the use of POMs as aggregation inhibitors of the amyloid-β peptide that are implicated in the development of amyloid plaques in the brains of those affected by Alzheimer's disease [38, 81]. The mechanism involved is believed to be POMs association with the positively charged motif of the amyloid-β peptide–$H^{13}HQK^{16}$—via electrostatic interaction [81]. This interaction seems even stronger in the case of metal-substituted POMs, which can be further chelated by the histidine residues of the peptide [38]. One of the simplest examples of Class I POM-peptide hybrid formation by self-assembly following the stepwise assembly process implies the use of the $A\beta_{15-20}$ peptide. This hexapeptide contains the "recognition element" KLVFF, a fragment of the native amyloid-β peptide, which can bind to the full-length peptide and prevent its aggregation [82, 83]. Two different types of POMs have been used to form POM@$A\beta_{15-20}$ hybrids: a Wells–Dawson [57] and a polyoxomolybdate [71]. Again, in the first step, the peptides and POMs bind by electrostatic interaction to form peptide-encapsulated clusters, which can further stack into supramolecular networks. Different forces take place in the second binding step, such as van der Waals forces, hydrogen bonding, and hydrophobic interactions. Both POM-$A\beta_{15-20}$ hybrids assemble into spheres of *ca.* 100 nm of diameter and reduce the fibrillation of the Aβ1-40 peptide and rescue the toxicity of the aggregates on PC-12 cells.

3.3.3.2 Adhesives

Inspiration for the development of wet adhesives comes, once again, from nature. Sessile animals such as mussels produce adhesive proteins, which allow them to stick to wet solid surfaces [7, 84]. Researchers are aiming at producing biomimetic coacervates and materials with adhesive properties, which could be used in diverse medical applications: soft and hard—bone and tooth—tissue adhesives, sutures and minimal invasive surgery, prosthesis, etc. [85, 86]. As seen in the previous section, it is the group of Wen Li that has done the most extensive work on adhesives formed by self-assembled hybrids of POMs and amino acids or peptides. Maybe the first idea came from the formation of hydrogels in 2016 with Keggin-type heteropolyacids and facial-like peptides in a 5:1 mixture of ethanol and water [63], which they further developed into coacervates and adhesives [64, 65, 70, 73]. As it has been seen in the previous section, cross-linking of POM and peptides clusters is the key for the establishment of 3D networks, which is the base to produce gels, coacervates, and adhesives. Since some of the works have been revised in section 3.2, we will not enter in much detail on the formation of the hybrids and their supramolecular structures in this part. A summary has been included for clarity (see Table 3.3).

In these materials, the role of the POM is the key in forming the 3D structure that will hold the gel and adhesive properties. POMs act here as double agents, not only facilitating the cross-linking of the peptides by electrostatic interactions, but also by acting as rigid fillers of the gel matrix. By this, they provide stiffness, greater adhesive properties to the materials and improve their mechanical strength [70]. In the cases of peptide hybrids, the supramolecular polymerization is mainly governed by electrostatic interaction between the POMs and the protonated residues of the peptides [65, 70]. In the case of hybrid formed with only one amino acid, its positively charged/protonated side chain also binds to the POM by electrostatic interaction and interestingly, intermolecular hydrogen bonding of the amino and carboxyl groups ($N^{+}H_2$-H ······· $^{-}$O-OC) help to support the 3D network [64, 73]. Varying the pH can affect the gel behavior and even disrupt the cross-linking due to the deprotonation of the side chains, leading to a loss of the adhesive properties. For example, this was observed when increasing the pH above 2.5 for HSiW@Ac-EEMQRRAD-NH_2 and above 9 for SiW_{11}@R. On the other hand, cross-linking density can be further improved by the inclusion of divalent metal cations such as Co^{2+}, as seen for the SiW_{11}@GHK hybrid [70]. These metal cations could bind by coordination to both main actors of the hybrids forming ternary complexes inside the matrix. Another modification that allows tuning the properties of the material is the choice of amino acids. For instance, the authors found that increasing the hydrophobicity of the peptide led to more rigid scaffolds, due to better POM-peptide stacking interactions [65].

Table 3.3 Summary of works on coacervates and adhesives of Class I POM-peptide hybrids

Main hybrid	HSiW@H	HSiW@Ac-EEMQRRAD-NH_2	SiW_{11}@GHK	SiW_{11}@R
Reference	[64]	[65]	[70]	[73]
Final POM-to-peptide molar ratio within the matrix	1:4	1:2	1:2.4	1:7
pH	3	2	6.5	6.5
[aa/peptide]	0.2 M	0.65 M	1.25 M	0.66 M
Behavior with ↑↓ pH	—	Loss of cross-linking at pH > 2.5	pH ≈ 4.5 from fluid coacervate to adhesive gel loss of cross-linking at pH < 2.5	Loss of cross-linking at pH > 9
Shear strength on glass	15.8 kPa	27.8 kPa	32.73 kPa	4.1 kPa

Physical studies of the behavior of these materials were performed and confirmed their adhesive properties with shear strength values between 15 and 30 kPa on glass for most of them. The strongest adhesive on glass was SiW_{11}@GHK, but HSiW@H showed to maintain its adhesiveness up to more than 400 kPa stretch pressure on the wood. All of them could self-heal within 5–10 minutes and presented good stability of the 3D matrix upon mechanical stress. The main characteristics of these POM-based adhesives reside on their underwater adhesive capacity, low toxicity, and shear thinning behavior, which allow them to be injected through a fine needle and gain stiffness once on the substrate or by varying conditions, making them good candidates as biomedical adhesives. They present some other added values as well. For example, the HSiW@H adhesive [64], as well as the HSiW@KFKFKFK gel [63], showed photo- and electro-chromic properties—turning from white to blue in the reduced state—due to the redox activity of tungsten. Another example of added values is the anticorrosive capacity of SiW_{11}@R to protect a copper sheet, which was increased by the adhesive properties of the hybrid avoiding being washed with rain [73].

3.3.3.3 Catalysis

POMs have a rich history of application in the field of catalysis [87, 88]. Their redox properties and thermal stability make POMs versatile catalytic materials for a variety of different types of reactions, and their high solubility in polar solvents even enables them to act as enzyme mimics. For example, Keggin-type POMs have shown peroxidase-like activity, facilitating the transfer of electrons and oxidizing the substrate 3,3',5,5'-tetramethylbenzidine (TMB). PW_{12} even showed higher activity than the enzyme horseradish peroxidase (HRP) [89].

The first example of Class I POM-peptide hybrids used as catalysts was described by Ma *et al.* [60] in 2015 (see Table 3.2, entry 5). In this case, hybrids incorporate a third precursor, graphene oxide (GO), and to go from a binary hybrid (FF@PW_{12}) to a ternary one (FF@PW_{12}@GO). Both hybrids were synthetized following the procedure described in section 3.3.2, and the assembly followed two steps: first, self-assembly by electrostatic interaction of the components, and further assembly into nanospheres. In the case of the ternary hybrid, the FF@PW_{12} nanospheres were dispersed on the surface of GO. The peroxidase-like activity of both materials was assayed using TMB as a chromogenic substrate in the presence of H_2O_2. A synergistically enhanced activity was found for both hybrids, FF@PW_{12} and FF@PW_{12}@GO. For the first one, the authors hypothesized that a suitable microenvironment would be generated upon encapsulation of FF on the surface of PW_{12}, which would facilitate the access of TMB and H_2O_2. Non-covalent interactions (H-bonding & π-π stacking) between TMB and FF favor the movement of TMB toward the spheres. It is also important to maintain a relatively low hydrophobicity of the surface of the composites to better capture H_2O_2 and increase the peroxidase-like activity.

For the optimum content of GO, 5% wt, FF@PW_{12}@GO showed 1.7 times higher performance than FF@PW_{12}. These hybrid materials would present some advantages of hybrids versus the use of POMs solutions [89]. For example, being heterogeneous catalysis, there would be fewer by-products and a better purification step.

Another example of a hybrid with interesting applications in catalysis is the EuW_{10}@K8 published in 2017 by Xin and co-workers [67]. The sub microspheres formed showed an improved photocatalytic behavior on the degradation of methylene blue in presence of H_2O_2 than the precursors. Again, synergy between the two components of the hybrids is the key for this improvement; the interactions between the POM and the peptide facilitate the exposure to the substrate, its adsorption to the hybrid and the electron transfer to EuW_{10} to activate H_2O_2 to $^{\bullet}OH$.

3.3.4 Future Perspectives

Class I POM-peptide hybrids are true multifunctional materials where rational design benefits from a large pool of possible combinations from commercially available or easy-to-prepare reagents. Furthermore, these versatile syntheses are often highly scalable and therefore present true potential for real-world needs. To date, the two main fields being explored are biotechnological applications (mainly antimicrobial, HPV-related, and anti-amyloid aggregation) and adhesives.

Even if Class I hybrid materials have been further developed, there is still room for preparing new sophisticated systems. Reviewing all the examples, it is clear that the complexity in the final aggregate arises from simple and universally accessible precursors—POM and peptides – which are either easily synthesized or commercially available. Following this mantra, the next level of sophistication would come from bridging proteins or enzymes with POMs. This strategy, although not new—since it has been applied in protein crystallization—would offer new tools to modulate protein activity. In this regard, further studies on fundamental biotechnological applications of Class I hybrids will be required. The application of hybrids in microbiology as antimicrobial agents and as sensors also offers potential although a proper understanding of the mechanisms of actions of the hybrids will be the key for the design of new and improved materials.

Acknowledgements

E. A. B. and S. G. M. acknowledge the European Union's Horizon 2020 research and innovation program for the Marie Skłodowska-Curie grant agreement No 845427. H. S. C. is grateful for FPU predoctoral contract from Ministerio de Educación, Cultura y Deporte (Spain). S.G.M and R.M.R acknowledge funding from Ministerio de Ciencia e Innovación (Spain) (RYC-2013-12570 and PID2019-109333RB-I00).

References

1. Anyushin, A. V., Kondinski, A. and Parac-Vogt, T.N. (2020). Hybrid polyoxometalates as post-functionalization platforms: From fundamentals to emerging applications, *Chem. Soc. Rev.*, **49**, pp. 382–432.
2. Proust, A., Matt, B., Villanneau, R., Guillemot, G., Gouzerh, P. and Izzet, G. (2012). Functionalization and post-functionalization: A step toward polyoxometalate-based materials, *Chem. Soc. Rev.*, **41**, pp. 7605–7622.
3. Dolbecq, A., Dumas, E., Mayer, C.R. and Mialane, P. (2010). Hybrid Organic–Inorganic Polyoxometalate Compounds: From Structural Diversity to Applications, *Chem. Rev.*, **110**, pp. 6009–6048.
4. Hamley, I.W. (2017). Small bioactive peptides for biomaterials design and therapeutics, *Chem. Rev.*, **117**, pp. 14015–14041.
5. Der Torossian Torres, M. and de la Fuente-Nunez, C. (2019). Reprogramming biological peptides to combat infectious diseases, *Chem. Commun.*, **55**, pp. 15020–15032.
6. Torres, M.D.T., Sothiselvam, S., Lu, T.K. and de la Fuente-Nunez, C. (2019). Peptide design principles for antimicrobial applications, *J. Mol. Biol.*, **431**, pp. 3547–3567.
7 Wei, W., Petrone, L., Tan, Y., Cai, H., Israelachvili, J.N., Miserez, A. and Waite, J.H. (2016). An underwater surface-drying peptide inspired by a mussel adhesive protein, *Adv. Funct. Mater.*, **26**, pp. 3496–3507.

8. Goyal, D., Shuaib, S., Mann, S. and Goyal, B. (2017). Rationally designed peptides and peptidomimetics as inhibitors of amyloid-β (Aβ) aggregation: Potential therapeutics of Alzheimer's disease, *ACS Comb. Sci.*, **19**, pp. 55–80.
9. Gao, P., Wu, Y. and Wu, L. (2016). Co-assembly of polyoxometalates and peptides toward biological applications, *Soft Matter.*, **12**, pp. 8464–8479.
10. Yvon, C., Surman, A.J., Hutin, M., Alex, J., Smith, B.O., Long, D.L. and Cronin, L. (2014). Polyoxometalate clusters integrated into peptide chains and as inorganic amino acids: Solution- and solid-phase approaches, *Angew. Chemie Int. Ed.*, **53**, pp. 3336–3341.
11. Ventura, D., Calderan, A., Honisch, C., Krol, S., Serratì, S., Bonchio, M. Carraro, M. and Ruzza, P. (2018). Synthesis and biological activity of an Anderson polyoxometalate bis-functionalized with a Bombesin-analog peptide, *Pept. Sci.*, **110**, p. e24047. doi:10.1002/pep2.24047.
12. Gao, N., Du, Z., Guan, Y., Dong, K., Ren, J. and Qu, X. (2019). Chirality-selected chemical modulation of amyloid aggregation, *J. Am. Chem. Soc.*, **141**, pp. 6915–6921.
13. Luo, J., Zhang, B., Yvon, C., Hutin, M., Gerislioglu, S., Wesdemiotis, C., Cronin, L. and Liu, T. (2019). Self-assembly of polyoxometalate–peptide hybrids in solution: Elucidating the contributions of multiple possible driving forces, *Eur. J. Inorg. Chem.*, **2019**, pp. 380–386.
14. Hu, X., Wang, H., Huang, B., Li, N., Hu, K., Wu, B., Xiao, Z., Wei, Y. and Wu, P. (2019). A new scheme for rational design and synthesis of polyoxovanadate hybrids with high antitumor activities, *J. Inorg. Biochem.*, **193**, pp. 130–132.
15. Wang, Y., Wang, F., Wang, D., Li, A.N., Chen, G.X., Xiong, H.W., Wei, Y.H. Wu, P.F. and Xiao, Z. (2020). Synthesis and structure studies of a new hexavanadate-glycine hybrid with high antitumor activities, *J. Mol. Struct.*, **1201**, p. 127138.
16. Huang, B., Xiao, Z., Wang, Y., Ke, D., Zhu, C., Zhang, S., H u, X. and Wu, P. (2019). Destroy the inherent symmetry of vanadium-based inorganic cluster through chiral organic ligand: Synthesis and characterization of a polyoxovanadate-derived amino acid ester hybrid, *J. Mol. Struct.*, **1195**, pp. 10–16.
17. Bareyt, S., Piligkos, S., Hasenknopf, B., Gouzerh, P., Lacôte, E., Thorimbert, M. and Malacria, M. (2003). Highly efficient peptide bond

formation to functionalized Wells–Dawson-type Polyoxotungstates, *Angew. Chemie Int. Ed.*, **42**, pp. 3404–3406.

18. Bareyt, S., Piligkos, S., Hasenknopf, B., Gouzerh, P., Lacôte, E., Thorimbert, S. and Malacria, M. (2005). Efficient preparation of functionalized hybrid organic/inorganic Wells–Dawson-type polyoxotungstates, *J. Am. Chem. Soc.*, **127**, pp. 6788–6794.
19. Micoine, K., Hasenknopf, B., Thorimbert, S., Lacôte, E. and Malacria, M. (2007). A general strategy for ligation of organic and biological molecules to Dawson and Keggin polyoxotungstates, *Org. Lett.*, **9**, pp. 3981–3984.
20. Boglio, C., Micoine, K., Derat, É., Thouvenot, R., Hasenknopf, B., Thorimbert, S., Lacôte, E. and Malacria, M. (2008). Regioselective activation of oxo ligands in functionalized Dawson polyoxotungstates, *J. Am. Chem. Soc.*, **130**, pp. 4553–4561.
21. Micoine, K., Hasenknopf, B., Thorimbert, S., Lacôte, E. and Malacria, M. (2009). Chiral recognition of hybrid metal oxide by peptides, *Angew. Chemie Int. Ed.*, **48**, pp. 3466–3468.
22. Vilona, D., Lachkar, D., Dumont, E., Lelli, M. and Lacôte, E. (2017). Elucidation of the conformation of polyglycine organo-polyoxotungstates: Evidence for zipper folding, *Chem. Eur. J.*, **23**, pp. 13323–13327.
23. Narasimhan, K., Micoine, K., Lacôte, E., Thorimbert, S., Cheung, E., Hasenknopf, B., Jauch, R. (2014). Exploring the utility of organo-polyoxometalate hybrids to inhibit SOX transcription factors, *Cell Regen.*, **3**, pp. 10.
24. Nikoloudakis, E., Karikis, K., Laurans, M., Kokotidou, C., Solé-Daura, A., Carbó, J.J., Charisiadis, A., Charalambidis, G., Izzet, G., Mitraki, A., Douvas, A.M., Poblet, J.M., Proust, A. and Coutsolelos, A.G. (2018). Self-assembly study of nanometric spheres from polyoxometalate-phenylalanine hybrids, an experimental and theoretical approach, *Dalton Trans.*, **47**, pp. 6304–6313.
25. Chen, Q. and Zubieta, J. (1990). Synthesis and structural characterization of a polyoxovanadate coordination complex with a hexametalate core: $[(n\text{-}C_4H_9)_4N]_2[V_6O_{13}\{O_2NC(CH_2O)_3\}_2]$, *Inorg. Chem.*, **29**, pp. 1456–1458.
26. Hasenknopf, B., Delmont, R., Herson, P. and Gouzerh, P. (2002). Anderson-type heteropolymolybdates containing tris(alkoxo) ligands: Synthesis and structural characterization, *Eur. J. Inorg. Chem.*, **2002**, pp. 1081–1087.

27. Yazigi, F.J., Wilson, C., Long, D.L. and Forgan, R.S. (2017). Considerations in the self-assembly of coordination polymers of pyridine-functionalized hybrid Mn-Anderson polyoxometalates, *Cryst. Growth Des.*, **17**, pp. 4739–4748. doi:10.1021/acs.cgd.7b00672.

28. He, Z., Li, B., Ai, H., Li, H. and Wu, L. (2013). A processable hybrid supramolecular polymer formed by base pair modified polyoxometalate clusters, *Chem. Commun.*, **49**, pp. 8039–8041.

29. Blazevic, A. and Rompel, A. (2016). The Anderson–Evans polyoxometalate: From inorganic building blocks *via* hybrid organic-inorganic structures to tomorrows "Bio-POM", *Coord. Chem. Rev.*, **307**, pp. 42–64.

30. Yan, J., Zheng, X., Yao, J., Xu, P., Miao, Z., Li, J., Lv, Z., Zhang, Q. and Yan, Y. (2019). Metallopolymers from organically modified polyoxometalates (MOMPs): A review, *J. Organomet. Chem.*, **884**, pp. 1–16.

31 Song, Y.-F., Long, D.-L., Kelly, S.E. and Cronin, L. (2008). Sorting the assemblies of unsymmetrically covalently functionalized Mn-Anderson polyoxometalate clusters with mass spectrometry, *Inorg. Chem.*, **47**, pp. 9137–9139.

32. Rosnes, M.H., Musumeci, C., Pradeep, C.P., Mathieson, J.S., Long, D.-L., Song, Y.-F., Pignataro, B., Cogdell, R. and Cronin, L. (2010). Assembly of modular asymmetric organic–inorganic polyoxometalate hybrids into anisotropic nanostructures, *J. Am. Chem. Soc.*, **132**, pp. 15490–15492.

33. Yvon, C., Macdonell, A., Buchwald, S., Surman, A.J., Follet, N., Alex, J., Long, D.L. and Cronin, L. (2013). A collection of robust methodologies for the preparation of asymmetric hybrid Mn–Anderson polyoxometalates for multifunctional materials, *Chem. Sci.*, **4**, pp. 3810–3817.

34. Liu, B., Yang, J., Yang, M., Wang, Y., Xia, N., Zhang, Z., Zheng P., Wang, W., Lieberwirth, I. and Kübel, C. (2011). Polyoxometalate cluster-contained hybrid gelator and hybrid organogel: A new concept of softenization of polyoxometalate clusters, *Soft Matter.*, **7**, pp. 2317–2320.

35. Hutin, M., Yvon, C. Yan, J., Macdonell, A. Long, D. and Cronin, L. (2013). Programming the assembly of carboxylic acid-functionalised hybrid polyoxometalates, *CrystEngComm.*, **15**, pp. 4422–4430.

36. Luo, J. and Liu, T. (2019). Competition and cooperation among different attractive forces in solutions of inorganic-organic hybrids containing macroionic clusters, *Langmuir.*, **35**, pp. 7603–7616.

37. Yan, X., Zhu, P. and Li, J. (2010). Self-assembly and application of diphenylalanine-based nanostructures, *Chem. Soc. Rev.*, **39**, pp. 1877–1890.

38. Gao, N., Sun, H., Dong, K., Ren, J., Duan, T., Xu, C. and Qu, X. (2014). Transition-metal-substituted polyoxometalate derivatives as functional anti-amyloid agents for Alzheimer's disease., *Nat. Commun.*, **5**, p. 3422.

39. Vilona, D., Lachkar, D., Dumont, E., Lelli, M. and Lacôte, E. (2017). Elucidation of the conformation of polyglycine organo-polyoxotungstates: Evidence for zipper folding, *Chem. Eur. J.*, **23**, pp. 13323–13327.

40. Zonnevijlle, F. and Pope, M.T. (1979). Attachment of organic groups to heteropoly oxometalate anions, *J. Am. Chem. Soc.*, **101**, pp. 2731–2732.

41. Chorghade, G.S. and M.T. Pope (1987). Heteropolyanions as nucleophiles. 1. Synthesis, characterization, and reactions of Keggin- and Dawson-type tungstostannates(II), *J. Am. Chem. Soc.*, **109**, pp. 5134–5138.

42. Matt, B., Moussa, J., Chamoreau, L.M., Afonso, C., Proust, A., Amouri, H. and Izzet, G. (2012). Elegant approach to the synthesis of a unique heteroleptic cyclometalated iridium(III)-polyoxometalate conjugate, *Organometallics*, **31**, pp.35–38.

43. Tomasello, G., Armenia, I. and Molla, G. (2020). The protein imager: A full-featured online molecular viewer interface with server-side HQ-rendering capabilities, *Bioinformatics*, **36**, pp. 2909–2911.

44. Kistler, M.L., Bhatt, A., Liu, G., Casa, D. and Liu, T. (2007). A complete macroion-"blackberry" assembly-macroion transition with continuously adjustable assembly sizes in $\{Mo_{132}\}$ water/acetone systems, *J. Am. Chem. Soc.*, **129**, pp. 6453–6460. doi:10.1021/ja0685809.

45. Zhu, Y., Yin, P., Xiao, F. Li, D., Bitterlich, E., Xiao, Z., Zhang, J., Hao, J., Liu, T., Wang, Y. and Wei, Y. (2013). Bottom-up construction of POM-based macrostructures: Coordination assembled paddle-wheel macroclusters and their vesicle-like supramolecular aggregation in solution, *J. Am. Chem. Soc.*, **135**, pp. 17155–17160.

46. Liu, T., Imber, B., Diemann, E., Liu, G., Cokleski, K., Li, H., Chen, Z. and Müller, A. (2006). Deprotonations and charges of well-defined $\{Mo_{72}Fe_{30}\}$ nanoacids simply stepwise tuned by pH allow control/variation of related self-assembly processes, *J. Am. Chem. Soc.*, **128**, pp. 15914–15920.

47. Yan, X., Zhu, P., Fei, J. and Li, J. (2010). Self-assembly of peptide-inorganic hybrid spheres for adaptive encapsulation of guests, *Adv. Mater.*, **22**, pp. 1283–1287.

48. Hasenknopf, B., Micoine, K., Lacôte, E., Thorimbert, S., Malacria, M. and Thouvenot, R. (2008). Chirality in polyoxometalate chemistry, *Eur. J. Inorg. Chem.*, **2008**, pp. 5001–5013. doi:10.1002/ejic.200800759.

49. Guan, Y., Li, M., Dong, K., Gao, N., Ren, J., Zheng, Y. and Qu, X. (2016). Ceria/POMs hybrid nanoparticles as a mimicking metallopeptidase for treatment of neurotoxicity of amyloid-β peptide, *Biomaterials*, **98,** pp. 92–102. doi:10.1016/j.biomaterials.2016.05.005.

50. Rubin, N., Perugia, E., Goldschmidt, M., Fridkin, M. and Addadi, L. (2008). Chirality of amyloid suprastructures, *J. Am. Chem. Soc.*, **130**, pp. 4602–4603.

51. Riordan, J.F. (1977). The role of metals in enzyme activity, *Ann. Clin. Lab. Sci.*, **7**, pp. 119–129.

52. Corbett, M.C., Hu, Y., Fay, A.W., Ribbe, M.W., Hedman, B. and Hodgson, K.O. (2006). Structural insights into a protein-bound iron-molybdenum cofactor precursor, *Proc. Natl. Acad. Sci. U.S.A.*, **103**, pp. 1238–1243.

53. Magalon, A., Fedor, J.G., Walburger, A. and Weiner, J.H. (2011). Molybdenum enzymes in bacteria and their maturation, *Coord. Chem. Rev.*, **255**, pp. 1159–1178.

54. Kowalewski, B., Poppe, J., Demmer, U., Warkentin, E., Dierks, T., Ermler, U. and Schneider, K. (2012). Natures polyoxometalate chemistry: X-ray structure of the Mo storage protein loaded with discrete polynuclear Mo-O clusters, *J. Am. Chem. Soc.*, **134**, pp. 9768–9774.

55. Beinert, H., Holm, R.H. and Münck, E. (1997). Iron-sulfur clusters: Nature's modular, multipurpose structures, *Science*, **277**, pp. 653–659.

56. Misra, A., Kozma, K., Streb, C. and Nyman, M. (2020). Beyond charge balance: Counter-cations in polyoxometalate chemistry, *Angew. Chemie Int. Ed.*, **59**, pp. 596–612.

57. Li, M. Xu, C., Wu, L., Ren, J., Wang, E. and Qu, X. (2013). Self-assembled peptide-polyoxometalate hybrid nanospheres: Two in one enhances targeted inhibition of amyloid β-peptide aggregation associated with Alzheimer's disease, *Small*, **9**, pp. 3455–3461.

58. Zhang, T., Li, H.W., Wu, Y., Wang, Y. and Wu, L. (2015). Self-assembly of an europium-containing polyoxometalate and the arginine/

lysine-rich peptides from human papillomavirus capsid protein L1 in forming luminescence-enhanced hybrid nanoparticles, *J. Phys. Chem. C.*, **119**, pp. 8321–8328.

59. Zhang, T., Li, H.-W., Wu, Y., Wang, Y. and Wu, L. (2015). The two-step assemblies of basic-amino-acid-rich peptide with a highly charged polyoxometalate, *Chem. Eur. J.*, **21**, pp. 9028–9033.

60. Ma, Z., Qiu, Y., Yang, H., Huang, Y., Liu, J., Lu, Y., Zhang, C. and Hu, P. (2015). Effective synergistic effect of dipeptide-polyoxometalate-graphene oxide ternary hybrid materials on peroxidase-like mimics with enhanced performance, *ACS Appl. Mater. Interfaces.*, **7**, pp. 22036–22045.

61. Li, J., Chen, Z., Zhou, M., Jing, J., Li, W., Wang, Y., Wu, L., Wang, L., Wang, Y. and Lee, M. (2016). Polyoxometalate-driven self-assembly of short peptides into multivalent nanofibers with enhanced antibacterial activity, *Angew. Chemie Int. Ed.*, **55**, pp. 2592–2595.

62. Li, J., Li, X., Xu, J.,Wang, Y., Wu, L., Wang, Y., Wang, L., Lee, M. and Li, W. (2016). Engineering the ionic self-assembly of polyoxometalates and facial-like peptides, *Chem. Eur. J.*, **22**, pp. 15751–15759.

63. Li, J., Xu, J., Li, X., Gao, W., Wang, L., Wu, L., Lee, M. and Li, W. (2016). Heteropoly acids triggered self-assembly of cationic peptides into photo- and electro-chromic gels, *Soft Matter.*, **12**, pp. 5572–5580.

64. Xu, J., Li, X., Li, J., Li, X., Li, B., Wang, Y., Wu, L. and Li, W. (2017). Wet and functional adhesives from one-step aqueous self-assembly of natural amino acids and polyoxometalates, *Angew. Chemie Int. Ed.*, **56**, pp. 8731–8735.

65. Xu, J., Li, X., Li, X., Li, B., Wu, L., Li, W., Xie, X. and Xue, R. (2017). Supramolecular copolymerization of short peptides and poly-oxometalates: Toward the fabrication of underwater adhesives, *Biomacromolecules*, **18**, pp. 3524–3530.

66. Li, J., Li, X., Jiang, F., Li, X., Xie, X., Wu, L., Wang, L., Lee, M. and Li, W. (2017). Short peptides directing 1D helical arrays of polyoxometalates with controllable pitches, *Chem. Eur. J.*, **23**, pp. 13510–1351.

67. Tong, L., Wang, Z., Xia, C., Yang, Y., Yuan, S., Sun, D. and Xin, X. (2017). Self-assembly of peptide-polyoxometalate hybrid sub-micrometer spheres for photocatalytic degradation of methylene blue, *J. Phys. Chem. B.*, **121**, pp. 10566–10573.

68. Gao, P.F., Liu, Y.X., Zhang, L., Zhang, S., Li, H.W.,Wu, Y. and Wu, L. (2018). Cell receptor screening for human papillomavirus invasion by using a polyoxometalate-peptide assembly as a probe, *J. Colloid Interface Sci.*, **514**, pp. 407–414.

69. Fu, D.-Y., Zhang, S., Qu, Z., Yu, X., Wu, Y. and Wu, L. (2018). Hybrid assembly toward enhanced thermal stability of virus-like particles and antibacterial activity of polyoxometalates, *ACS Appl. Mater. Interfaces*, **10**, pp. 6137–6145.

70. Li, X., Zheng, T., Liu, X., Du, Z., Xie, X., Li, B., Wu, L. and Li, W. (2019). Coassembly of short peptide and polyoxometalate into complex coacervate adapted for pH and metal ion-triggered underwater adhesion, *Langmuir*, **35**, pp. 4995–5003.

71. Liu, Y., Sun, J., Gong, Y., Zhou, H., Chen, X., Zhu, X., Zhao, Y., Wen, Y., Qin, X. and Liu, J. (2019). Peptide-modified Mo polyoxometalate nanoparticles suppress Zn^{2+}-induced Aβ aggregation, *ChemNanoMat.*, **5**, pp. 897–910.

72. Zhang, S., Peng, B., Xue, P., Kong, X., Tang, Y., Wu, L. and Lin, S. (2019). Polyoxometalate-antioxidant peptide assembly materials with NIR-triggered photothermal behavior and enhanced antibacterial activity, *Soft Matter.*, **15**, pp. 5375–5379.

73. Liu, X., Xie, X., Du, Z., Li, B., Wu, L. and Li, W. (2019). Aqueous self-assembly of arginine and $K_8SiW_{11}O_{39}$: Fine-tuning the formation of a coacervate intended for sprayable anticorrosive coatings, *Soft Matter.*, **15**, pp. 9178–9186.

74. Kaba, M.S., Song, I.K., Duncan, D.C., Hill, C.L. and Barteau, M.A. (1998). Molecular shapes, orientation, and packing of polyoxometalate arrays imaged by scanning tunneling microscopy, *Inorg. Chem.*, **37**, pp. 398–406.

75. Kozhevnikov, I.V. (1998). Catalysis by heteropoly acids and multicomponent polyoxometalates in liquid-phase reactions, *Chem. Rev.*, **98**, pp. 171–198.

76. Fjell, C.D., Hiss, J.A., Hancock, R.E.W. and Schneider, G. (2012). Designing antimicrobial peptides: Form follows function, *Nat. Rev. Drug Discov.*, **11**, pp. 37–51.

77. Mahlapuu, M., Håkansson, J., Ringstad, L. and Björn, C. (2016), Antimicrobial peptides: An emerging category of therapeutic agents, *Front. Cell. Infect. Microbiol.*, **6**, p. 194.

78. Xie, X., Wang, L., Liu, X., Du, Z., Li, Y., Li, B., Wu, L. and Li, W. (2020). Light-powered and transient peptide two-dimensional assembly driven by trans-to-cis isomerization of azobenzene side chains, *Chem. Commun.*, **56**, pp. 1867–1870.

79. Paavonen, J., Naud, P., Salmerón, J., Wheeler, C.M., Chow, S.N., Apter, D., Kitchener, H., Castellsague, X., Teixeira, J.C., Skinner, S.R., Hedrick, J., Jaisamrarn, U., Limson, G., Garland, S., Szarewski, A.,

Romanowski, B., Aoki, F.Y., Schwarz, T.F., Poppe, W., Bosch, F.X., Jenkins, D., Hardt, K., Zahaf, T., Descamps, D., Struyf, F., Lehtinen, M. and Dubin, G. (2009). Efficacy of human papillomavirus (HPV)-16/18 AS04-adjuvanted vaccine against cervical infection and precancer caused by oncogenic HPV types (PATRICIA): Final analysis of a double-blind, randomized study in young women, *Lancet*, **374**, pp. 301–314.

80. Soscia, S.J., Kirby, J.E., Washicosky, K.J., Tucker, S.M., Ingelsson, M., Hyman, B., Burton, M.A., Goldstein, L.E., Duong, S., Tanzi, R.E. and Moir, R.D. (2010). The Alzheimer's disease-associated amyloid β-protein is an antimicrobial peptide, *PLoS ONE*, **5**, pp. 1–10.

81. Geng, J., Li, M., Ren, J., Wang, E. and Qu, X. (2011). Polyoxometalates as inhibitors of the aggregation of amyloid β peptides associated with Alzheimer's disease, *Angew. Chemie Int. Ed.*, **50**, pp. 4184–4188.

82. Tjernberg, L.O., Näslundt, J., Lindqvist, F., Johansson, J., Karlström, A.R., Thyberg, J., Tereniust, L. and Nordstedt, C. (1996). Arrest of β-amyloid fibril formation by a pentapeptide ligand, *J. Biol. Chem.*, **271**, pp. 8545–8548.

83. Lowe, T.L., Strzelec, A., Kiessling, L.L. and Murphy, R.M. (2001). Structure-function relationships for inhibitors of β-amyloid toxicity containing the recognition sequence KLVFF, *Biochemistry*, **40**, pp. 7882–7889.

84. Waite, J.H. (2017). Mussel adhesion: Essential footwork, *J. Exp. Biol.*, **220**, pp. 517–530.

85. Park, K.H., Seong, K.Y., Yang, S.Y. and Seo, S. (2017). Advances in medical adhesives inspired by aquatic organisms' adhesion, *Biomater. Res.*, **21**, pp. 1–9.

86. Kim, H.J., Yang, B., Park, T.Y., Lim, S. and Cha, H.J. (2017). Complex coacervates based on recombinant mussel adhesive proteins: Their characterization and applications, *Soft Matter*, **13**, pp. 7704–7716.

87. Gumerova, N.I. and Rompel, A. (2018). Synthesis, structures and applications of electron-rich polyoxometalates, *Nat. Rev. Chem.*, **2**, p. 0112.

88. Roy, S., Crans, D.C. and Parac-Vogt, T.N. (2019). Editorial: Polyoxometalates in catalysis, biology, energy and materials science, *Front. Chem.*, **7**, pp. 882–885.

89. Wang, J., Han, D., Wang, X., Qi, B. and Zhao, M. (2012). Polyoxometalates as peroxidase mimetics and their applications in H_2O^2 and glucose detection, *Biosens. Bioelectron.*, **36**, pp. 18–21.

Chapter 4

Polyoxometalate–Polymer Hybrid Materials

Leire Ruiz-Rubio,[a,b] Isabel Moreno,[c] and José Luis Vilas-Vilela[a,b]

[a]*Macromolecular Chemistry Group (LQM), Department of Physical Chemistry, Faculty of Science and Technology, University of the Basque Country, 48940 Leioa, Spain*
[b]*BCMaterials, Basque Center for Materials, Applications and Nanostructures, UPV/EHU Science Park, 48940 Leioa, Spain*
[c]*Macromolecular Chemistry Group (LQM), Department of Organic and Inorganic Chemistry, Faculty of Science and Technology, University of the Basque Country, 48940 Leioa, Spain*

leire.ruiz@ehu.eus

The great variety of different structures and compositions of both polyoxometalates (POMs) and polymers eases that the hybrid materials formed from them could be used in a diverse range of applications, such as sensor, catalysts, electronic, or photochromic materials among others. In addition, their versatility means that they can be adapted seeking a continuous improvement of their properties. In this chapter, the state of art of hybrid POM/polymer hybrid materials is analyzed. The main synthetic strategies employed on the development of these hybrid materials are summarized: POM/polymer blends, composites assembled by non-covalent interactions, and

Polyoxometalates: Advances, Properties, and Applications
Edited by Leire Ruiz Rubio, José Luis Vilas Vilela, Beñat Artetxe, and Juan Manuel Gutiérrez-Zorrilla

ISBN 978-981-4968-14-0 (Hardcover), 978-1-003-27744-6 (eBook)
www.jennystanford.com

covalently linked POM/polymer hybrids. Finally, some of the most relevant applications in which these kinds of hybrid materials have been used are summarized.

4.1 Introduction

Polyoxometalates (POMs) are a class of polyanionic clusters. They are discrete assemblies of early-transition, commonly Mo, W, or V, in high oxidation states. The high diversity regarding chemical composition, structure, and electronic properties gives rise to materials that find applications in very diverse fields of materials science, such as catalysis, gas adsorption, energy storage, biological processes [1]. Moreover, POM-based hybrid materials, in which the outstanding properties of the polyanions are improved by their incorporation on matrixes, have gained the attention of many researchers due to the wide potential applications [1–3]. There are many examples of materials used as matrixes to incorporate POMs inside them, such as carbon nanotubes, metallic nanoparticles, among others [4, 5]. However, polymeric materials provide many interesting properties that have aroused them as one of the most viable options.

In fact, polymers present good mechanical properties, such as excellent toughness and ductility, and they are easy to process. In addition, the wide amount of available polymers could improve their use as matrixes in many applications. For example, conductive polymers could be used in several applications related to flexible electronic, sensors, or batteries, among others. Other polymers, such as poly(methyl methacrylate) (PMMA), present excellent optical properties and have been used for the development of polymeric optic fibers. On the other hand, many examples of biocompatible polymers have been used in the fabrication of biomedical devices or as drug delivery systems. This diversity makes polymeric materials highly appropriate counterparts for POM-based hybrid materials.

The wide variety of polymers available as a matrix for these kinds of materials, with different chemical nature, structures, and properties, provides many huge amount synthetic pathways to fabricate these kinds of hybrids. POMs/polymer hybrid materials could be obtained by different methods like physical blending, non-covalent interactions (electrostatic, hydrogen bond, or van der Waals), supramolecular modifications of the clusters (e.g., micelles), or covalent bonding [6–8]. In this context, the combination of the properties of the POMs with the versatility of polymeric materials could result in new materials with unique and highly advantageous functionalities that could be used in many applications.

In this chapter, the state of art of hybrid POM/polymer hybrid materials is analyzed. The main synthetic strategies employed on the development of these hybrid materials are summarized: POM/polymer blends, composites assembled by non-covalent interactions, and covalently linked POM/polymer hybrids. Finally, some of the most relevant applications in which these kinds of hybrid materials have been used are summarized.

4.2 Development of Hybrid POM/Polymer Materials

4.2.1 Physical Blends of POM/Polymers

Many examples of POM/polymer physical mixtures or blends have been reported up to date, being three main strategies employed in their fabrication: physical mixture on common solvents, by in situ polymerization of the polymer in presence of POM or physical mixture with melted polymers.

When hybrid materials are obtained by physical blending in a common solvent, the most common used solvent used is water. However, due to the spread used of this method, these kinds of materials are going to be analyzed in section 2.2., describing each type of physical interactions formed between POMs and polymer chains. Although water constitutes the main solvent used in those systems that POMs are involve, some works have described the employment of other solvents [7, 9]. For example, Peng *et al.* [10] reported the fabrication of photochromic composite films based on physical mixtures of poly(methacrylate-co-vinylpyrrolidone) and Keggin-type phosphotungstic acid (PWA) employing ethanol as solvent for both components. After vigorous agitation, the films were casted in several substrates. It is important to notice that this system initially transparent turned in to blue after being exposed to ultraviolet (UV)-light, being this phenomenon related to intervalence charge transfer (IVCT) ($W^{6+} \rightarrow W^{5+}$). Similarly, in a recent study, Espindola and co-workers described the fabrication of POM/PMMA composites in a dichloromethane/tetrahydrofuran (CH_2Cl_2/THF) solution [11]. The films are highly transparent due to the excellent optical properties of PMMA, but they also present a photochromic behavior of polyoxotungstoeuropate, $K_{12}[EuP_5W_{30}O_{110}] \cdot 54H_2O$ (EPW).

However, POM/polymer hybrid composites obtained in organic solvents present poorer compatibility compared to those obtained in water [12]. In order to overcome this drawback, some authors have incorporated polyoxometales modified with surfactant improving the stability of these hybrid materials. The use of surfactant allows the replacement of the small counterions of the POMs, commonly potassium cations, for hydrophobic cations, such as alkyl chains, that present better miscibility both with organic solvents and hydrophobic polymers. These POMs modified with surfactants are also known as surfactant-encapsulated POM (SEP) [13, 14]. As an example, Sun *et al.* [15] developed a POM/poly(styrene) hybrid film by using an SEP complex, $(DODA)_9[EuW_{10}O_{36}]$. Authors reported that an optimum ratio between SEP and PS matrixes added to adequate humidity and concentration conditions allowed the formation of highly ordered hybrid films with a characteristic honeycomb arrays on their surface.

Composites based on physical mixtures of POM/polymers could also be fabricated by in situ polymerization of selected monomers in the presence of POMs. Commonly, the polymerization processes used in this approach are radical (co)polymerizations or electropolymerizations depending on the nature of the monomers involved or the potential applications of the material. POM/polymer composite formation could also be obtained by the electropolymerization of a conductive polymer, such as polypyrrole (PPy) or polyaniline (PANI), among others, in presence of POMs [16, 17]. As an example, Liu and Dong reported a successful formation composite films of molibdosilicic heteropoly complex with dysprosium $[Dy(SiMo_{11})_2]$ immobilized on a PPy matrix that was electropolymerized on a glassy carbon electrode [18]. This electrochemical immobilization on a poly(3-methylthiophene) of several Keggin-type POMs, such as $PW_{12}O_{40}^{-3}$, $SiW_{12}O_{40}^{4-}$ and $PMo_{12}O_{40}^{3-}$, has also been reported [19]. More recently, a Keggin-type $PW_{12}O_{40}^{-3}$ was immobilized by electrochemical polymerization of polydiphenylamine (PDPA) [20].

In addition to radical polymerization (chain polymerization) and electropolymerizations, a step polymerization can be used to obtain POM/polymer composites of polyurethanes or epoxy resins, among others. Küçük *et al.* [21–23] reported several studies involving hybrid POM/polymer composites with different polyurethanes as a matrix. Different PU were synthetized in situ by using several diisocyantes, such as dicyclohexylmethane-4,4′-diisocyanate (DHDI) and 1,6-diisocyanatehexane (HMDI) [23], and ethylene glycol. Cobalt ion-based polyoxoanions (Co-POM) were added to the reaction mixture in order to be immobilized during the polymerization step (Fig. 4.1). Authors studied the influence of the Co-POMs on the dielectric constant of the composites reporting a decrease on the property that they related to the increase on the free volume due to the presence of the POMs.

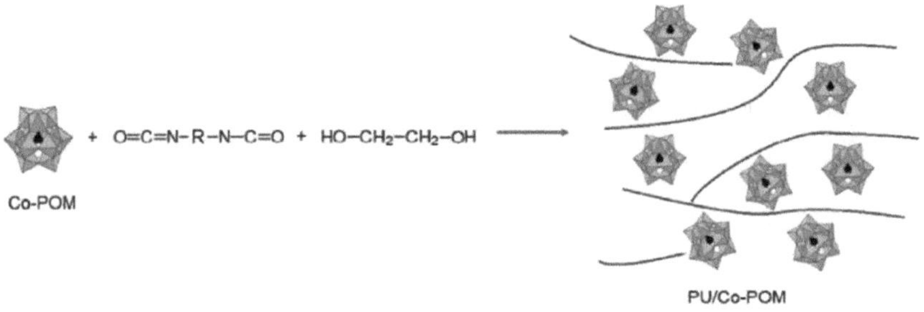

Figure 4.1 Scheme of the Co-POM/PU composite formation by in situ polymerization of polyurethanes with POMs. Reproduced with permission from reference [23]. Copyright 2020 Taylor & Francis.

Pinto and co-workers fabricated photochromic fibers of polypropylene (PP)/silica@POMs hybrids [24]. In this study, two types of Keggin-type phosphomolybdates hybrids electrostatically bonded to SiO_2 nanoparticles were mixed with

PP by melt compounding using a twin-screw extruder. Then, a mixture of the hybrid PP/silica@POMs composite with a pristine PP was melted spinning, being the obtained fibers are shown in Fig. 4.2. These fibers were used to produce photochromic meshes presenting a reversible respond, that is, the fibers present a dark blue color under UV irradiation at $\lambda = 254$ nm) and they could be recessively bleach at temperature range of 60–100 °C.

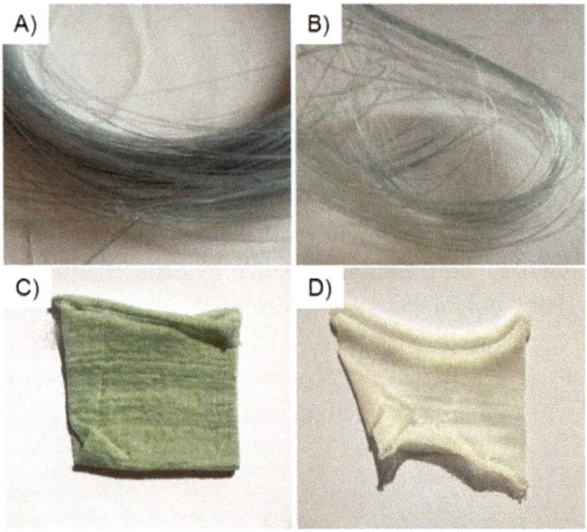

Figure 4.2 (A) Hybrid fibers obtained from PP/SiO_2@ PMo_{12} fibers, (B) PP/SiO_2@ PMo_{12} fibers dried at 60°C, (C) PP/SiO_2@PMo_{12} mesh, and (D) mesh obtained from pristine PP. Reproduced with permission from reference [24]. Copyright 2018 Elsevier.

4.2.2 Hybrid Composites Formed by Non-Covalent Interactions

The miscibility problems that could arise from the materials obtained by physical blends could be overcome both by promoting the non-covalent interactions between the polymer chains and the POMs or covalently anchoring the POMs to the polymers. In this section, some examples of non-covalent interactions used as a driving force for the hybrid composite formation are studied. Considering the water affinity of the polyoxianions, the main physical interactions in these kinds of systems are the hydrogen bond and the electrostatic interactions.

Many examples of POMs/polymer hybrid materials formed due to the hydrogen bonded formation in aqueous media have been reported until date. Considering the water-soluble polymers, the most used polymers in this approach have been poly(ethylene oxide) (PEO), poly(vinyl pyrrolidone) (PVP), polyacrylamide (PAM), and poly(vinyl alcohol) (PVA) [25–30].

Several authors have studied POM and poly(ethylene glycol) (PEG) or PEO composites. Buchecker *et al.* [31] studied the interactions between the POM and PEG in water using a Keggin-type POM, $PW_{12}O_{40}^{3-}$. This $PW_{12}O_{40}^{3-}$/PEG system could form crystalline domains under mild conditions (aqueous solution at room temperature. These crystalline domains are formed due repetition of the unit formed by the spontaneous self-assembly of POM with ethylene oxide, being 4 the minimum number of repeated units. Taking advantage of this self-assembly capability, Jing *et al.* [29] fabricated POMS/poly(ethylene oxide-*b*-ε-caprolactone) (PEG-*b*-PCL) polymersomes formed by hydrogen bond. In this case, the POM used presented wheel-like structure, $Na_{14}[Mo_{154}O_{462}H_{14}(H_2O)_{70}] \cdot 400H_2O$ ({Mo_{154}}). A scheme of the interactions presented in the formed polymersome is depictured in Fig. 4.3. As can be seen the hydrogen bonds established in the block copolymer between the repeating units formed by the POM and the ethylene oxide are responsible of the formation of this self-assembly structures. So, the presence of this POM enables the physical crosslinking of the copolymer by hydrogen bonding. Moreover, this complex also provides a stable shapeand mechanical reinforcement to the polymersomers.

Similarly to the composites fabricated by physical blends, there are many examples devoted to the fabrication of photochromic POM/polymer composites favored by the physical interactions, such as electrostatic or hydrogen bonding to obtain highly homogenous hybrid materials. The majority of the reported examples, Layer-by-Layer (LbL) approach is used to fabricate electrochromic and/photochromic films. For example, Xu *et al.* [32] employed Dawson-type phosphotungstate anion $[P_2W_{18}O_{62}]^{6-}$

(P_2W_{18}) hydrogen bonded with a polyvinyl alcohol (PVA) and poly(allylamine hydrochloride) (PAH) to form electrochromic multilayers [PVA–P_2W_{18}/PAH] (Fig. 4.4).

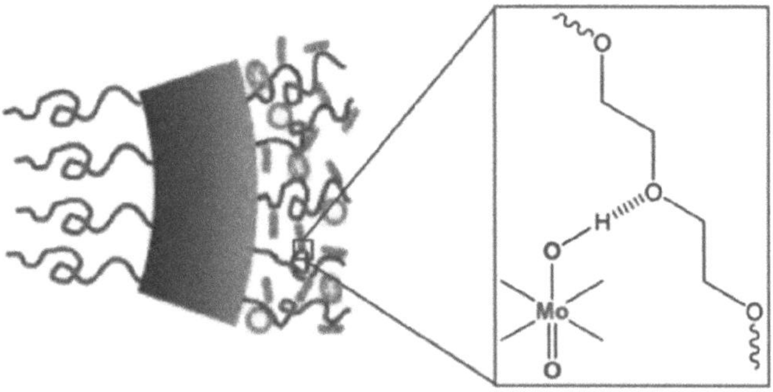

Figure 4.3 Scheme of the hydrogen bond formed between the $\{Mo_{154}\}$ and ethylene oxide groups of the PEG-*b*-PCL copolymer. Reproduced with permission from reference [29]. Copyright 2017 Elsevier.

PAM, due to its intensive use in many applications over decades [33], is another widely used polymer matrix for the formation of hybrid POM nanocomposites [34, 35]. An interesting study on these kinds of hybrid materials was described by Chen *et al.* [36]. In this study, the influence of the Fe^{2+} cations on the properties of molybdenumphosphoric acid (PMoA)/PAM hybrid films was analyzed. Authors reported changes both on microstructure and photochromic properties due to the Fe^{2+} doping. The presence of the cations varied the nanoparticles shape from well-defined spheres to irregular conglomeration. In addition, this Fe^{2+} doping also decreased the intensity of the color change observed in PMoA/PAM film. The effect of the Fe^{2+} doping could be related to the variation in the number of the hydrogen bonds between POM/polymer systems since many of them could be disrupted due to the incorporation of the metal.

The luminescent properties of the europium-based POMs constitute another aspect that it is important to notice. For example, Wang *et al.* [37] reported the fabrication of $[Eu(SiW_{10}MoO_{39})_2]^{13-}$/agarose composite films, that turned out capable of switching their luminescence recessively. This film was obtained by mixing both components in water and, then, by solvent casting the mixture on indium tin oxide (ITO) substrate. The so-obtained film presented good transparency and flexibility. Considering the scanning electron microscopy (SEM) and transmission electron microscopy (TEM) images, the authors proposed that the hydrogen bond formed between the $[Eu(SiW_{10}MoO_{39})_2]^{13-}$ and agarose could ease the homogeneity of the films, favoring the described luminescence, transparence, and mechanical properties.

The variety of polyelectrolytes available, both synthetic and biopolymers, has arisen the fabrication route based on electrostatic interactions between POMs and polymers. Many of these polymers present a high water-affinity, therefore, using this approach, physical hydrogels could be obtained in which, the crosslinking between the polymer chains is commonly induced by the POMs [38–41]. In addition, this kind of interaction has been used for the development of thin films based on LbL methodology [42, 43].

As an example, Feng *et al.* [44] developed hybrid POM/polymer thin films by LbL methodology. In their study, they evaluated the LbL formation by using two different Keggin-type POMs, $[SiW_{12}O_{40}]^{4-}$ (SiW_{12}), $[PMo_{12}O_{40}]^{3-}$ (PMo_{12}), and chitosan as a positively charge polyelectrolyte. The study reported that the LbL approach constitutes a highly reproducible and regular approach to fabricate hybrid thin films.

In a recent study, Liang and co-workers fabricated hybrid POM/polymer composites by using polymerizable imidazole-type zwitterion 3-(1-vinyl-3-imidazolio)propanesulfonate (VIPS) [40]. The POM used was $Na_9EuW_{10}O_{36}$ (EuW_{10}), a Weakley-type POM. In this case, the electrostatic interactions formed between the POMS and the VIPS established a secondary network, since the principal hydrogel network was formed by covalent crosslinking with crosslinker *N,N′*-methylene-bis(2-propenamide) (MBAA). This secondary network provided improved mechanical and luminous properties to the hydrogels. In addition, authors also proposed that the incorporation of free zwitterions, such as 3-(1-methyl-3-imidazolio)propanesulfonate (MIPS) or 3-(1-decyl-3-imidazolio)propanesulfonate (C10IPS), to the hydrogel/POM systems could suppose an additional improvement of the mentioned properties.

The electrostatic interactions between the POMs and some polysaccharides, such as chitosan, have been used as driving forces for the formation of physical hydrogel based on the interaction between the positively charged amine groups present in the chitosan and the polyoxoanions [45–48]. Pandya *et al.* [49] fabricated a pH reversible hydrogel based in the interaction between Wells–Dawson type polyoxoanions and carboxymethyl chitosan (CMC). In this study, two different isostructural POMs, $[P_2W_{18}O_{62}]^{6-}$ and $[P_2W_{15}V_3O_{62}]^{9-}$ were used. They reported a pH sensitive behavior on the hydrogel formation, being the hydrogel formed with $P_2W_{15}V_3$ only obtained at pH 7.4, whereas the one with P_2W_{18} was obtained in a pH range between 7 and 9.

In another study devoted to POM/polysaccharide hydrogels, Wang *et al.* [50] studied the formation of hydrogels and chiroptical properties of two families of POMs in three natural polysaccharides, agarose, ι-carrageenan and κ-carrageenan (Fig. 4.4). It is noteworthy that achieving chirality in this type of structures is normally hampered by the high symmetry of the POMs. The most commonly used approach to solve this problem is to make a covalent modification in the POM structure; however, this could limit the number of possible POMs structures that could be fabricated. On the other hand, it is important to notice that polysaccharides present not only the chirality associated to glucose groups, but also those of them that present a high number of repeating units in the polymer chains, such as agarose and carrageenans, could present a helical conformation that can also imply asymmetry and consequently chirality. Authors described the chiroptical properties obtained due to the physical interactions between the studied polyoxomolybdates and polyoxotungstates with these hydrogels. Agarose forms hydrogen bonds with the POMs, while both κ/ι-carrageenan form electrostatic interactions due to their ionic nature. These different types of interactions induce variations on the chiroptical properties. The extrinsic chiral factors are dominated in electrostatic interactions. On the other hand, the hydrogen bond interactions present in agarose vary the alternating bond length (ABL), being the dominant effect on induced circular dichroism for this hydrogel. In addition, since these interactions are sensitive to several stimuli, such as ionic strength or UV irradiation, the optical properties that are dependent of these interactions also change in response to them.

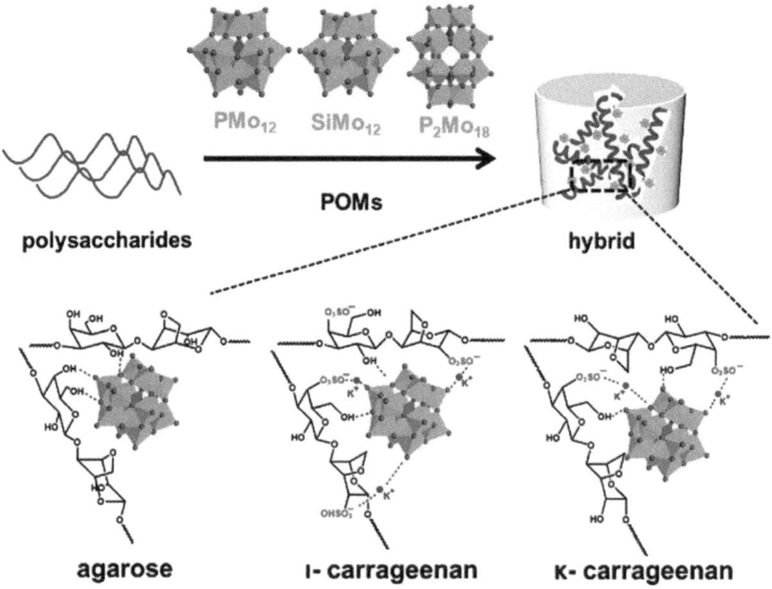

Figure 4.4 Scheme of the main interactions formed between the agarose and carrageenan the study reported by Wang *et al.* Reproduced with permission from reference [50]. Copyright 2018 Royal Society of Chemistry.

Hybrid materials based on biopolymers and POMs have also been studied. Carn and co-workers described a complex coacervation behavior between gelatin and decavanadate POMs [51]. This system presents three different regions (Fig. 4.5) depending on the different concentration ratios of decavanadate and gelatin present in the aqueous solution.

In Region 1, attractive electrostatic interactions are established between decavanadates and individual gelatin chains that are in their protonated state. In this case, the low POM concentration induces the formation of water-soluble inter-complexes by non-covalent interactions. However, when the decavanadate concentration increases, the non-covalent crosslinking increases until a critical concentration in which the previous structures collapsed forming micro-gel aggregates (Region 2). Finally, in Region 3, the low electrostatic repulsion between aggregates allows the aggregate growth forming a macroscopic phase separation. Considering the phase behavior of this system, the control over each region could allow the generation of hybrid nanocomposites with different size and functional properties.

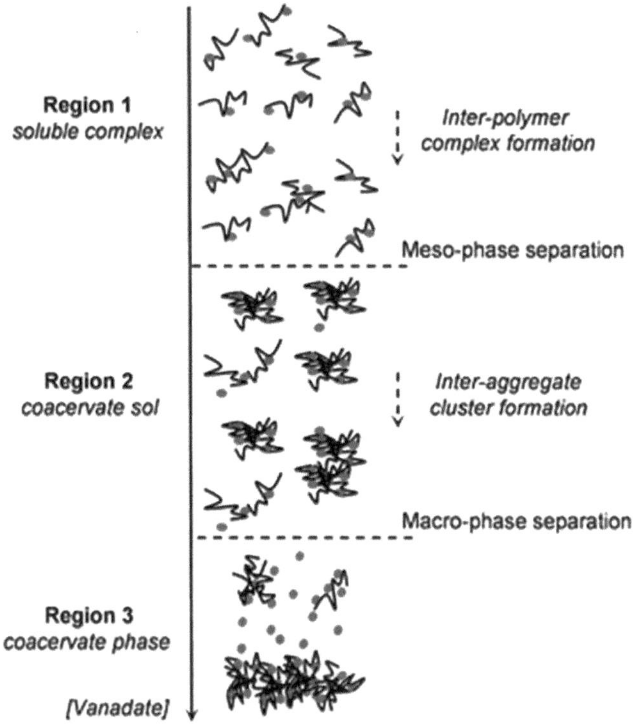

Figure 4.5 Schematic summary of the different phase separations observed for decavanadate/gelatin system. Reproduced with permission from reference [51] Copyright 2018 Elsevier.

4.2.3 Covalently Linked POM/Polymer Hybrids

The capability of the POMs to be functionalized opens wide potential possibilities to be incorporated in to polymeric materials due to the structural diversity of these compounds [3, 6, 52]. The use of covalent bond usually provides stronger POMs/

polymers integration and improves the stability of the hybrid system [53, 54]. Covalently bonded POM/polymer hybrid (PPHs) can be prepared by many synthetic approaches: (a) the previous modification of the monomers with POMs and subsequent polymerization or (b) post-polymerization modification of the polymer structure with POMs [8].

4.2.3.1 Hybrid materials formed by monomer modification

In the first approach, the radical polymerization by using POM-modified vinylic monomers [55, 56] represents the most common methodology. As an examples, Kalyani *et al.* [57] reported the synthesis of two hybrid POM-based monomers and their subsequent copolymerization with (methacryloyloxy) phenyldimethylsulfonium triflate (MAPDST) and/or methylmethacrylate (MMA). The prepared monomers were based on two types of POMs, Wells–Dawson-type cluster anions $[P_2V_3W_{15}O_{62}]^{9-}$ and Keggin-type cluster anions $[PV_2Mo_{10}O_{40}]^{5-}$. The synthetic paths used for the fabrication of Wells–Dawson-based monomer $(Bu_4N)_5(H)[P_2V_3W_{15}O_{59}\text{-}\{(OCH_2)_3CNHCO(CH_3)C{=}CH_2\}]$ are summarized in Fig. 4.6, being the yield of the synthesis of 44%. A similar procedure was used for the synthesis of the Keggin-type monomer reported in this work being the yield of the synthesis of 51%.

The obtained monomers were used as comonomers of the copolymerization with MAPDST and MMA by radical polymerization with 2,2′-azobis(2-methylpropio-nitrile) (AIBN) as initiator. The yields of the copolymers based on MMA with Wells–Dawson type monomer and Keggin type monomer were 45 and 64%, respectively. On the other hand, when MAPDST was used as a comonomer, the polymerization yield for Wells–Dawson type was 67%. These hybrid POM/polymers were also tested as photoresists using electron beam (E-beam)/ extreme ultraviolet (EUV) lithographic techniques.

In addition to traditional radical polymerizations, POMs covalently linked to monomers have also been employed in more demanding living polymerizations, such as ring-opening metathesis polymerization (ROMP) [58], atom transfer radical polymerization (ATRP) or reversible addition fragmentation chain transfer (RAFT) [59].

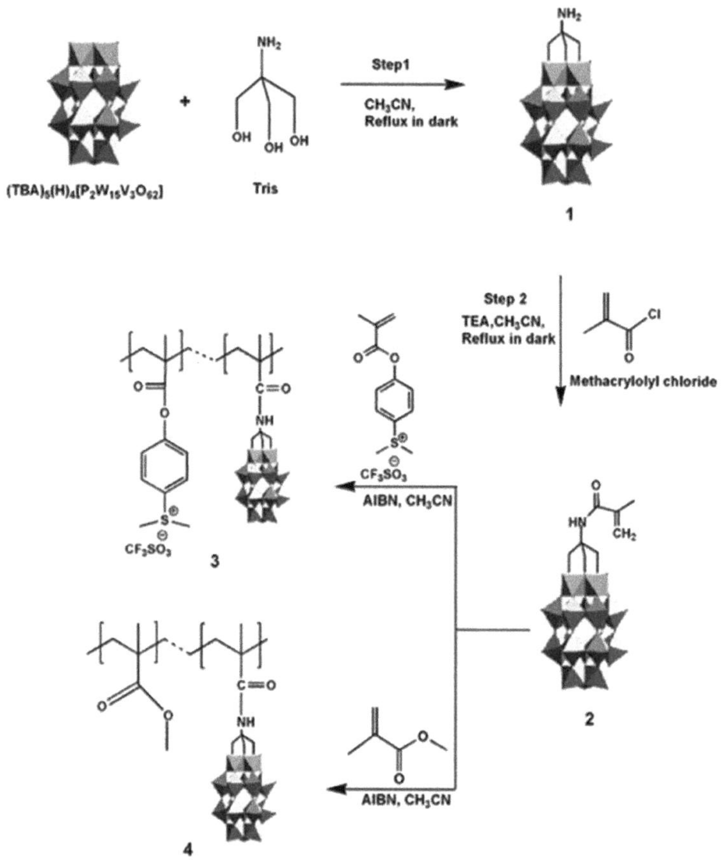

Figure 4.6 Summary of the synthetic route developed by Kalyani *et al.* [57] for the synthesis of Wells–Dawson-based monomer and its copolymerization with methylmetacrylate monomers. Reproduced with permission from reference [57]. Copyright 2014 Wiley.

Miao *et al.* [58] described the incorporation of a Wells–Dawson type POM with catalytic activity as a pendant functional group of the monomer and the successful polymerization. So, they synthesized a norbornene functionalized POM from trivanadium-substituted Wells–Dawson type polyoxotungstate $[(Bu_4N^+)_6H_3(P_2W_{15}V_3O_{62})^{9-}]$ to be used as a monomer. The norbornene-substituted POM was polymerized by ROMP polymerization using a Grubbs catalyst (Fig. 4.7) obtaining almost quantitatively the corresponding polymer.

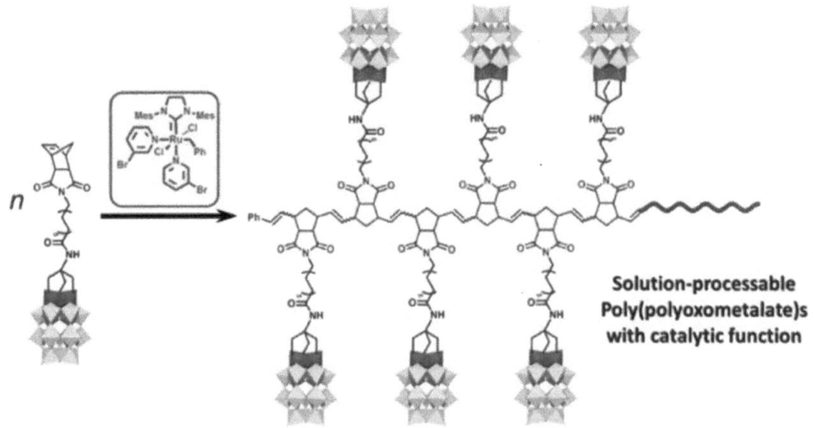

Figure 4.7 Scheme of the synthetic pathway carried out by Miao *et al.* to obtain a Wells–Dawson functionalized polymer. Reproduced with permission from reference [58]. Copyright 2014 American Chemical Society.

Authors also reported that the obtained polymer with POM pendant groups presented good processability and catalytic properties. On the other hand, this synthetic route could be applied to other similar POMs capable to be coordinated with three hydroxyl pendant groups.

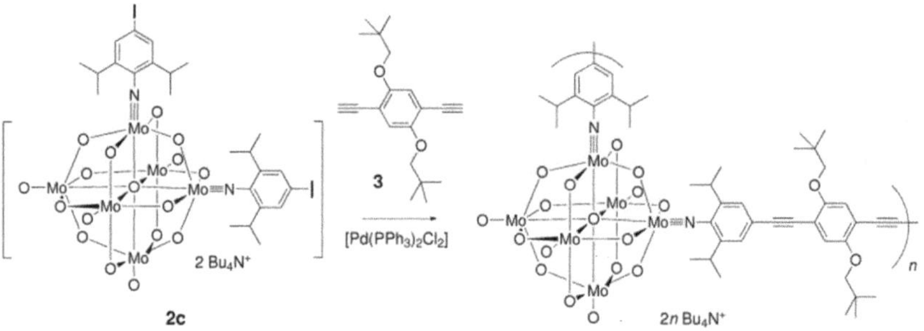

Figure 4.8 Scheme of the POM-based monomer and its subsequent polymerization. Reproduced with permission from reference [62]. Copyright 2002 Wiley.

The fabrication of hybrid monomers has not only been carried out through the synthesis of monomers with POMs in their side chain, but some examples of POM-substituted monomers that could include these inorganic entities on the polymer backbone have been described [60, 61]. However, the steric hindrance that these clusters induce in the main chain of the polymer reduces significantly the potential application of these kinds of structures. Xu and co-workers synthesized bifunctionalized organoimido polyoxomolybdates and they successfully tested their polymerization obtaining hybrid polymers with POMs in their backbone (Fig. 4.8) [62].

4.2.3.2 Covalently linked hybrid POM/polymers formed after polymerization

There are many examples of post-polymerization covalently immobilized POMs. These approaches could take advantages of well-established modification reactions commonly used in bioconjugation and chemistry modifications [63, 64].

For example, click reaction between azides and alkynes has been used for the incorporation of polyoxometales onto a solid resin [65]. The Wells–Dawson-type POM ($(NBu_4)_6[\alpha_2\text{-}P_2W_{17}O_{61}(SiC_6H_4CH_2N_3)_2O]$) was previously modified with two azido groups [66]. The catalytic properties presented by this POM make its incorporation in macroporous resins with a high surface area can reveal a suitable approach for its industrial application for oxidative catalysis. On the other hand, D380 benzylamine resin with a styrene–divinylbenzene matrix was modified with pent-4-ynoic acid and subsequently the azido-functionalized POMs were immobilized in it by click reaction. The obtained hybrid materials presented good catalytic properties tested in the oxidation of tetrahydrothiophene (THT). The findings provide a valuable protocol for practical catalytic oxidation.

In another example, Geng *et al.* [67] reported the functionalization of poly(ether ether ketone) (PEEK) with phosphotungstic acid (PWA). This hybrid material presented a ultra-low dielectric constant (κ = 1.96 at the frequency of 1 MHz when the PWA content reached 10 wt %), so it could be used in several electric applications. They used a previously modified poly(ether ether ketone) containing (3-trifluoromethyl)phenyl groups and carboxyl groups (PEEK–CF_3–COOH) which enables the immobilization of PWA as a side group.

4.3 Applications

4.3.1 Drug Delivery System for Breast Cancer Therapies

Menon *et al.* [68] developed hybrid nanocarriers based on nanocomplexes formed between chitosan and $[Cs\subset Eu_6As_6W_{63}O_{218}(H_2O)_{14}(OH)_4]^{25-}$ (EuWAs) POM. These nanocarriers were fabricated by ionotropic gelation, which takes advantages of the non-covalent electrostatic interactions between the POM and the chitosan chains in order to obtain nanosized particles. Authors reported that the nanoparticles presented a crosslinking efficiency of 81%, which allowed a slow and sustained POMS release at the physiological media. In addition, the cytotoxicity assays carried out on several host of cancer cell lines had proved the anticancer activity of these nanoparticles.

Recently, nanocarriers based on POM/chitosan were developed for their potential used for breast cancer applications. The Wells–Dawson-type polyanion used for this development, $[P_2W_{18}O_{62}]^{6-}$, can inhibit the action of Sox2 and Ap-2 gamma transcription factors [69, 70]. It is important to notice that, frequently, cancer disease is caused by an incorrect expression of transcription factors; however, there is a significant lack on drug capable to modulate these transcription factors, so the successful use of these types of POMs for this aim has opened a highly interesting and promising field. Nevertheless, their high toxicity represents an important limitation in the use of these materials as active ingredients, and so, limits their application. Thus, the manufacture of chitosan-based nanocarriers can prevent the liberation of POMs before reaching the target cells [71].

4.3.2 Self-Healing Hybrid Hydrogels

Self-healing materials are those, which in contract to classical materials, are capable to restore their functionality after suffering damage [72]. The self-healing process could be carried out due to the presence of non-covalent bonds (electrostatic interactions, hydrogen bonds, etc.) or dynamic covalent bonds, capable to autonomously form new bonds after the damage. Among the potential interactions capable to reversibly form these materials, some authors have reported POMs/polymer interactions as good candidates for these self-healable materials [73, 74]. In this context, Wei *et al.* [75] successfully fabricated an hydrogel based on electrostatic interactions between $Na_9EuW_{10}O_{36}$ (EuW_{10}) POM and ABA type triblock copolymer, poly(2-(2-guanidinoethoxy)ethyl methacrylate)-*b*-poly(ethylene oxide)-*b*-poly(2-(2-guanidinoethoxy)ethyl methacrylate) (PGn-*b*-PEO230-*b*-PGn) (Fig. 4.9).

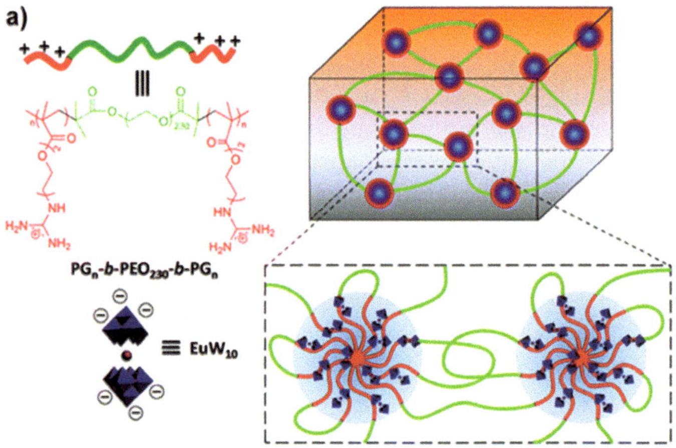

Figure 4.9 Scheme of the self-healing hybrid hydrogels reported by Wei *et al.* Reproduced from reference [75]. Copyright 2014 Royal Society of Chemistry.

The presence of europium-based POMs not only provides the self-healing capability to the material due to POM/polymer electrostatic interaction, but also gives luminescent properties to this material. In addition, this hydrogel presents high mechanical strength that could be tailor by variation on the composition of triblock copolymer or the ionic strength of the media, among other possible factors considering the nature of the polymer involved in the hydrogel.

4.3.3 Moisture Responsive Sensors

The combination of the capability of the hydrogels for water-absorption with some of the interesting properties of the POMs, such as the luminescence of the europium containing POMs could be very advantageous for the development of moisture responsive sensors. Considering this, Qiu and co-workers developed a moisture responsive film based on $Na_9[EuW_{10}O_{36}]\cdot 32H_2O$ (EuW_{10}) and agarose as a potential moisture sensor [76]. It is important to notice that EuW_{10} could reversibly switch into a non-luminescent by photoreduction and being recovered by oxidation (Fig. 4.10). This oxidation process capable to recover the luminescent property of this POM takes place in a humid media at relative humidity (RH) oscillating from 43 to 78% depending on the exposure time.

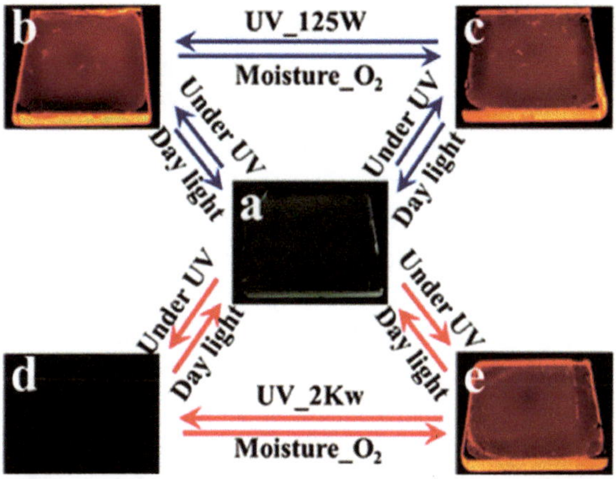

Figure 4.10 Scheme of the EuW10@Agarose films response process. (a) Film under daylight, (b) the luminescent image hybrid film under a UV lamp (12 W); (c) film after 10 min exposure to UV lamp (125 W), (d) after 10 min exposure under a UV lamp (2 kW), and (e) films after being store at 78% RH for 10 min. Reproduced with permission from reference [76]. Copyright 2015 Royal Society of Chemistry.

4.3.4 Solar UV Sensor

Continuing with the applications in the field of sensors and actuators, Liu and co-workers reported a solar UV sensitive composite film based on POM/polymer materials and viologen [77]. The films were fabricated by casting a mixture of fluorescent POM, EuW_{10}, PVP, poly(ethylene imine) (PEI) and a diaminopropyl viologen of *N,N′*-bis(δ-aminopropyl)-4,4′-

bipyridine bromide hydrobromide (AV^{2+}). These films presented photoswitchable emission that could be observed by fluorescence intensity returning. The fast response photochromism and switchable fluorescence reported for this system suggested that it could have great potential as a sensor in portable devices for solar UV detection.

4.3.5 Intumescent Flame Retardant (IFR)

One of the most common problems of the polymers or plastics is their flammability. In order to overcome this problem many flame retardant formulations have been used. In this context, the use of some POMS as IFR has been described by some authors. These IFR agents present several highly advantageous characteristics, such as the low smoke formation, non-toxicity, or capability to avoid the formation of corrosive or halogenated gasses. In this context, Chen *et al.* [80] described and evaluated three different POM/PP composites by using different POMs-based ionic liquid (PIL). Three commercially available POMs ($H_3PW_{12}O_{40}\cdot nH_2O$ (PWA), $H_3PMo_{12}O_{40}\cdot nH_2O$ (PMoA), and $H_4SiW_{12}O_{40}\cdot nH_2O$ (SiWA)) were modified with 1-butyl-3-methylimidazolium ([BMIm]) chloride, and these hybrid materials were subsequently melt-mixed with PP. All the formulations presented an improved limiting oxygen index (LOI). However, among all the studied formulations only the one formed with $[BMIm]_3PMo$/PP presented improvements in the UL-94 vertical burning tests and in thermal stability.

In another example of flame retardant materials, Peng *et al.* [79] described the efficiency of the Anderson-type POM, $[MnMo_6O_{18}]^{3-}$, functionalized with 9,10-dihydro-9-oxa-10-phosphaphenanthrene-10-oxide (DOPO) added to epoxy resin with an improved flame retardancy. With the aid of the organic modification, the addition of P–$MnMo_6$ not only maintained the transparency of EP but also enhanced the dynamic mechanical properties. It is important to highlight that the epoxy resin formulation with 8 wt % of POM passed the vertical burning grade (UL-94) V-0 rating with LOI value of 30.4%. In addition, the cone calorimeter test also indicated an LOI value of 41% from the pristine epoxy resin.

4.3.6 Electrochemical Capacitors

POM/polymer composite multilayers have also been reported as electrochemical capacitators [80–82]. Akter *et al.* [83] described the incorporation of different pseudocapacitive POMs, such as $SiMo_{12}O_{40}^{-4}$ ($SiMo_{12}$) and $PMo_{12}O_{40}^{-3}$ (PMo_{12}), to polymer multilayers. The studied POMs were deposited onto multi-walled carbon nanotubes (MWCNTs) forming multilayers with poly(diallyldimethylammonium chloride) (PDDA). The authors reported that the combination of these two POMs exhibited continuous overlapping oxidation/reduction reactions through LbL providing an easy and efficient approach to design and deposit active layers in order to achieve the desired electrochemical behavior.

4.3.7 Solid-State Proton Conductors (SSPCs)

SSPCs are considered key materials in the development of new environment-friendly energy storage devices [84–88]. Until date, the highly interesting use of some hybrid composite films based on POM/polymer interactions as SSPCs has been described. In this context, recently, Zheng *et al.* [89] reported hybrid $H_3PW_{12}O_{40}\cdot nH_2O$/PEO nanocomposite films with promising proton conductivities at ambient conditions (Fig. 4.11).

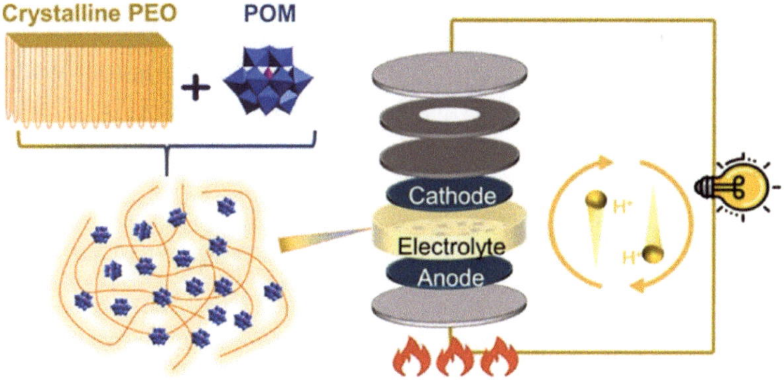

Figure 4.11 Scheme of the hybrid composite material developed by Zheng *et al.* Reproduced with permission from reference [89]. Copyright 2021 American Chemical Society.

The strong interactions formed between POMs and PEO, semicrystalline polymer, could inhibit its crystallization forming a highly stable composite, which eases fast proton transportation through the polymer matrix and improves the proton conductivity of this hybrid composite. Authors reported a conductivity of 6.86×10^{-3} S cm^{-1} at 100°C in anhydrous conditions for the composites formed with 70 wt % of POM. In addition, they proved that the conductivities could be maintained in a temperature range from −20 to 100°C in anhydrous conditions. The reported results indicated that these composite materials could be successfully used in proton exchange membranes (PEMs) in fuel cells and electrolytes in supercapacitors.

4.3.8 Near-Infrared and Visible Light Modulated Electrochromic Devices

Gu *et al.* [90] reported a modulated electrochromic device based on hybrid POM/polymer materials that could be potentially applied in the development of smart windows. In their study, they used two POMs, $K_{28}Li_5H_7P_8W_{48}O_{184}\cdot 92H_2O$ (P_8W_{48}) and $W_{18}O_{49}$ nanowires, which were sequentially incorporated to poly(ethylene imine) (PEI). These POM/PEI were anchored to ITO substrates by using PEI and poly(sodium 4-styrenesulfonate) (PSS) multilayers to ensure de-adhesion. The obtained results indicated that this system could successfully modulate the transmittance both near-infrared (750–1360 nm) and visible light (400–750 nm) regions by only varying the applied potential (Fig. 4.12).

4.4 Conclusion

In this chapter, the main approaches used to in the synthesis of POM/polymer hybrids have been depicted summarizing some of the most relevant examples of hybrid materials described until date. In addition, the state of art of many of the most relevant applications involving these kinds of hybrid materials has been described.

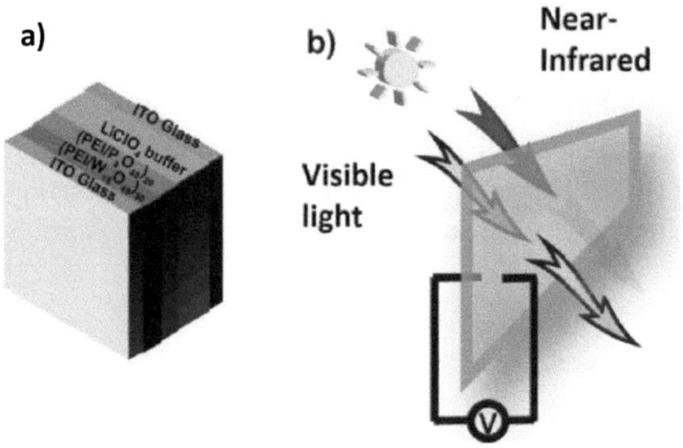

Figure 4.12 (a) Scheme of the sandwich configuration developed in Gu *et al.*'s [90] study. (b) Description of how the device operates. Reproduced with permission from reference [90]. Copyright 2014 American Chemical Society.

It is foreseeable that considering the wide range of potential combinations of POMs and polymers that could be designed, tailoring a huge amount of properties, such as mechanical, electrical, or photochromic, among others, the number of studies devoted to the fabrication of these types of materials or their applications will increase in the following years. Among all the applications described, the arising necessity for highly efficient energy storage devises could be one of the most interesting applications of these materials that could gain more attention. Nevertheless, as described throughout the text, there are many other areas that could benefit from the improvement in the progress of these materials, such as in optics, electronics, biology, or catalysis.

Acknowledgment

The authors acknowledge MINECOG within the project "Polímeros autorreparables con memoria de forma basados en interacciones orgánicas-inorgánicas reversibles" (MAT2017-89553-P).

References

1. Pope, M.T. and Müller, A. (eds) (1994). *Polyoxometalates: From Platonic Solids to Anti-Retroviral Activity*. Springer, The Netherlands.

2. Li, Q., Zhang, L., Dai, J., Tang, H., Li, Q., Xue, H. and Pang, H. (2018). Polyoxometalate-based materials for advanced electrochemical energy conversion and storage, *Chem. Eng. J.*, **351**, pp. 441–461.
3. Proust, A., Matt, B., Villanneau, R., Guillemot, G., Gouzerh, P. and Izzet, G. (2012). Functionalization and post-functionalization: a step towards polyoxometalate-based materials, *Chem. Soc. Rev.*, **41**, pp. 7605–7622.
4. Wang, Y. and Weinstock, I.A. (2012). Polyoxometalate-decorated nanoparticles, *Chem. Soc. Rev.*, **41**, pp. 7479–7496.
5. Toma, F.M., Sartorel, A., Iurlo, M., Carraro, M., Parisse, P., Maccato, C., Rapino, S., Gonzalez, B.R., Amenitsch, H., Da Ros, T., Casalis, L., Goldoni, A., Marcaccio, M., Scorrano, G., Scoles, G., Paolucci, F., Prato, M. and Bonchio, M. (2010). Efficient water oxidation at carbon nanotube–polyoxometalate electrocatalytic interfaces, *Nat. Chem.*, **2**, pp. 826–831.
6. Dolbecq, A., Dumas, E., Mayer, C.R. and Mialane, P. (2010). Hybrid organic–inorganic polyoxometalate compounds: from structural diversity to applications, *Chem. Rev.*, **110**, pp. 6009–6048.
7. Qi, W. and Wu, L. (2009). Polyoxometalate/polymer hybrid materials: fabrication and properties, *Polym. Int.*, **58**, pp. 1217–1225.
8. Wu, H., Yang, H.K. and Wang, W. (2016). Covalently-linked polyoxometalate–polymer hybrids: optimizing synthesis, appealing structures and prospective applications, *New J. Chem.*, **40**, pp. 886–897.
9. Zukowska, G., Stevens, J.R. and Jeffrey, K.R. (2003). Anhydrous gel electrolytes doped with silicotungstic acid, *Electrochim. Acta*, **48**, pp. 2157–2164.
10. Peng, X.J., Zhang, Y., Feng, W., Ai, L.M., Chen, J. and Zhang, F.-J. (2013). The characterization and photochromism of copolymer composite films containing phosphotungstic acid, *J. Mol. Struct.*, **1041**, pp. 139–143.
11. Espindola, A., Gonçalves, N.S., Nalin, M., Ribeiro, S.J.L., Barud, H.S. and Molina, C. (2019). Casting and inkjet printable photochromic films based on polymethylmethacrylate—phosphotungstic acid, *Opt. Mater.*, **96**, p. 109345.
12. Glezos, N., Argitis, P., Velessiotis, D. and Diakoumakos, C.D. (2003). Tunneling transport in polyoxometalate based composite materials, *Appl. Phys. Lett.*, **83**, pp. 488–490.

13. Li, H., Sun, H., Qi, W., Xu, M. and Wu, L. (2007). Onionlike hybrid assemblies based on surfactant-encapsulated polyoxometalates, *Angew. Chemie—Int. Ed.*, **46**, pp. 1300–1303.

14. Nisar, A. and Wang, X. (2012). Surfactant-encapsulated polyoxometalate building blocks: controlled assembly and their catalytic properties, *Dalt. Trans.*, **41**, pp. 9832–9845.

15. Sun, H., Li, H., and Wu, L. (2009). Micro-patterned polystyrene surfaces directed by surfactant-encapsulated polyoxometalate complex via breath figures, *Polymer*, **50**, pp. 2113–2122.

16. Yang, H., Song, T., Liu, L., Devadoss, A., Xia, F., Han, H., Park, H., Sigmund, W., Kwon, K. and Paik, U. (2013). Polyaniline/polyoxometalate hybrid nanofibers as cathode for lithium ion batteries with improved lithium storage capacity, *J. Phys. Chem. C*, **117**, pp. 17376–17381.

17. Lan, Y., Wang, E., Song, Y., Song, Y., Kang, Z., Xu, L. and Li, Z. (2006). An effective layer-by-layer adsorption and polymerization method to the fabrication of polyoxometalate–polypyrrole nanoparticle ultrathin films, *Polymer*, **47**, pp. 1480–1485.

18. Liu, M. and Dong, S. (1995). Electrochemical behavior of molibdosilicic heteropoly complex with dysprosium and its doped polypyrrole film modified electrode, *Electrochim. Acta*, **40**, pp. 197–200.

19. Bidan, G., Geniés, E.M., and Lapkowski, M. (1989). One-step electrochemical immobilization of Keggin-type heteropolyanions in poly(3-methylthiophene) film at an alectrode surface: electrochemical and electrocatalytic properties, *Synth. Met.*, **31**, pp. 327–334.

20. Baibarac, M., Baltog, I., Smaranda, I. and Magrez, A. (2015). Photochemical processes developed in composite based on highly separated metallic and semiconducting SWCNTs functionalized with polydiphenylamine, *Carbon N. Y.*, **81**, pp. 426–438.

21. Küçük, İ., Vural, S., Kıvılcım, N., Adıgüzel, İ., Köytepe, S. and Seçkin, T. (2020). Preparation of the copper-based polyoxometalate/polyurethane composites and their dielectric properties, *Polym. Polym. Compos.*, **28**, pp. 473–483.

22. Küçük, İ., Vural, S., Köytepe, S. and Seçkin, T. (2019). Synthesis, characterization and dielectric properties of nickel-based polyoxometalate/polyurethane composites, *Polym. Technol. Mater.*, **58**, pp. 1445–1460.

23. Küçük, İ., Vural, S., Kıvılcım, N., Adıgüzel, I., Köytepe, S. and Seçkin, T. (2020). Hydrothermal synthesis of polyoxometalate containing cobalt ion, and its polyurethane composites for dielectric material applications, *Polym. Technol. Mater.*, **59**, pp. 1822–1841.

24. Pinto, T.V., Fernandes, D.M., Guedes, A., Cardoso, N., Durães, N.F. Silva, C., Pereira, C. and Freire, C. (2018). Photochromic polypropylene fibers based on UV-responsive silica@phosphomolybdate nanoparticles through melt spinning technology, *Chem. Eng. J.*, **350**, pp. 856–866.
25. Yin, R., Guan, X.-H., Gong, J. and Qu, L.-Y. (2007). Evaluation of swelling capacity of poly(vinyl alcohol) fibrous mats dealt with polyoxometalate containing vanadium, *J. Appl. Polym. Sci.*, **106**, pp. 1677–1682.
26. Lu, X., Liu, X., Wang, L., Zhang, W. and Wang, C. (2006). Fabrication of luminescent hybrid fibers based on the encapsulation of polyoxometalate into polymer matrices, *Nanotechnology*, **17**, pp. 3048–3053.
27. Zhang, T.R., Lu, R., Liu, X.L., Zhao, Y.Y., Li, T.J. and Yao, J.N. (2003). Photochromic polyoxotungstoeuropate $K_{12}[EuP_5W_{30}O_{110}]$/polyvinylpyrrolidone nanocomposite films, *J. Solid State Chem.*, **172**, pp. 458–463.
28. Feng, W., Zhang, T.R., Wei, L., Lu, R., Bai, Y.B., Li, T.J., Zhao, Y.Y. and Yao, J.N. (2002). Photochromic behavior and mechanism of thin films in $H_3PW_{12}O_{40}$/polyacrylamide system, *Mater. Lett.*, **54**, pp. 309–313.
29. Jing, B., Wang, X., Wang, H., Qiu, J., Shi, Y., Gao H. and Zhu, Y. (2017). Shape and mechanical control of poly(ethylene oxide) based polymersome with polyoxometalates via hydrogen bond, *J. Phys. Chem. B*, **121**, pp. 1723–1730.
30. Tsuboi, M., Hibino, M., Mizuno, N. and Uchida, S. (2016). Crystalline polyoxometalate (POM)–polyethylene glycol (PEG) composites aimed as non-humidified intermediate-temperature proton conductors, *J. Solid State Chem.*, **234**, pp. 9–14.
31. Buchecker, T., Le Goff, X., Naskar, B., Pfitzner, A., Diat, O. and Bauduin, P. (2017). Polyoxometalate/polyethylene glycol interactions in water: from nanoassemblies in water to crystal formation by electrostatic screening, *Chem.– A Eur. J.*, **23**, pp. 8434–8442.
32. Xu, B., Xu, L., Gao, G. and Jin, Y. (2007). Nanosized multilayer films with concurrent photochromism and electrochromism based on Dawson-type polyoxometalate, *Appl. Surf. Sci.*, **253**, pp. 3190–3195.
33. Bai, B., Zhou, J. and Yin, M. (2015). A comprehensive review of polyacrylamide polymer gels for conformance control, *Pet. Explor. Dev.*, **42**, pp. 525–532.

34. Tonkushina, M.O., Alekseeva, O.V., Agafonov, A.V. and Ostroushko, A.A. (2016). Viscosity and electrophysical characteristics of solutions containing nanocluster polyoxometalates and polyvinylpyrrolidone, *Russ. J. Phys. Chem. A*, **90**, pp. 838–842.

35. Feng, W., Zhang, T.R., Liu, Y., Lu, R., Zhao, Y.Y., Li, T.J. and Yao, J.N. (2002). Sonochemical preparation of photochromic nanocomposite thin film based on polyoxometalates well dispersed in polyacrylamide, *J. Solid State Chem.*, **169**, pp. 1–5.

36. Chen, J. Liu, S.L., Feng, W., Bao, X.J. and Yang, F.L. (2013). The influence of Fe^{2+} doping on the microstructure and photochromic behavior of polyoxometalates acid/polyacrylamide hybrid films, *Opt. Mater.*, **35**, pp. 973–977.

37. Wang, Z., Ma, Y., Zhang, R., Peng, A., Liao, Q., Cao, Z., Fu, H. and Yao, J. (2009). Reversible luminescent switching in a $[Eu(SiW_{10}MO_{39})_2]^{13-}$-agarose composite film by photosensitive intramolecular energy transfer, *Adv. Mater.*, **21**, pp. 1737–1741.

38. Fang, Y., Liu, T., Xing, C., Chang, J. and Li, M. (2020). A blend hydrogel based on polyoxometalate for long-term and repeatedly localized antibacterial application study, *Int. J. Pharm.*, **591**, p. 119990.

39. Wu, A., Sun, P., Sun, N., Yu, Y. and Zheng, L. (2018). Coassembly of a polyoxometalate and a zwitterionic amphiphile into a luminescent hydrogel with excellent stimuli responsiveness, *Chem. –A Eur. J.*, **24**, pp. 16857–16864.

40. Liang, L., Sun, N., Yu, Y., Ren, S., Wu, A. and Zheng, L. (2020). Photoluminescent polymer hydrogels with stimuli-responsiveness constructed from Eu-containing polyoxometalate and imidazolium zwitterions, *Soft Matter*, **16**, pp. 2311–2320.

41. Haider, A., Kortz, U., Ullah, S. and Sohail, M. (2017). Novel pH responsive supramolecular hydrogels of chitosan hydrochloride and polyoxometalate: in-vitro, in-vivo and preliminary safety evaluation, *Int. J. Pharm.*, **533**, pp. 125–137.

42. Li, H., Pang, S., Wu, S., Feng, X., Müllen, K. and Bubeck, C. (2011). Layer-by-Layer assembly and UV photoreduction of graphene–polyoxometalate composite films for electronics, *J. Am. Chem. Soc.*, **133**, pp. 9423–9429.

43. Wang, B., Vyas, R.N. and Shaik, S. (2007). Preparation parameter development for layer-by-layer assembly of Keggin-type polyoxometalates, *Langmuir*, **23**, pp. 11120–11126.

44. Feng, Y., Han, Z., Peng, J., Lu, J., Xue, B., Li, L., Ma, H. and Wang, E. (2006). Fabrication and characterization of multilayer films based on Keggin-type polyoxometalate and chitosan, *Mater. Lett.*, **60**, pp. 1588–1593.

45. Draget, K.I., Värum, K.M., Moen, E., Gynnild, H. and Smidsrød, O. (1992). Chitosan cross-linked with Mo(VI) polyoxyanions: a new gelling system, *Biomaterials*, **13**, pp. 635–638.

46. Fiorani, G., Saoncella, O., Kaner, P., Altinkaya, S.A., Figoli, A. Bonchio, M. and Carraro, M. (2014) Chitosan-polyoxometalate nanocomposites: synthesis, characterization and application as antimicrobial agents, *J. Clust. Sci.*, **25**, pp. 839–854.

47. Geisberger, G., Paulus, S.,. Gyenge, E.B, Maake, C. and Patzke, G.R. (2011). Targeted delivery of polyoxometalate nanocomposites, *Small*, **7**, pp. 2808–2814.

48. Pamin, K., Jachimska, B., Onik, K., Połtowicz, J. and Grabowski, R. (2009). Electrostatic self-assembly of polyoxometalates on chitosan as catalysts of oxidation of cyclic hydrocarbons, *Catal. Letters*, **127**, pp. 167–174.

49. Pandya, V.M., Kortz, U. and Joshi, S.A. (2014). Encapsulation and stabilization of polyoxometalates in self-assembled supramolecular hydrogels., *Dalton Trans.*, **44**, pp. 58–61.

50. Wang, R., Wan, X. and Zhang, J. (2019). Multi-stimuli-responsive induced chirality of polyoxometalates in natural polysaccharide hydrogels, *Chem. Commun.*, **55**, pp. 4711–4714.

51. Carn, F., Steunou, N., Djabourov, M. Coradin, T., Ribot, F. and Livage, J. (2008). First example of biopolymer–polyoxometalate complex coacervation in gelatin–decavanadate mixtures, *Soft Matter*, **4**, p. 735–738.

52. Anyushin, A.V., Kondinski, A. and Parac-Vogt, T.N. (2020). Hybrid polyoxometalates as post-functionalization platforms: from fundamentals to emerging applications, *Chem. Soc. Rev.*, **49**, pp. 382–432.

53. Ren, Y., Wang, M. Chen, X., Yue, B. and He, H. (2015). Heterogeneous catalysis of polyoxometalate based organic–inorganic hybrids, *Materials*, **8**, pp. 1545–1567.

54. Lan, Y.-Q., Li, S.-L., Shao, K.-Z., Wang, X.-L. and Su, Z.-M. (2008). Construction of different dimensional inorganic–organic hybrid materials based on polyoxometalates and metal–organic units via changing metal ions: from non-covalent interactions to covalent connections, *Dalt. Trans.*, **29**, p. 3824–3835.

55. Mayer, C.R., Thouvenot, R. and Lalot, T. (2000). New hybrid covalent networks based on polyoxometalates: part 1. Hybrid networks based on poly(ethyl methacrylate) chains covalently cross-linked by heteropolyanions: synthesis and swelling properties, *Chem. Mater.*, **12**, pp. 257–260.

56. Carraro, M., Fiorani, G., Mognon, L., Caneva, F., Gardan, M., Maccato, C. and Bonchio, M. (2021). Hybrid polyoxotungstates as functional comonomers in new cross-linked catalytic polymers for sustainable oxidation with hydrogen peroxide, *Chem.–A Eur. J.*, **18**, pp. 13195–13202.

57. Kalyani, V., Satyanarayana, V.S.V., Singh, V., Pradeep, C.P., Ghosh, S., Sharma, S.K. and Gonsalves, K.E. (2015). New polyoxometalates containing hybrid polymers and their potential for nano-patterning, *Chem.–A Eur. J.*, **21**, pp. 2250–2258.

58. Miao, W.K., Yan, Y.K., Le Wang, X., Xiao, Y., Ren, L.J., Zheng, P., Wang, C.H., Ren, L.X. and Wang, W. (2014). Incorporation of polyoxometalates into polymers to create linear poly(polyoxometalate)s with catalytic function, *ACS Macro Lett.*, **3**, pp. 211–215.

59. Lesage De La Haye, J., Beaunier, P., Ruhlmann, L., Hasenknopf, B., Lacôte, E. and Rieger, J. (2014). Synthesis of well-defined Dawson-type poly(*N,N*-diethylacrylamide) organopolyoxometalates, *Chempluschem*, **79**, pp. 250–256.

60. Wang, R., Li, Y., Shetye, K., Dutta, T., Jin, L., Li, S. and Peng, Z. (2015). Luminescent polythiophene-based main-chain polyoxometalate-containing conjugated polymers with improved solar-cell performance, *Eur. J. Inorg. Chem.*, **2015**, pp. 656–663.

61. Wei, Y., Lu, M., Cheung, C.F., Barnes, C.L. and Peng, Z. (2001). Functionalization of $[MoW_5O_{19}]^{2-}$ with aromatic amines: synthesis of the first arylimido derivatives of mixed-metal polyoxometalates, *Inorg. Chem.*, **40**, pp. 5489–5490.

62. Xu, L., Lu, M., Xu, B., Wei, Y., Peng, Z. and Powell, D.R. (2002). Towards main-chain-polyoxometalate-containing hybrid polymers: a highly efficient approach to bifunctionalized organoimido derivatives of hexamolybdates, *Angew. Chemie Int. Ed.*, **41**, pp. 4129–4132.

63. Hermanson, G.T. (2013). *Bioconjugate Techniques*. Academic Press, San Diego, CA

64. Theato, P. and Klok, H.-A. (eds). (2002). *Functional Polymers by Post-Polymerization Modification*. Wiley-VCH Verlag, Germany.

65. Xiao, Y., Chen, D., Ma, N., Hou, Z., Hu, M., Wang, C. and Wang, W. (2013). Covalent immobilization of a polyoxometalate in a porous polymer

matrix: a heterogeneous catalyst towards sustainability, *RSC Adv.*, **3**, pp. 21544–21551.

66. Odobel, F. Séverac, M., Pellegrin, Y., Blart, E., Fosse, C., Cannizzo, C., Mayer, C.R., Elliott, K.J. and Harriman, A. (2009). Coupled Sensitizer-catalyst dyads: electron-transfer reactions in a perylene-polyoxometalate conjugate, *Chem.–A Eur. J.*, **15**, pp. 3130–3138.

67. Geng, Z., Ba, J., Zhang, S., Luan, J., Jiang, X., Huo, P. and Wang, G. (2012). Ultra low dielectric constant hybrid films *via* side chain grafting reaction of poly(ether ether ketone) and phosphotungstic acid, *J. Mater. Chem.*, **22**, pp. 23534–23540.

68. Menon, D., Thomas, R.T., Narayanan, S., Maya, S., Jayakumar, R., Hussain, F., Lakshmanan, V.K. and Nair, S.V. (2011). A novel chitosan/polyoxometalate nano-complex for anti-cancer applications, *Carbohydr. Polym.*, **84**, pp. 887–893.

69. Narasimhan, K., Pillay, S., Bin Ahmad, N.R., Bikadi, Z., Hazai, E., Yan, L., Kolatkar, P.R., Pervushin, K. and Jauch, R. (2011). Identification of a polyoxometalate inhibitor of the DNA binding activity of Sox2, *ACS Chem. Biol.*, **6**, pp. 573–581.

70. Hu, J., Pillay, S., Bin Ahmad, N.R., Bikadi, Z., Hazai, E., Yan, L., Kolatkar, P.R., Pervushin, K. and Jauch, R. (2018). Identification of a Wells–Dawson polyoxometalate-based AP-2γ inhibitor with pro-apoptotic activity. *Biochem. J.*, **475**, pp. 1965–1977.

71. Pérez-Álvarez, L., Ruiz-Rubio, L., Artetxe, B., Vivanco, M. dM., Gutiérrez-Zorrilla, J.M. and Vilas-Vilela, J.L. (2019). Chitosan nanogels as nanocarriers of polyoxometalates for breast cancer therapies, *Carbohydr. Polym.*, **213**, pp. 159–167.

72. Hager, M.D. (2017). Self-healing materials. *Handbook of Solid State Chemistry*, Wiley-VCH Germany. pp. 201–225.

73. Wei, X., Ma, K., Cheng, Y., Sun, L., Chen, D., Zhao, X., Lu, H., Song, B., Yang, K. and Jia, P. (2020). Adhesive, conductive, self-healing, and antibacterial hydrogel based on chitosan–polyoxometalate complexes for wearable strain sensor, *ACS Appl. Polym. Mater.*, **2**, pp. 2541–2549.

74. Yang, J., Chen, M., Li, P., Cheng, F., Xu, Y., Li, Z., Wang, Y. and Li, H. (2018). Self-healing hydrogel containing Eu-polyoxometalate as acid-base vapor modulated luminescent switch, *Sensors Actuators B Chem.*, **273**, pp. 153–158.

75. Wei, H., Du, S., Liu, Y., Zhao, H., Chen, C., Li, Z., Lin, J., Zhang, Y., Zhang, J. and Wan, X. (2014). Tunable, luminescent, and self-healing

hybrid hydrogels of polyoxometalates and triblock copolymers based on electrostatic assembly, *Chem. Commun.*, **50**, pp. 1447–1450.

76. Qiu Y.-F., Liu, H., Liu, J.-X., Zhang, C., Ma, Z., Hu, P.-A. and Gao, G.-G. (2015). Moisture-responsive films consisting of luminescent polyoxometalates and agarose, *J. Mater. Chem. C*, **3**, pp. 6322–6328.

77. Liu, H., Lv, Y., Li, S., Yang, F., Liu, S., Wang, C., Sun, J.-Q., Meng, H. and Gao, G.-G. (2017). A solar ultraviolet sensor based on fluorescent polyoxometalate and viologen, *J. Mater. Chem. C*, 5 (36), pp. 9383–9388.

78. Chen, S., Li, J., Zhu, Y. and Su, S. (2014) Roles of anion of polyoxometalate-based ionic liquids in properties of intumescent flame retardant polypropylene, *RSC Adv.*, **4**, pp. 32902–32913.

79. Peng, C., Chen, T., Zeng, B., Chen, G., Yuan, C., Xu, Y. and Dai, L. (2020). Anderson-type polyoxometalate-based hybrid with high flame retardant efficiency for the preparation of multifunctional epoxy resin nanocomposites, *Compos. Part B Eng.*, **186**, p. 107780.

80. Genovese, M. and Lian, K. (2015). Polyoxometalate modified inorganic–organic nanocomposite materials for energy storage applications: a review, *Curr. Opin. Solid State Mater. Sci.*, **19**, pp. 126–137.

81. Yang, M., Choi, B.G., Jung, S.C., Han, Y.-K., Huh, Y.S. and Lee, S.B. (2014). Polyoxometalate-coupled graphene *via* polymeric ionic liquid linker for supercapacitors, *Adv. Funct. Mater.*, **24**, pp. 7301–7309.

82. Martel, D., Nguyen Cong, H., Molinari, M., Ebothé, J. and Kityk, I.V. (2008). Adsorption of polyanions on nanostructured polypyrrole submonolayer grafted on semiconducting transparent support, *J. Mater. Sci.*, **43**, pp. 3486–3490.

83. Akter, T., Hu, K. and Lian, K. (2011). Investigations of multilayer polyoxometalates-modified carbon nanotubes for electrochemical capacitors, *Electrochim. Acta*, **56**, pp. 4966–4971.

84. Zheng, Z., Zhou, Q., Li, M. and Yin, P. (2019) Poly(ethylene glycol) nanocomposites of sub-nanometer metal oxide clusters for dynamic semi-solid proton conductive electrolytes, *Chem. Sci.*, **10**, pp. 7333–7339.

85. Lin, G., Bai, Z., Han, M., Jia, K., Huang, Y. and Liu, X. (2020). Tungstophosphoric acid-doped sulfonated poly(arylene ether nitriles) composite membranes with improved proton conductivity and excellent long-term stability, *Solid State Ionics*, **357**, p. 115487.

86. Wen, T., Qiu, L., Zheng, Z., Gong, Y., Yuan, J., Wang, Y., Huang, M. and Yin, P. (2020). Inclusion crystallization of silicotungstic acid

and poly(ethylene oxide) and its impact on proton conductivities, *Macromolecules*, **53**, pp. 1415–1421.

87. He, H., Wang, G., Chai, S., Li, X., Zhai, L., Wu, L. and Li, H., (2021). Self-assembled lamellar nanochannels in polyoxometalate-polymer nanocomposites for proton conduction, *Chinese Chem. Lett.*, In press (February 2021).

88. Zhai, L. and Li, H. (2019). Polyoxometalate–polymer hybrid materials as proton exchange membranes for fuel cell applications, *Molecules*, **24**, p. 3425.

89. Zheng, Z., Li, M., Zhou, Q., Cai, L., Yin, J.-F., Cao, Y. and Yin, P. (2021). Polyoxometalate-poly(ethylene oxide) nanocomposites for flexible anhydrous solid-state proton conductors, *ACS Appl. Nano Mater.*, **4**, pp. 811–819.

90. Gu, H., Guo, C., Zhang, S., Bi, L., Li, T., Sun, T. and Liu, S. (2018). Highly efficient, near-infrared and visible light modulated electrochromic devices based on polyoxometalates and $W_{18}O_{49}$ nanowires, *ACS Nano*, **12**, pp. 559–567.

Chapter 5

Polyoxometalates in Catalysis

Juan Alcañiz Monge[a] and Santiago Reinoso[b]

[a]*Departamento de Química Inorgánica, Facultad de Ciencias, Universidad de Alicante, 03080 Alicante, Spain*
[b]*Departamento de Ciencias and Institute for Advanced Materials and Mathematics (InaMat2), Universidad Pública de Navarra, 31006 Pamplona, Spain*

jalcaniz@ua.es, santiago.reinoso@unavarra.es

This chapter focuses on providing a basis for understanding the mechanisms involved in the catalysis carried out by selected polyoxometalate (POM)-based compounds. The catalysts discussed herein are either insoluble POM salts or POM clusters heterogenized on porous solid supports, including activated carbon materials and metallic oxides, such as zirconia. The influence on the catalytic activity of both the POM catalytic species and the active porous support is the main aspect to be commented and analyzed in detail within the chapter.

5.1 Introduction

The current framework of the chemical industry demands that production processes must comply with restrictive environmental

Polyoxometalates: Advances, Properties, and Applications
Edited by Leire Ruiz Rubio, José Luis Vilas Vilela, Beñat Artetxe, and Juan Manuel Gutiérrez-Zorrilla

ISBN 978-981-4968-14-0 (Hardcover), 978-1-003-27744-6 (eBook)
www.jennystanford.com

standards to avoid the use of depleting feedstocks. Moreover, the technologies implemented to such end should also be sustainable from an energetic and environmental point of view. All of these premises have led to the emergence of the so-called Green Chemistry. In this context, the use of catalytic processes has become an essential part of the chemical industry, as they are present in the production of as much as the 80% of all industrial chemicals ranging from fine and specialty chemicals, cosmetics, and pharmaceuticals to agrochemicals and foods, petrochemicals and even polymers [1].

Catalytic processes are classified into homogeneous, when catalyst and substrate coexist in the same phase, or heterogeneous, when the catalytically active species is in a phase separate from that of the substrate. In most cases, homogeneous catalysts are more active and selective than heterogeneous counterparts, and they usually profit from milder reaction conditions. Homogeneous catalysts include inorganic acids (HF, H_2SO_4), alkali/alkaline earth hydroxides and salts, and classical Lewis acids (e.g., $AlCl_3$, BF_3, $SnCl_4$, or $TiCl_4$). These are among the cheapest, most widely available, and most extensively used of such catalysts [2–4]. Their main drawback, however, is the separation of the homogeneous catalyst from the reaction products, since both components are found in the same phase. Associated with this fact is the generation of corrosive waste products related with the catalyst and unreacted raw materials.

This drawback can be overcome by making the molecular homogeneous catalysts insoluble in the reaction medium, thereby paving the way for the development of new heterogeneous solid catalysts. Heterogeneous catalytic processes involve a solid catalyst in contact with a fluid medium containing the precursors and products, thereby allowing for an easy separation and recovery of the former. Despite being less active and selective, this ease of separation endows solid catalysts with main advantages over conventional homogeneous catalysts because it lowers the amount of wastes, avoids potential contamination of the reaction products, and allows for the catalyst to be largely recyclable and reusable in consecutive reaction runs [5–7], thus making large-scale production processes more economical than those carried out in the homogeneous phase. One elegant approach to design such type of solid catalysts consists in the so-called heterogenization, that is, the immobilization of a soluble species (homogeneous catalyst) as the catalytically active functionality on a solid support that provides the heterogeneous character to the resulting material. A critical aspect in the design of such type of supported catalysts is the selection of a suitable support because this can strongly enhance or reduce the selectivity of the catalytically active functionality, as well as its activity, or even both at the same time. This fact has been theoretically analyzed in the framework of the concept of strong metal–support interactions (SMSI) [8, 9] and, more recently, also on the basis of the term electronic metal–support interaction (EMSI) [10]. The basic background behind both theories is that the chemical bonding of the catalytic functionality to the support, which is required for its suitable anchoring, can also change its electronic and/or molecular structure, hence greatly affecting its catalytic behavior.

In this context, there is an ongoing research effort devoted to the development of more economic and environmentally benign materials that could replace catalysts, such as expensive noble-metal species or hazardous mineral acids in industrial processes. The most promising among such a type of catalytic materials are those that incorporate polyoxometalates (POMs), a class of anionic metal-oxo clusters with dual properties as strong acids and efficient oxidants [11]. Based on their chemical composition, POMs can be classified into two large families: isopolyoxometalates of general formula $[M_mO_n]^{p-}$, which only contain transition metals from groups 5 or 6 and bridging/terminal oxygen atoms, and heteropolyoxometalates of general formula $[X_yM_mO_n]^{q-}$, which contain additional elements (X) besides the transition metals and oxygen. In the heteropoly case, X is the heteroatom and is usually located in the center of the cluster. The elements playing the X role are mainly non-metals of the p-block (e.g., P or Si), but there are numerous examples for over 70 different elements. The addenda metals M composing the POM framework are usually molybdenum, tungsten, or vanadium, but they can be partially replaced with other transition metals, such as Mn, Fe, Co, etc. This broad range of combinations between addenda metals, substituting metals, and heteroatoms allows for the chemical (acid-base and redox behavior) and physical (solubility in different liquid media, thermal stability) properties of POM clusters to be tuned by choosing the most suitable constituent elements. The literature contains a thorough body of reports on the preparation and catalytic applications of POM species [12], among which the large subclass of Keggin-type $[XM_{12}O_{40}]^{n-}$ heteropolyanions constitute the most relevant representatives. The protonated forms of such species, known as heteropolyacids (HPAs), are Brønsted acids stronger than mineral acids and can undergo reversible uptake of several electrons in fast, multistep redox processes coupled with protonation. These processes proceed without any substantial structural modification, and hence these POM clusters have been long regarded as electron/proton reservoirs.

The catalytic properties of POMs have been mainly inspected in homogeneous conditions because they can be solubilized in manifold solvents from water to alkanes depending on the type of counterion. For example, alkaline POM salts are soluble in water and other aqueous solvents, but insoluble in organic medium. In turn, (organo)ammonium salts are less water-soluble, but their solubility in polar protic organic solvents like alcohols improves. Increasing the organic character of the counterion leads to water-insoluble salts that can be dissolved in other common organic solvents, such as acetonitrile or acetone (e.g., tetra-*n*-butylammonium salts). The use of cationic surfactants (e.g., DODA, di-*n*-octadecylammonium) allows for POMs to be transferred into nonpolar organic media, such as *n*-hexane.

Thanks to their solubilities being highly tunable depending on the nature of its counterions, POMs have traditionally been transformed into heteropolysalts insoluble in the reaction medium for their use in heterogeneous conditions [5, 13]. Insoluble heteropolysalts can be readily achieved through cation metathesis or proton-cation exchange, but unfortunately, these approaches lead to non-porous solids in almost all cases. The exception are some alkaline POM salts obtained from the replacement of protons in HPAs with bulky monovalent cations (e.g., Cs^+ and NH_4^+), which results in solids with significant specific surface areas among heteropolysalts (> 100 m^2g^{-1}) [14]. Porosity is a key aspect in the design of heterogeneous catalysts because it improves the accessibility of reagent molecules toward catalytically active sites and prevents deactivation by facilitating reaction products to leave the solid material. Attempts to improve the specific surface areas of POM-based solid catalysts have been made by supporting such type of catalytically active clusters on highly porous solids [15, 16]. Supported POMs are usually prepared by aqueous impregnation [17, 18] and several different types of porous solids have been used for this purpose, including mesoporous MCM-41 silica [19], zirconia [16], alumina [20], or activated carbonaceous (AC) materials [21]. Using AC as support for metallic catalysts is long known to provide with high surface areas and pH stability [22], and thus, several studies on AC-supported POM materials can be found in the literature [23–25]. Many of them focus on the catalytic application of such solid materials in liquid-phase reactions of interest, such as esterification, hydrolysis, etc. Others intend to analyze variations in the acid strength of bulk POMs when supported in AC, but as indicated by Swhwegler *et al.* [25], it is not even clear whether the mechanism of supporting POMs in AC is physical or chemical in nature.

The main drawback of supported catalysts is the loss of weakly bound catalytically active species through transfer into the reaction medium with consequent solubilization, denoted as leaching. In the case of POMs, the surface of the support is often derivatized with functional groups at which POMs can become tightly anchored to avoid leaching of clusters as the reaction proceeds [26]. This approach can be exemplified by the immobilization of 12-phosphotungstic acid in MCM-41 silica displaying protonated amino groups at the surface [27]. Other methods for the firm integration of POM clusters into mesoporous matrices include the use of nanocasting techniques to prepare fully inorganic frameworks built up from transition-metal oxides and POMs building blocks [28], or the direct copolymerization of metal–organic frameworks and POMs via sol–gel techniques [29].

The two parameters that are most commonly taken into account to evaluate the performance of a given catalyst are the activity, concerning to the reaction rate, and the selectivity, when it favors the synthesis toward the product of interest over other possible products. In the case of heterogeneous catalysts, however, other key factors beyond the activity/selectivity dualism must be considered for a given catalyst to succeed from the industrial viewpoint, such as the thermal and chemical stability, and the reactant accessibility to the active centers. Considering these and all of the other aspects commented above, this chapter will present an overview of our main recent findings in the application of POMs as acidic or oxidation catalysts in heterogeneous phase. The catalysts discussed herein will be either insoluble POM salts or supported catalysts based on POM clusters heterogenized on porous solid supports, including activated carbon materials and metallic oxides, such as zirconia. The main aspects to be commented will show the reach of our experimental results about the stability and accessibility of our catalyst, as well as provide a basis for understanding the mechanisms involved in the catalysis carried out by the selected POM species (i.e., values of conversion and selectivity obtained for some specific model reactions). Another focus of interest that will be analyzed in detail is the influence of both POM clusters and active supports on the catalytic activity of our solid materials. The catalytic role of POMs will be illustrated with their application to selected environmentally friendly reactions, such as the esterification of palmitic acid to generate biodiesel and the organic-phase oxidation of aniline with hydrogen peroxide to obtain nitrosobenzene selectively.

5.2 Stability of POMs

The thermal stability of POM clusters is of supreme relevance in regard to their performance as heterogeneous catalysts because they can irreversibly lose any catalytic activity if thermal decomposition takes place during the reaction process [12]. The thermal stability of POM-based solid materials must be considered from two different points of view: (1) the structural viewpoint, from which the catalytic material must remain solid without any undesired phase transition in the temperature range of interest;

and (2) the compositional viewpoint, from which both POM clusters and supports must retain their respective chemical nature and identity during any thermal treatment. A change in the chemical identity is usually, but not always, accompanied by decomposition involving the release of light gas products. Thus, thermogravimetric analysis (TGA) becomes a very helpful tool to determine the thermal stability range of a given solid catalytic material.

From the compositional viewpoint, HPAs usually display lower thermal stability than ionic salts based on the same POM cluster. For example, the TGA curve of $H_4PMo_{11}VO_{40}$ (VPMo) shown in Fig. 5.1 indicates that the thermal decomposition of this HPA species takes place through four different stages: (1) an initial mass loss that extends up to 200 °C and originates from partial release of crystallization water molecules (24 out of the 32 per POM cluster present in the starting HPA material [30]); (2) a second mass loss in the temperature range from 200 to 500 °C that originates from release of the remaining water molecules strongly retained at the acidic centers of the partially dehydrated $H_4PMo_{11}VO_{40}\cdot 8H_2O$ phase, as well as from partial decomposition with associated reduction of V^{V}/Mo^{VI} centers to V^{IV}/Mo^{V}; (3) a subsequent mass gain that takes place between 500 and 600 °C and can be attributed to reoxidation of the previously reduced metal atoms [31]; and (4) a final mass loss above 600 °C, which is related to the full breakdown of the Keggin-type framework with consequent release of water molecules originating from the protons of the HPA species [18, 30].

HPAs are usually immobilized on porous solid supports for their use as heterogeneous catalysts and this heterogenization process can affect the thermal stability of the HPA species, as illustrated by the TGA curve of the 50VPMo/Z100 solid catalyst (Fig. 5.1) with a 50 wt % load of the VPMo species supported on a zirconia $Zr(OH)_4$ matrix (Z) treated at 100 °C. This curve displays also two main mass loss stages that somehow resemble those found separately in the Z support and the VPMo starting material, but the overall mass loss in the catalyst is much lower than those observed separately for the individual Z and VPMo components. This fact indicates a strong interaction between components, which must be established during the impregnation procedure and must involve the loss of hydroxyl groups of the $Zr(OH)_4$ hydrogel upon condensation with the protons of the HPA species, as well as a decrease in the hydration sphere of the VPMo clusters to maximize the contact with the surface of the support.

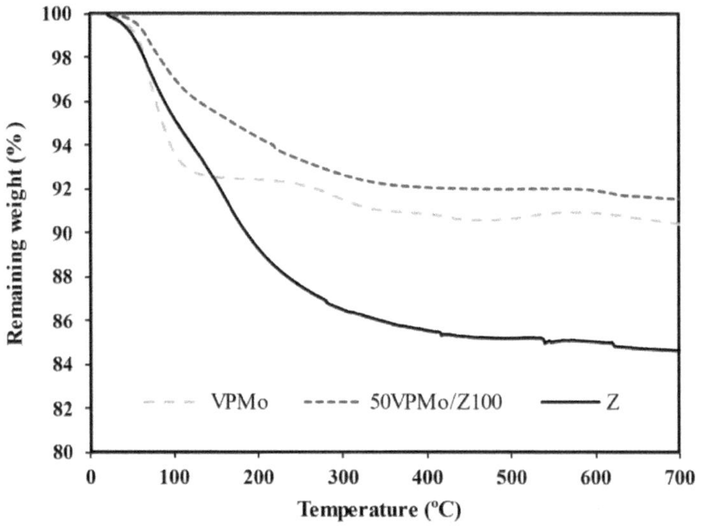

Figure 5.1 Comparison of the TGA curves of the zirconia support (Z), the starting HPA material (VPMo), and the 50VPMo/Z100 catalyst (50 wt %).

The increase of thermal stability of HPAs upon heterogenization can also be related to confinement effects when HPA clusters result entrapped within the porosity of the supporting material. Two of the HPAs most commonly used in catalysis, namely, $H_3PMo_{12}O_{40}$ (HPMo) and $H_3PW_{12}O_{40}$ (HPW), are highly hydrated species known to lose significant amounts of water of crystallization at temperatures even lower than VPMo, with those molecules strongly retained at the acidic centers being released at 150 °C according to TGA analyzes [21]. Figure 5.2 shows, however, that neither of these two HPAs undergo any mass loss up to temperatures above 200 °C when immobilized within the microporous domain of an activated carbon fiber support (A20). This fact reflects a gain in the thermal stability of HPAs upon entrapment within the porous structure of the support.

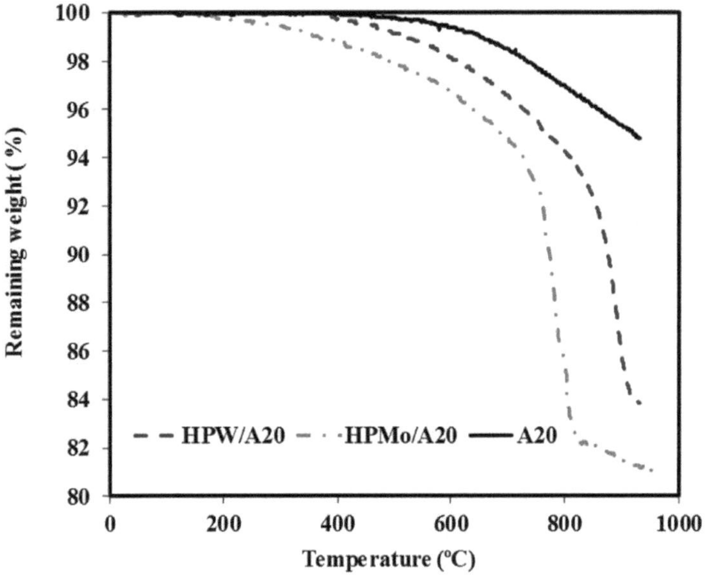

Figure 5.2 TGA experiments corresponding to the activated carbon fiber support (A20) and the fresh HPW/A20 and HPMo/A20 supported catalysts.

As pointed out at the beginning of this section, TGA alone is not sufficient for establishing the thermal stability of a given catalytic material. Powder X-ray diffraction (PXRD) becomes also essential to determine whether a given solid catalyst undergoes any phase transition when thermally treated and to monitor such structural modifications. This aspect is nicely illustrated in Fig. 5.3, which shows the results of PXRD experiments performed on pristine Z supports and 50VPMo/Z materials calcined in air for 2 h at different temperatures between 100 and 400 °C to inspect the potential effect of thermal treatments on the structure of the Z-supported VPMo solid catalysts.

Our results show that the Z support is initially an amorphous material and gains significant crystallinity upon thermal treatment at 400 °C and. At this temperature, it is partially transformed into a monoclinic ZrO_2 crystal phase [JCPDS 7-3430M] as evidenced from the diffraction maxima observed at 2θ ca. 24, 28, 32, 42, 46, 54, and 56°. The contribution of a fraction of tetragonal ZrO_2 can also be identified from the diffraction maxima located at 2θ ca. 30, 35, and 50°. In the case of the 50VPMo/Z catalysts, calcination up to 300 °C does not appear to induce any remarkable structural modification in the materials, but the diffraction pattern of 50VPMo/Z400 calcined at 400 °C is indeed substantially modified. Additional new diffraction maxima are observed, and they can only be associated in part with the crystallization of the Z support, which proceeds fully toward the tetragonal phase in this case, as the presence of monoclinic ZrO_2 cannot be detected. The appearance of a highly intense diffraction maximum at 2θ = 23° does not relate to any structural modification of the support, but corresponds to the formation of the β-type crystal phase of MoO_3 [32]. This fact reveals that a significant fraction of the VPMo clusters, if not all, decompose at 400 °C, which is consistent with literature reports addressing that the VPMo component alone decomposes at temperatures above 300 °C [33].

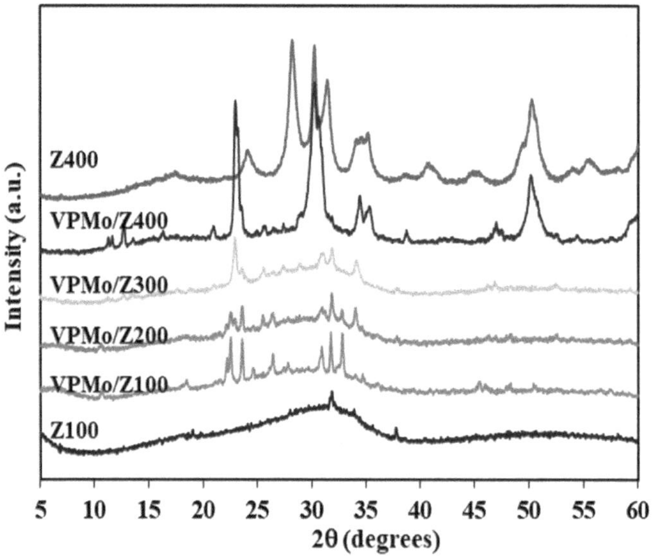

Figure 5.3 PXRD patterns of Z and 50VPMo/Z samples calcined at temperatures T = 100, 200, 300, and 400 °C.

The thermal stability of this type of catalytic materials can also be monitored by using diffuse reflectance infrared Fourier transform (DRIFT) spectroscopy. Following with the above supported catalyst as example, we show in Fig. 5.4 a comparison of the DRIFT spectra of the VPMo starting material, the pristine Z support, and two 50VPMo/Z catalysts calcined at 100 and 300 °C [34]. The spectrum of the starting VPMo precursor is dominated by the fingerprint of the primary structure of a Keggin-type phosphomolybdate cluster [30]: $\upsilon_{as}(P–O_a)$ at 1070 cm^{-1}, $\upsilon_{as}(Mo=O_d)$ at 980 cm^{-1}, $\upsilon_{as}(Mo–O_b–Mo)$ at 912 cm^{-1}, and $\upsilon_{as}(Mo–O_c–Mo)$ at 840 cm^{-1} (O_a: central oxygen atom; O_d: terminal oxido ligand; O_b: bridging oxygen atom between corner-sharing MoO_6 octahedra; and O_c: bridging oxygen atom between edge-sharing MoO_6 octahedra). The bands of strong intensity located at 3500 and 1600 cm^{-1} are due to water molecules in the secondary structure of the hydrated VPMo and assigned to $\upsilon(O–H)$ and $\delta(H–O–H)$ vibration modes [35, 36]. The representative signals originating from vanadium (V) mono-substitution in the Keggin-type metal framework are partially masked as they appear at 1076 and 990 cm^{-1} ($\upsilon_{as}(V–O_a)$ and $\upsilon_{as}(V=O_d)$, respectively).

The Z-supported VPMo samples display in turn significant changes in the fingerprint of the primary structure of Keggin-type clusters, which suggests a good interaction between VPMo species and Z support. Two additional signals can be observed at 997 and 1037 cm^{-1}, and these can respectively be related to the presence of MoO_3 and V_2O_5 in our samples [31, 37]. This fact reveals that a fraction of VPMo clusters result partially degraded with extrusion of vanadium (V) during the impregnation procedure followed to prepare our 50VPMo/Z catalysts [38, 39]. This partial VPMo degradation can be explained on the basis of the synthetic method of the Z support, which involves its precipitation as a $Zr(OH)_4$ hydrogel at basic conditions (pH = 10), because Keggin-type phosphomolybdate HPAs are known to be water-stable only in highly acidic conditions (pH < 2) and to gradually degrade into $[PMo_{11}O_{39}]^{7-}$ and $[HP_2Mo_5O_{23}]^{5-}$ species through alkaline hydrolysis of M–O–M bonds as the pH increases to 3 and then to 5, in such a way that full degradation into simple oxoanions is achieved at pH > 6 [40]. The spectrum of 50VPMo/Z300 suggests an increase in the content of MoO_3 and V_2O_5 upon thermal treatment at 300 °C, and in this regard, it has been suggested that the structural disorder caused by the introduction of vanadium (V) in the primary structure of the phosphomolybdate Keggin-type anion lowers its decomposition temperature by facilitating the output of VO_x units toward the surface sites of the secondary structure with consequent partial decomposition of the cluster framework into lacunary fragments and oxide phases [30, 32].

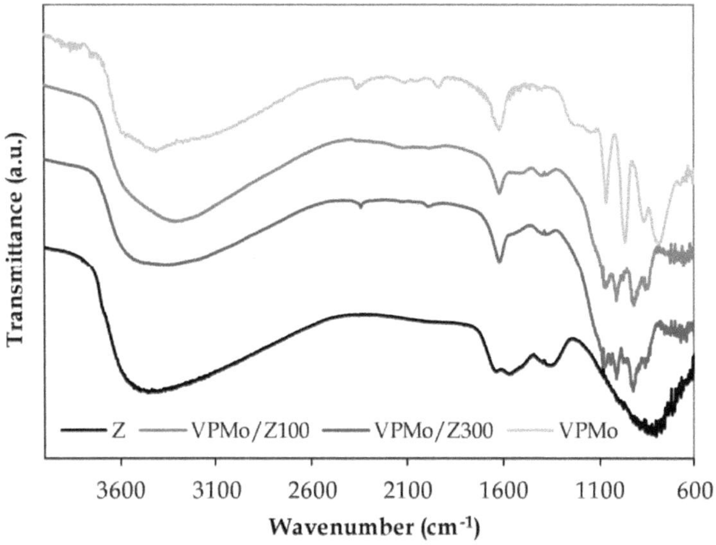

Figure 5.4 DRIFT spectra of the VPMo HPA, the Z support, and two 50VPMo/Z samples calcined at 100 and 300 °C.

Supported HPA materials have gathered great interest mainly as solid acid catalysts. The very strong acidity shown by HPAs is due to the presence of mobile protons in the secondary structure, which are bound to or shared between water molecules to form oxonium cations. Thus, the calcination treatments that the preparative procedures of supported catalysts usually involve can affect the acidity of HPAs due to the release of crystallization water molecules. DRIFT spectroscopy is a powerful technique to inspect the thermal stability of HPAs from the point of view of their acidity because it allows to monitor the evolution of the water of crystallization associated with protons during the calcination treatments.

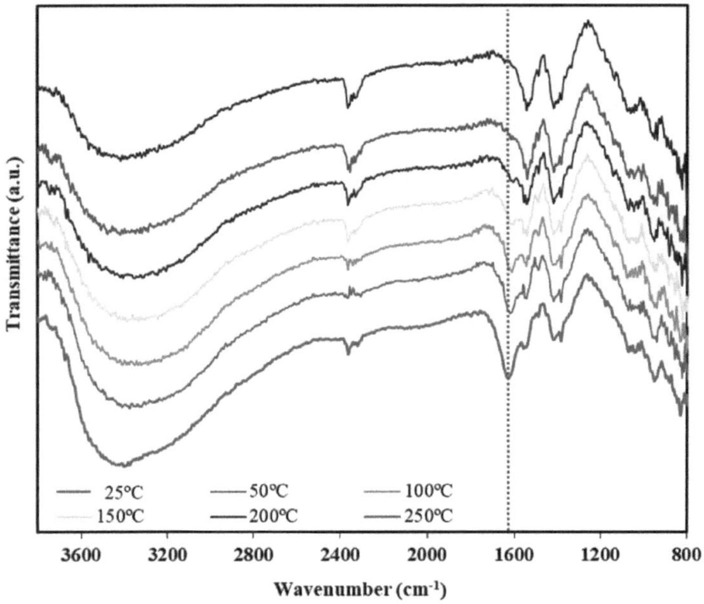

Figure 5.5 Evolution of the DRIFT spectra of the 30HPW/Z sample with the calcination temperature in He atmosphere.

For example, Fig. 5.5 shows the evolution with increasing calcination temperature of the DRIFT spectra of the zirconia-supported HPW catalyst 30HPW/Z (HPW load of 30 wt %). The main features of the spectra are a broad band of very strong intensity in the 2500–3700 cm^{-1} range that originates from stretching O–H vibrations of surface ≡Zr–OH groups, a peak of medium-to-strong intensity in the 1500–1700 cm^{-1} domain that associates with bending vibrations in physisorbed water molecules, and a second broad band of medium intensity at ca. 820 cm^{-1} that corresponds to stretching vibrations of Zr–O bonds [41]. One of most intense peaks at room temperature is observed at ca. 1620 cm^{-1} and corresponds to the in-plane bending H–O–H vibration [35, 36] of those water molecules forming the $H_5O_2^+$ ions in the secondary structure of the hydrated HPW species, rather than of those being merely physisorbed. Thermal treatments at temperatures equal to or above 150 °C remarkably lower the intensity of this peak at ca. 1620 cm^{-1}.

This fact is indicative of a significant reduction in the number of $H_5O_2^+$ ions in the secondary structure of the Keggin-type clusters, and hence in the acidity of the hydrated HPW species. Thus, DRIFT monitoring shows that the use of such type of zirconia-supported HPA materials as solid acid catalysts would be limited to operation temperatures below ca. 200 °C.

POM-based salts obtained from HPAs through partial or full proton-cation exchange display higher thermal stability than the parent HPA species. Their thermal stability can be modulated depending on the extent to which protons are exchanged with a given cation, and also on the specific nature of such cation. The TGA curves of the series of salts $Cs_xH_{3-x}PW_{12}O_{40}\cdot nH_2O$ (Fig. 5.6) clearly illustrates the dependence of the thermal stability on the extent of the proton-cation exchange. For this series, the temperature at which the second mass loss corresponding to the release of water molecules strongly retained at acidic centers takes place increases significantly as the Cs content becomes higher in the salts obtained upon exchanging protons of the parent HPW with Cs cations.

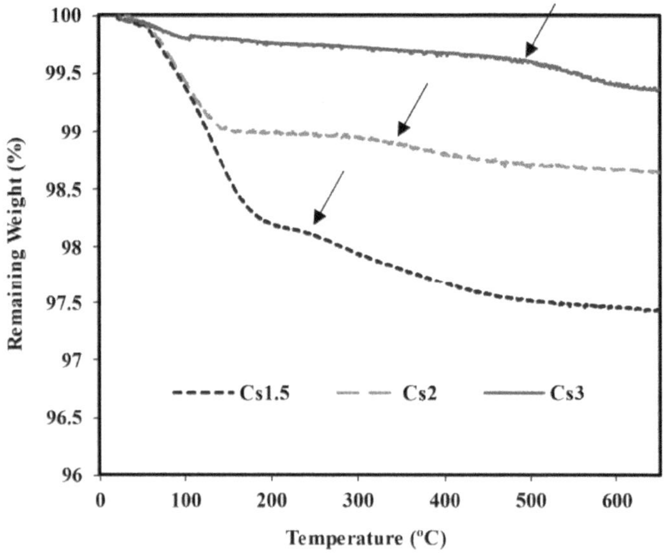

Figure 5.6 TGA curves of the $Cs_xH_{3-x}PW_{12}O_{40}\cdot nH_2O$ series with x = 1.5, 2, and 3.

In regard to the type of counterion selected for replacing protons in HPA species, we found in the course of our studies that large metallic cations, such as Cs^+ or Ba^{2+}, provides the structure formed by Keggin-type anions with higher thermal stability than other smaller metal cations (e.g., Fe^{3+}, Co^{2+}, and Cu^{2+}). For example, the salt resulting from replacing all protons in the $H_5(CuPMo_{11}O_{39})$ HPA with Cs^+ is thermally more stable than that obtained when using Cu^{2+} cations instead, as shown by the comparison of the TGA curves gathered in Fig. 5.7. This latter $Cu_2H(CuPMo_{11}O_{39})$ salt is noteworthy because of melting at 180 °C, which exemplifies how proton-cation exchange can result in a significant loss of thermal stability from the structural viewpoint, which is an essential aspect for solid materials to be used in heterogeneous catalytic reactors. The use of differential thermal analysis (DTA) coupled to TGA allowed us to detect such behavior from a signal of endothermic nature at 180 °C originating from the melting process.

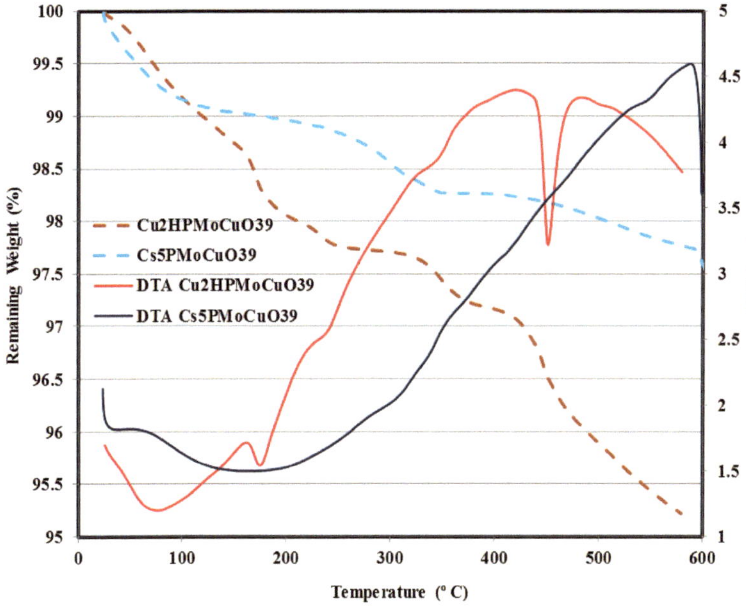

Figure 5.7 TGA and DTA curves of the salts $Cu_2H(CuPMo_{11}O_{39})$ and $Cs_5(CuPMo_{11}O_{39})$.

The thermal properties of the cation itself can greatly influence the stability of the resulting salt as well. Ammonium, organo-ammonium derivatives, and other organic cations are widely used to prepare POM-based salts relevant for catalysis, but they often result in compounds the thermal stability of which is limited by that of the specific cationic species. For example, the TGA curves in Fig. 5.8 show how the ammonium salts derived from the HPMo and HPW HPAs display lower thermal stability than their Cs-containing counterparts, which is due to NH_4^+ and other organic cations undergoing thermal decomposition in contrast to the metallic analogs [5, 12d].

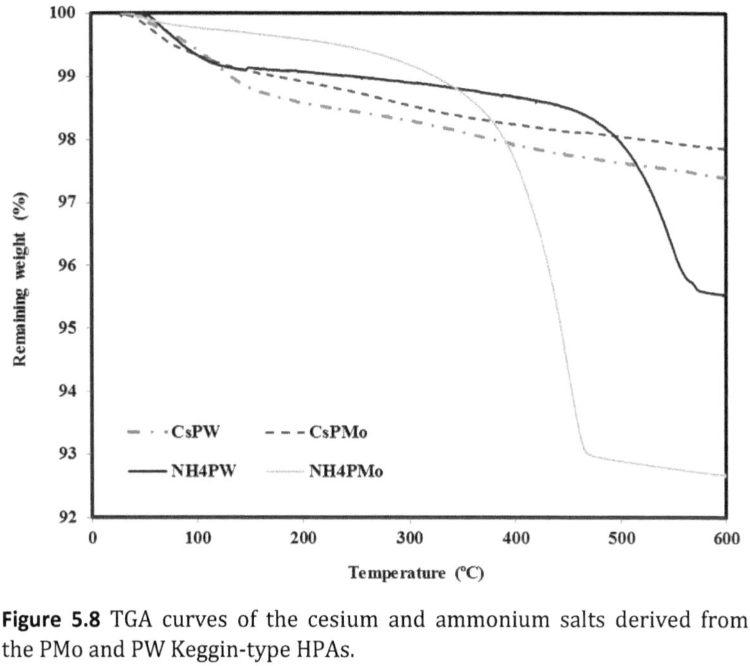

Figure 5.8 TGA curves of the cesium and ammonium salts derived from the PMo and PW Keggin-type HPAs.

In summary, the design of supported heterogeneous catalysts must be directed at obtaining solid materials able to effectively catalyze a given reaction of interest at temperatures below thermal decomposition. Our results above constitute a strong indication of the fact that the use of POM-based salts, and cesium-containing salts in particular, would most likely be of great help to avoid any thermal stability problem related with structural or compositional aspects when operating at temperatures below 350 °C. In the case of supported HPA catalysts, however, operation temperatures should be lower than 200 °C to retain the chemical identity of the HPA species and its acid properties intact, according to our experimental findings.

5.3 Porosity-Accessibility in POMs

5.3.1 HPAs Supported on Porous Solids: Impregnation and Sol–Gel Methods

As mentioned above, HPAs behave as very strong Brønsted acids close to the super-acid region, and they are significantly less harmful than common mineral acids in terms of their environmental effects. HPAs show very high solubility in a range of polar solvents, and therefore, they need to be immobilized on porous solid supports for being applied under heterogeneous catalytic conditions. In recent years, we have developed a number of studies on the process through which HPAs are supported on several different types of porous solids, including activated carbon materials or zirconia, with the aim at analyzing where the clusters are adsorbed in the support and whether the accessibility of the substrate toward the catalytically active centers inside the porosity remains suitable enough for catalytic prospects upon heterogenization.

Table 5.1 Characterization of the porous texture of AC-supported HPMo catalysts

	BET (m^2/g)	V_{N2} ($cm^3\ g^{-1}$) micropore (<2 nm)	V_{CO2} ($cm^3\ g^{-1}$) micropore (<0.8 nm)	V_{meso} ($cm^3\ g^{-1}$)
A5	650	0.28	0.29	–
A10	1220	0.53	0.40	–
A15	1290	0.57	0.33	–
A20	1990	0.93	0.52	–
West	1750	0.68	0.38	0.54

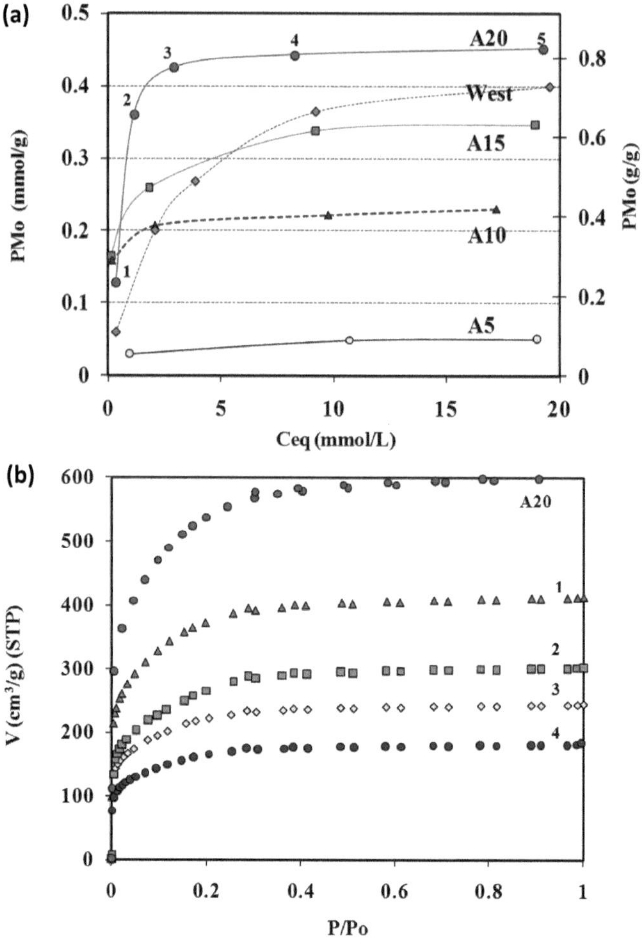

Figure 5.9 (a) Variation of the amount of adsorbed HPMo with the concentration of the impregnate solution; (b) N_2 adsorption isotherms at −196 °C of the AC support A20 and the A20-supported HPMo samples impregnated with HPMo mass percentages increasing from sample 1 to 4.

Our investigations on the heterogenization of HPMo on different types of activated carbons (ACs) through the impregnation method demonstrates that such HPA species can be supported in large amounts up to loads of 45 wt % depending on the specific AC material [16a]. The adsorption of HPMo on ACs follows a trend similar to that of the physical adsorption of N_2: it takes place at relatively low concentrations and reaches saturation quickly as the concentration of the impregnate solution increases, in such a way that the higher the AC porosity, the larger the adsorbed HPMo amount (Fig. 5.9a). Despite the large HPMo amounts with which ACs can be loaded, the resulting materials still retain a remarkable degree of open and accessible porosity as shown in Table 5.1. This aspect constitutes an essential requirement for further using AC-supported HPA materials as heterogeneous catalysts.

Figure 5.9b contains the N_2 isotherms of HPMo-impregnated samples of the microporous activated carbon A20. Their comparison shows that the N_2 adsorption capacity of A20 decreases upon impregnation with HPMo and that the latter species adsorbs within the microporous domain of the AC support. The micropore volume of A20 decreases upon impregnation with HPMo, but this reduction is located mainly in the supermicroporous structure, whereas the VCO_2 narrow microporosity (pore size < 0.8 nm) and mesoporosity (2 nm < pore size < 50 nm) only show slight variations. These results demonstrate that HPMo is adsorbed preferentially within a very limited range of porosity. This range corresponds to the supermicropores only (1 < pore size < 2 nm), which is consistent with the average size of Keggin-type POMs. Such type of clusters displays diameters of 0.8–1.0 nm regardless of the addenda metal or heteroatom, and hence their incorporation into the narrow ultramicropores (pore size < 0.8 nm) is hindered due to size exclusion [42]. PXRD and transmission electron microscopy (TEM) experiments performed on our AC-supported HPMo samples evidenced also that such HPA species is in all cases immobilized with a high degree of dispersion regardless of the HPMo amount with which the support is loaded, which can be as high as 45 wt % [16a].

When zirconia is used as support instead of activated carbons, HPA species are also immobilized within the supermicroporous domain by following the impregnation method. In this case, however, the HPA amount with which the Z support can be loaded is very much limited because high amounts of 50 wt % result in the porosity of the resulting material being fully blocked [34]. As shown by the N_2 adsorption–desorption isotherms in Fig. 5.10, impregnation of the Z support with VPMo leads to a progressive decrease in the porosity with increasing the HPA load. The adsorption capacity remains nearly unmodified when the material is loaded with just 5 wt % of VPMo, but drops substantially with loads as low as 10 wt % (ca. 30% compared to the Z support) and disappears completely for the material 50VPMo/Z with the highest HPA load achievable (50 wt %). The open shape of the isotherm elbow in the low-pressure region indicates a wide distribution of micropore sizes for the pristine Z support. Monitoring the shape of such elbow reveals that impregnation tends to homogenize the micropore sizes in the support as the HPA load becomes higher because the isotherm elbow narrows gradually with increasing the amount of supported VPMo.

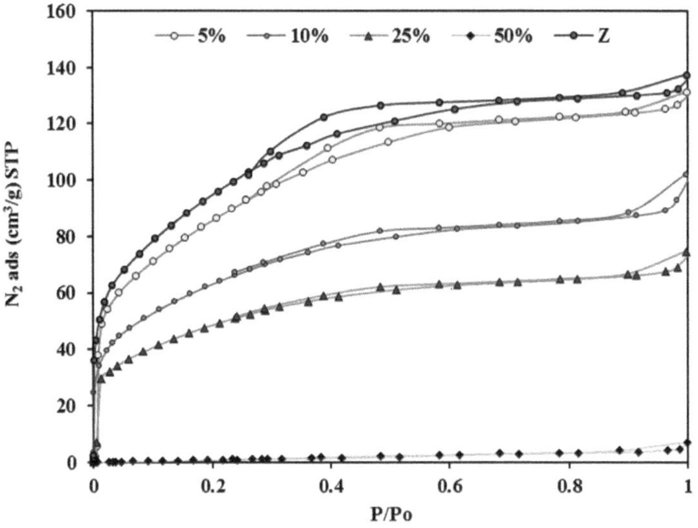

Figure 5.10 N_2 adsorption–desorption isotherms at −196 °C of the Z support and *x*VPMo/Z100 catalysts obtained by the impregnation method with VPMo loads in the range *x* = 0–50 wt %.

Due to the limited amount of HPA that can be immobilized on zirconia through impregnation without significantly affecting the porous texture of the support, we have developed an alternative one-pot route for preparing Z-supported HPA catalysts based on the hydrothermal sol-gel synthesis of the zirconia support in the presence of the HPA species to be immobilized in the final material [34]. Table 5.2 lists the textural properties of a series Z-supported HPW catalysts prepared by a sol–gel method and subsequent hydrothermal treatment. This series exemplifies how changing the preparative procedure can lead to materials with analogous composition but very different porous textures, and it consists in four *x*HPW/Z solid catalysts with HPW loads of x = 10, 20, 30, and 50 wt %, together with the corresponding pristine Z support obtained under similar synthetic conditions [16b].

Table 5.2 Textural properties of a series of *x*HPW/Z supported catalysts with HPW loads in the x = 10–50 wt % range obtained from the immobilization of the $H_3PW_{12}O_{40}$ HPA over zirconia through a combined sol–gel/hydrothermal approach

Catalyst	S_{BET}(m^2 g^{-1})	V_{N2}(cm^3 g^{-1})	V_{CO2}(cm^3 g^{-1})	V_{meso} (cm^3 g^{-1})
Pristine ZrO_2 (Z)	164	0.08	0.05	0.17
10HPW/Z (x = 10%)	129	0.06	0.04	0.18
20HPW/Z (x = 20%)	270	0.13	0.09	0.19
30HPW/Z (x = 30%)	365	0.18	0.12	0.11
50HPW/Z (x = 50%)	321	0.16	0.11	0.09
$H_3PW_{12}O_{40}$ (HPW)	0	0	0	0

The data above show that the porous texture of the final solid materials undergoes significant development as higher amounts of HPW are embedded into the Z support. Such development of porosity along the *x*HPW/Z series reaches its maximum extent when the Z support is loaded with an HPW mass percentage as high as x = 30%. The 30HPW/Z sample displays the highest microporous volume and the largest specific surface area in the series with an SBET value that more than doubles that of the Z support alone. This fact is almost exclusively due to a substantial increase of ultramicropores in the material, which develops the narrow microporosity but retaining the mesoporous domain almost unaffected. All of the results above evidence that Keggin-type clusters can have a key role on modifying the textural properties of zirconia when this support is prepared via hydrolysis of zirconium(IV) alkoxides in the presence of HPA species.

Hydrolysis of Zr^{IV} molecular precursors, such as alkoxides results in $Zr(OH)_4$ hydrogels, into which the Keggin-type units (HPW in the series above) are embedded as gelation proceeds. Such types of HPA-loaded inorganic hydrogels are known to age into the corresponding inorganic oxides upon hydrothermal conditions are applied [43]. In our case, HPW clusters embedded in the $Zr(OH)_4$ initial hydrogel are immobilized in the hydrothermally-generated ZrO_2 final matrix through both electrostatic interactions and extensive hydrogen-bonding. Some ≡Zr–OH groups in the ZrO_2 matrix protonate due to the acidic environment provided by the HPA species, and lead to $[\equiv Zr\text{–}OH_2]^+[H_2PW_{12}O_{40}]^-$ pairing through an acid-base-like reaction [6, 14]. These electrostatic pairs are reinforced by extensive hydrogen bonding interactions established between the surface O atoms of the HPW clusters and other neighboring ≡Zr–OH non-protonated groups. This mechanism for immobilizing HPW units on ZrO_2 is consistent with the large development of the microporous structure observed for our *x*HPW/Z series as the HPW load increases. This development takes place mainly in the domain of the ultramicropores with sizes equal to or lower than 0.8 nm, which are comparable to the diameter of Keggin-type clusters. As higher amounts of HPW are loaded, an increasing number of such units are grafted at the surface of ZrO_2 nanoparticles through the interactions discussed above, and hence they start acting as spacers between adjacent nanoparticles hampering their aggregation effectively. Adjacent nanoparticles would then be separated by a distance analogous to the diameter of the cluster, resulting in the formation of interparticle void spaces corresponding to the ultramicroporous domain.

It is worth noting, however, that the thermal stability of the porous structure is very much limited for Z-supported HPA catalysts obtained from the combined sol–gel/hydrothermal approach. This type of material shows a substantial decrease of the accessible porosity when the hydrothermal treatment is carried out at temperatures above 100 °C, which is largely due to the thermal stability of the zirconia support itself. The

N_2 adsorption-desorption isotherms displayed in Fig. 5.11 finely illustrate this behavior. They correspond to a series of 50VPMo/ZT catalysts prepared by the sol-gel synthetic procedure and calcined at temperatures T = 100, 200, 300, and 400 °C. Overall, this type of sample displays larger porosity than analogs obtained from the impregnation method, as shown by the comparison of the textural properties of the sol-gel 50VPMo/Z100 sample with those of the impregnation 50VPMo/Z100 material, which did not show any type of porosity according to the isotherm in Fig. 5.10. On the other hand, calcination of sol-gel samples at temperatures above 100 °C does not significantly modify the shape of the isotherm, but produces a remarkable reduction of the adsorption capacity mainly at the relative low-pressure region (P/P_0 < 0.2). This fact indicates that the samples undergo a decrease in the microporous volume while retaining the mesoporous domain almost intact.

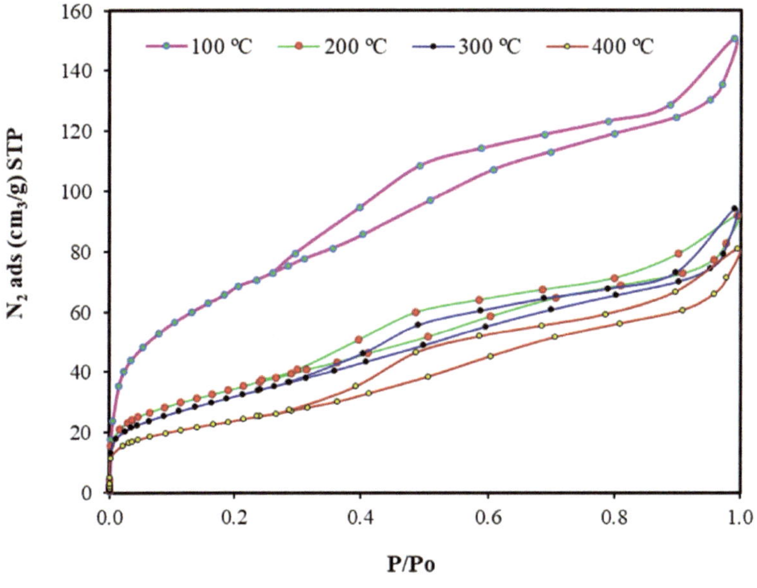

Figure 5.11 N_2 adsorption-desorption isotherms at −196 °C of 50VPMo/ZT catalysts (VPMo load of 50 wt %) obtained by the sol-gel method and hydrothermally treated at temperatures in the range T = 100–400 °C.

5.3.2 Development of Tailored Porosity in Nanostructured Heteropolysalts

In contrast to non-porous HPA solids, POM-based salt derivatives obtained through proton-cation exchange can display remarkable porous texture when bulky monovalent cations, Cs^+ and NH_4^+ in particular, are used. It is worth mentioning in this context that highly-porous POM-based salts can only be accessed by using clusters with the Keggin-type framework and cations with diameters in the 0.27–0.34 nm range.

Figure 5.12 and Table 5.3 illustrate the observation above for a series of (organo)ammonium salts derived from HPMo, the developed porosity of which substantially decreases when going from NH_4^+ to bulkier tetra-alkylammonium species [5]. Moreover, the use of the smaller Cs^+ leads to higher porosity than any of the latter tetra-alkyl cations, and the resulting porous structure displays remarkable thermal stability according to the slight differences found when comparing the N_2 adsorption–desorption isotherm of CsPMo with that obtained upon calcination at 375 °C (CsPMo-375).

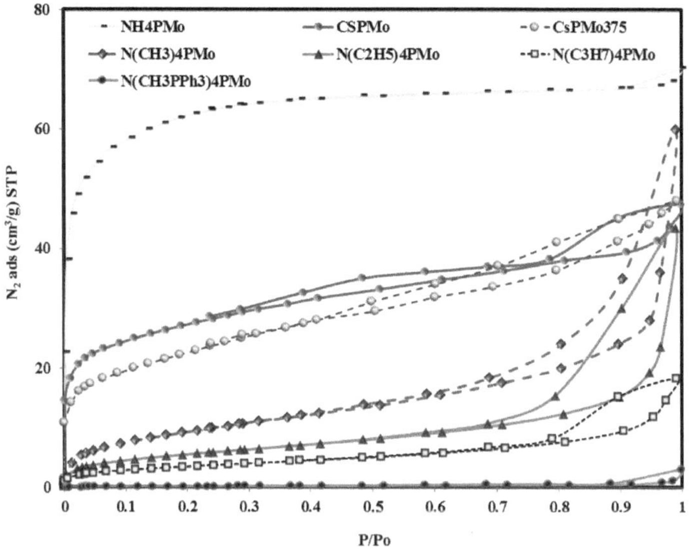

Figure 5.12 N_2 adsorption–desorption isotherms at −196 °C of representative POM-based salts derived from HPMo.

Table 5.3 Textural properties of representative POM-based salts derived from HPMo

Salt	S_{BET} (m^2 g^{-1})	V_{N2}	V_{CO2}	V_{Meso}	V_{Total}
		(cm^3 g^{-1})			
NH_4PMo	225	0.113	0.095	0.005	0.120
CsPMo	102	0.053	0.042	0.020	0.073
CsPMo-375	80	0.042	0.041	0.030	0.073
$(N(CH_3)_4)PMo$	37	0.017	0.015	0.075	0.093
$(N(C_2H_5)_4)PMo$	24	0.010	0.011	0.057	0.067
$(N(C_3H_7)_4)PMo$	14	0.006	0.011	0.022	0.028
$(Ph_3PC_3H_7)PMo$	2	0	0	0.004	0.004

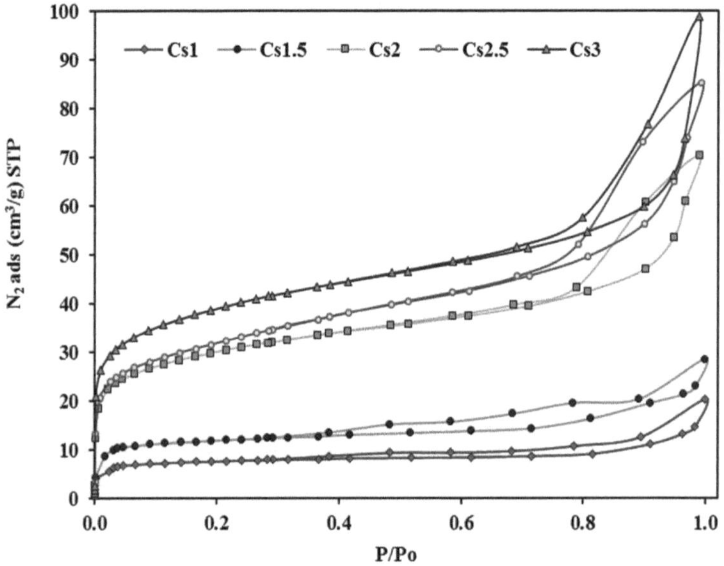

Figure 5.13 N_2 adsorption–desorption isotherms at −196 °C of a series of salts with formula $Cs_xH_{3-x}PW_{12}O_{40}$ (x = 1–3).

When POM-based salts are aimed at being used as solid acid catalysts, an agreement must be reached between the development of porosity upon cation exchange and the presence of protons required for the acid catalysis to proceed. Not all protons in the HPA solid must be replaced and this fact affects greatly the porosity of the resulting salt. Figure 5.13 exemplifies through a series of $Cs_xH_{3-x}PW_{12}O_{40}$ salts that the development of porosity is heavily dependent on the extent to which protons in the parent HPA solid are replaced with cations in the final salt. For the most common HPAs, namely, HPMo and HPW with three protons each, a replacement of 2/3 of the original protons is required to obtain salts with suitable porous structures. Replacing 1/2 of the protons leads to salts with better acidic properties, but with scarce development of the porosity, which hampers their application as solid acid catalysts. Therefore, optimal relations between acidity and accessibility to acidic sites must be found for each POM-based salt.

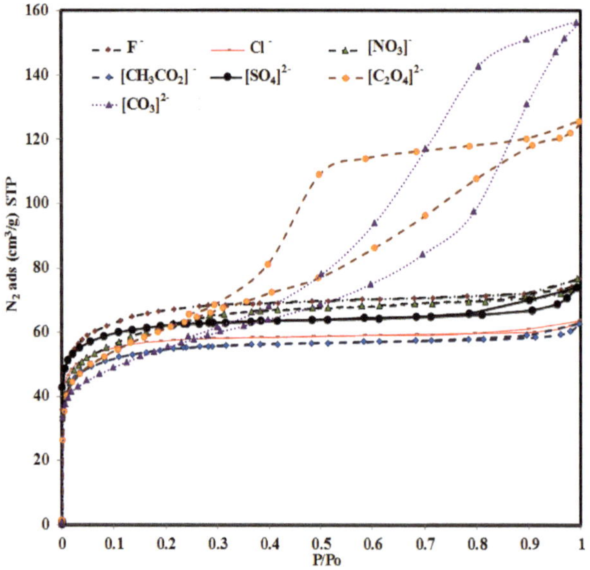

Figure 5.14 N_2 adsorption–desorption isotherms at −196 °C of $(NH_4)_3[PMo_{12}O_{40}]$ salts synthesized using different ammonium precursors.

In regard to synthetic aspects related to the preparation of ammonium heteropolysalts through proton-cation exchange, the nature of the anion in the ammonium precursor has proven to play a key role as directing agent for the development of tailored porosity in $(NH_4)_3[PMo_{12}O_{40}]$ despite being absent from the final solid samples. Figure 5.14 evidences that the development of porosity in $(NH_4)_3[PMo_{12}O_{40}]$ can be controlled using ammonium precursors bearing different inorganic or organic anions, and that the obtained solid samples are essentially microporous with different adsorption capacities as a function of the anion or even with some remarkable contribution of mesoporosity for oxalate and carbonate.

The observations above allow for reaching some conclusions about how the anion in the ammonium precursor affects the final textural properties of $(NH_4)_3[PMo_{12}O_{40}]$ solid samples. Our studies have demonstrated that the development of mesoporosity in $(NH_4)_3[PMo_{12}O_{40}]$ is mainly determined by the pH of the solution from which the samples are precipitated upon proton-cation exchange. Solutions with pH values higher than 3 are essential for materials with significant contribution of mesopores to be isolated, and this parameter depends mainly on the acid-base properties of the anion in the ammonium precursor. Therefore, the more basic this anion, the higher the solution pH, and consequently, the greater the extent to which the mesoporous domain of the final $(NH_4)_3[PMo_{12}O_{40}]$ material develops. For the examples in Fig. 5.14, anions with negligible acid-base properties (e.g., chloride and nitrate) result in essentially microporous samples, whereas basic anions (e.g., carbonate and oxalate) leads to substantial development of mesoporosity. In addition, the local concentration of NH_4^+ during precipitation has also been found to significantly affect the pore texture of the resulting $(NH_4)_3[PMo_{12}O_{40}]$ samples, in such a way that the higher the local NH_4^+ concentration, the more amount of porosity in the final material.

During the past decade, a great interest in the design of porous materials with bimodal distribution of the porosity has emerged. Such materials combine large microporous domains, hosting the active centers at which the catalytic reaction takes place, with a substantial mesoporous contribution that allows for: (1) avoiding any diffusional problem related with the access of reagents toward and the release of reaction products from the active centers; and (2) minimizing the catalyst deactivation as a result of the deposition of the reaction products and consequent porous blockage [16, 21]. As mentioned above, certain heteropolysalts can show an essentially microporous texture, but the controlled development of mesoporosity in such heteropolysalts is a topic yet to be explored to the best of our knowledge. The existence of mesoporosity as a result of voids in between nanoparticles has been described for a few nanostructured heteropolysalts [14], but this phenomenon was identified as merely incidental because no synthetic route could ensure formation of such voids, much less exert control over the size of the mesopores.

Based on the observations above about the facts that the choice of cationic source can determine the textural properties of heteropolysalt powders prepared from HPAs through proton-cation exchange [44], and that the basicity of the anion in ammonium precursors has a crucial role in developing mesoporosity in nanostructured $(NH_4)_3[PMo_{12}O_{40}]$ salts prepared from HPMo by using titration methods, we have been able to develop a very simple approach to implement tailored mesoporosity in any heteropolysalt with control over both the mesoporous volume and pore size. This method consists in treating an M_x[POM] heteropolysalt at a given pH value determined by the concentration of a M[B] salt, in which B is a basic anion, and it constitutes a new procedure for the preparation of mesoporous solids with no reported precedents [45].

We selected the essentially microporous $(NH_4)_3[PMo_{12}O_{40}]$, the porosity of which originates from accessible voids amongst nanocrystals that aggregate into dodecahedral particles, for our investigation and this model nanostructured heteropolysalt allowed for two main conclusions to be drawn: (1) the micro- and meso-porous structures present in the heteropolysalt nanoparticles are independent from each other; and (2) the development of mesoporosity must be related to the nanocrystals forming the dodecahedral particles undergoing alkaline degradation upon interaction with the basic agent during the preparation procedure (Fig. 5.15).

From our viewpoint, the only suitable explanation for the independent trends observed in the variation of the micro- and meso-pore volumes with the loss of mass that the heteropolysalt undergoes upon alkaline degradation is to consider that the latter takes place preferentially in the innermost part of the nanocrystals, as the TEM images suggest (Fig. 5.16). Thus, implicit mesoporosity develops gradually within the nanocrystal cores as the alkaline degradation proceeds, while the explicit microporosity remains nearly unaltered because the outermost shell of the nanocrystals, and the voids amongst them in the epitaxial aggregates as a result, are essentially preserved up to mass losses of around 40% during this process.

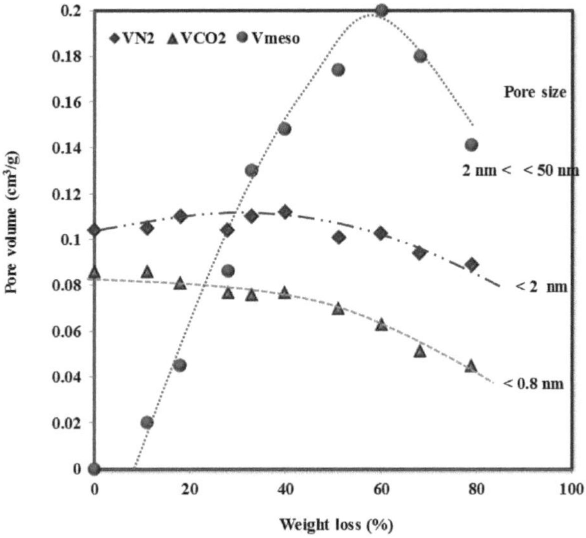

Figure 5.15 Evolution of the narrow microporous ($V(CO_2)$), total microporous ($V(N_2)$), and mesoporous (V(meso)) volumes with the mass loss that the $(NH_4)_3[PMo_{12}O_{40}]$ precursor undergoes upon alkaline degradation.

The results commented up to this point highlight the great possibilities of developing POM-based salts with any type of porosity distribution, but whether the pore volumes and specific surface areas (SBET values typically around 100 m^2/g) achieved in such type of materials were enough to be comparable to other commonly used porous materials remained an open question. This aspect was inspected in detail through the comparative analysis of the textural properties in a wide collection of representative porous materials of very different nature [46]. Besides $(NH_4)_3[PMo_{12}O_{40}]$ heteropolysalt samples, the adsorbents in this collection included: the commercial West powdered activated carbon, the activated carbon fiber A20, the zeolite NaA, the mesoporous silica MCM-41, and the two porous metal organic frameworks MOF-5 and HKUST-1. This study brought to light that the most suitable method to quantify and compare the adsorption capacity of porous materials with so different characteristics is to express such parameter on a volumetric basis rather than a gravimetric one (Table 5.4). This method is also applied when the results obtained from different heterogeneous catalytic reactions need to be compared as they are usually referred to the space velocity (or reactant flow rate/catalyst bed volume). Due to their remarkable skeletal density, POM-based salts are the materials with the highest packing density among all adsorbents analyzed, and hence a given bed volume would encompass larger amounts of heteropolysalt than of any other material. This fact results in such salts showing high porosity values on a volumetric basis that are comparable to those afforded by the rest of adsorbents on a gravimetric basis.

(a)

(b)

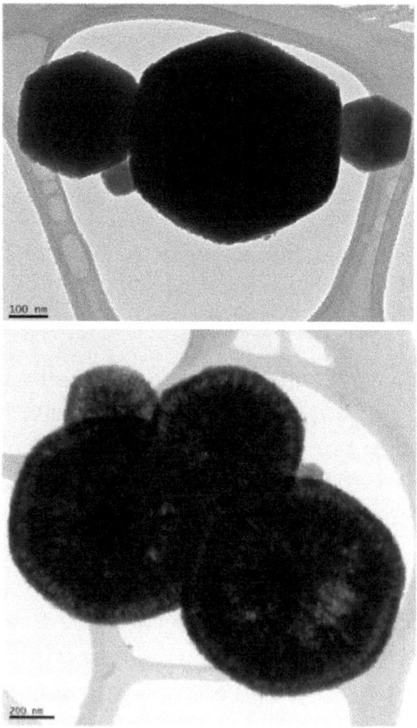

Figure 5.16 TEM images of the $(NH_4)_3[PMo_{12}O_{40}]$ precursor (left) compared with the sample in which 60% of the initial mass is lost upon treatment (right).

Table 5.4 Porous textural properties on a gravimetric ($cm^3\ g^{-1}$) and a volumetric ($cm^3\ cm^{-3}$) basis of selected adsorbents before and after compaction at 180 Mpa

Sample	$V_{CO2}{}^1$ $cm^3\ g^{-1}$	$V_{CO2}{}^1$ $cm^3\ cm^{-3}$	$V_{N2}{}^2$ $cm^3\ g^{-1}$	$V_{N2}{}^2$ $cm^3\ cm^{-3}$	V_{Meso} $cm^3\ g^{-1}$	V_{Meso} $cm^3\ cm^{-3}$
$(NH_4)_3[PMo_{12}O_{40}]$	0.088	0.11	0.088	0.11	0.082	0.10
$(NH_4)_3[PMo_{12}O_{40}]$ 180	0.087	0.19	0.083	0.18	0.091	0.20
Zeolite NaA	0.171	0.11	0.028	0.02	0.058	0.04
Zeolite NaA 180	0.122	0.14	0.022	0.02	0.045	0.05
MCM-41	0.27	0.06	0.48	0.12	0.39	0.09
MCM-41 180	0.27	0.17	0.48	0.31	0.31	0.20
West	0.33	0.11	0.71	0.25	0.49	0.17
West 180	0.34	0.17	0.70	0.35	0.47	0.24
A20	0.56	0.13	1.01	0.24	–	–
A20 180	0.54	0.32	0.99	0.59	–	–
MOF-5	0.25	0.08	1.06	0.34	–	–
MOF-5 180	0.07	0.12	0.14	0.22	–	–
HKUST-1	0.66	0.36	0.55	0.30	–	–
HKUST-1 180	0.34	0.39	0.24	0.27	–	–

Notes: [1]Narrow micropore volume (< 0.7 nm); [2]Total micropore volume (< 2 nm).

5.4 POMs as Catalysts

This last section will be devoted to showing that the different POM-based materials commented above (supported HPAs and porous heteropolysalts) can be applied as effective catalysts in a wide range of heterogeneous reactions. It is worth reminding in this sense that POMs can act as acid catalysts due to their very strong Brønsted acidity, as well as oxidation catalysts thanks to undergoing reversible uptake of several electrons in fast multistep redox processes without substantial structural changes, and to their capability of transferring oxygen atoms to the catalyst–substrate adduct.

5.4.1 HPAs as Acid Catalysts

This first sub-section summarizes our studies on the applicability of a series of HPA-supported porous solids as heterogeneous acid catalysts for a process of industrial and environmental relevance, such as the preparation of biodiesel. Palmitic acid as the free fatty acid and its esterification with methanol into de corresponding mono-alkyl ester as the model reaction are selected for such analysis. In our activated carbon- and zirconia-supported catalysts, the active HPA species are immobilized within the porous structure of the supporting material, which might bring up the uncertainty of whether such HPAs: (1) still display the high acidity required for acting as an effective catalyst; and (2) contain enough number of acid sites accessible to the reagents.

To solve the two questions above, the total amount of acid sites present in the supported catalysts must be first evaluated. Table 5.5 shows the results of such evaluation through temperature-programmed desorption (TPD) of NH_3 for our series of AC- (activated carbon fibers A20, A15, and A10) and Z-supported HPW materials [21]. As expected, the number of acid sites increases with increasing the HPW load and the values obtained are close to those expected on the basis of the amount of HPW. This trend develops up to a certain HPW mass percentage (e.g., 20–25 wt % in the case of the HPW/A20 series), above which starts decreasing gradually with higher HPA loads. According to the literature, different factors could explain this latter decrease in the surface acidity: (1) protons at high HPA loads become progressively inaccessible due to HPA clusters blocking the pores [16]; (2) protons are partially neutralized by the basic groups of the support [47]; and (3) protons become gradually inaccessible due to the formation of a multilayer HPA coverage on the support [48]. In the case of the HPW/Z series, it is worth remarking that the pristine Z support can also catalyze the esterification reaction to the extent of achieving conversions of ca. 30%, which can be attributed to the existence of Brønsted acidic sites at the zirconia surface due to its synthetic procedure involving the use of hydrochloric acid to set the pH conditions. The main conclusion of TPD analyses is that the acidity of the

HPA species is in all cases partially retained upon immobilization and that all catalysts should in principle show catalytic activity for the esterification reaction according to the recount of their acid sites.

Table 5.5 Quantification of acid sites in a series of AC- and Z-supported HPW catalysts

Sample	HPW load (wt %)	Acid sites[a] (μmol g^{-1})
HPW	100	1120
12HPW/A20	11.7	184
24HPW/A20	24.9	296
52HPW/A20	50.0	254
27HPW/A15	27.5	301
28HPW/A10	28.1	297
ZrO_2	0	578
10HPW/Z	10.3	720
20HPW/Z	20.5	952
50HPW/Z	49.7	1134

Note: [a]Total acidity calculated by desorption of NH_3 at temperatures higher than 120 °C.

Figure 5.17 shows the conversions achieved in the esterification of palmitic acid with methanol at 60 °C by using the homogeneous HPW commercial catalyst and the heterogeneous AC- and Z-supported HPW catalysts as a function of the amount of HPW present in the methanolic reaction medium. Commercial HPW is soluble in methanol and affords high conversions of ca. 70% with amounts as low as 10 μmols (30 mg) when acting as a homogeneous acid catalyst in the esterification of palmitic acid. Upon immobilization, the HPW species can still catalyze such reaction. For low HPA amounts, the supported HPW catalysts exhibit lower activity than the corresponding unsupported sample, which evidences that HPAs partially lose their activity as acid catalysts upon heterogenization. However, the activity among homogeneous and heterogeneous catalysts becomes more similar as the HPA amount increases, and it reaches comparable values for ca. 50 and 60 mg of HPW in the case of the HPW/A20 and HPW/ZrO_2 series, respectively. The activity remains mostly unaffected by further increases in the amount of the catalytically active HPW species added to the reaction medium.

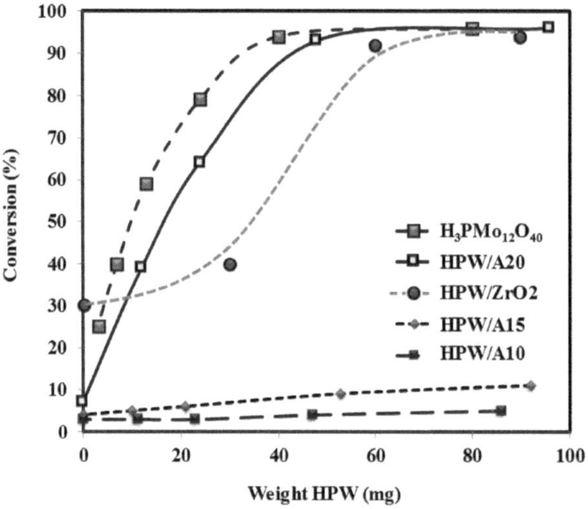

Figure 5.17 Conversions afforded by unsupported HPW (homogeneous) and AC- or Z-supported HPW (heterogeneous) acid catalysts in the esterification of palmitic acid vs the amount of the HPW species.

The similar activities shown by our heterogeneous catalysts regardless of the supporting material, together with the results obtained from the homogeneous phase, evidences the facile accessibility of palmitic acid toward the active sites located in the supermicroporosity of the A20 fiber and ZrO_2 material. Considering its large dimensions (a linear alkyl chain of 16 °C atoms and a COOH head accounting for 1.9 nm in length and a diameter of 0.6 nm), the ease with which palmitic acid reaches the active sites of the supported catalysts would be unexpected if the HPW clusters were located only within the microporous structure with pore sizes lower than 0.8 nm. In this sense, the experiments carried out with other HPW-impregnated activated carbon fibers (A10 and A15) at the same experimental conditions demonstrates that the activity of the latter catalysts is much lower than that of the HPW/A20 series even for materials with the same HPA loads and similar amounts of acid sites (Table 5.5). The microporous structure of the A10 and A15 samples is less developed than that of A20 and the narrowest ultramicroporous domain is more predominant according to Table 5.1, and therefore, the results above confirm the influence of the pore size on the kinetic control of the reaction by allowing or restricting the accessibility of the reagents [49].

A key feature of heterogeneous catalysts is their reusability in consecutive reaction cycles while retaining their catalytic properties mostly unmodified. The recyclability is of great relevance not only from an economic viewpoint, but also because of operational aspects related to the charge and discharge of catalytic reactors [1, 50]. The recyclability of our supported catalysts was evaluated through their application in five successive esterification runs without any intermediate regeneration stage, and the conversions achieved are shown in Fig. 5.18. In general, the catalytic activity decreases after each cycle, but this decrease depends on the specific nature of the support. For example, HPW/A20 affords the highest conversion in the first cycle, but loses catalytic activity much faster throughout the five consecutive cycles, which makes this catalyst less efficient than HPW/Z in terms of recyclability.

Table 5.6 Mass percentage of HPW leached over five consecutive esterification cycles based on the initial amount of HPW present in the fresh supported catalysts

Catalyst	Cycle 1	Cycle 2	Cycle 3	Cycle 4	Cycle 5
52HPW/A20	54	15	11	7	5
30HPW/Z	4.6	4.2	4.1	4.3	4.2

The decrease of activity in heterogeneous catalysts throughout successive reaction runs can be caused by factors [49] like: (i) loss of the supported catalytically active species through transfer into the bulk of solution (leaching); (ii) inaccessibility of reagents toward the active sites present in the microporous domain (fouling); or (iii) deactivation of active sites. For heterogenized catalysts, leaching of the supported active species toward the reaction medium usually represents the most relevant among such factors, which results in poor recyclability associated with a sequential loss of catalytic activity and the potential contamination of the reaction products. This process is particularly common for supported POM-based catalysts and it is long known for several materials based on HPW immobilized on different solid matrices that operate under conditions analogous to those of our model esterification reaction [19, 21]. The leaching percentages with respect to the initial amount of fresh catalyst (80 mg HPA) found for two of our HPW-supported catalysts are summarized in Table 5.6. Their values are clearly aligned with the trends observed in Fig. 5.18 for the sequential reduction of catalytic activity. The HPW/A20 material shows the highest amount of leached catalyst, which agrees well with its highest loss of activity particularly upon the first cycle, whereas the amount of leached HPW species in HPW/Z is very low and nearly constant throughout the reaction runs, which is consistent with its loss of activity displaying a much smoother trend. These results highlight the suitability of our sol–gel/hydrothermal synthetic protocol for immobilizing HPW species over zirconia supports, which leads to supported catalysts with strongly anchored POM clusters and hinders significant leaching of the catalytically active species under the reaction conditions applied in the esterification of palmitic acid.

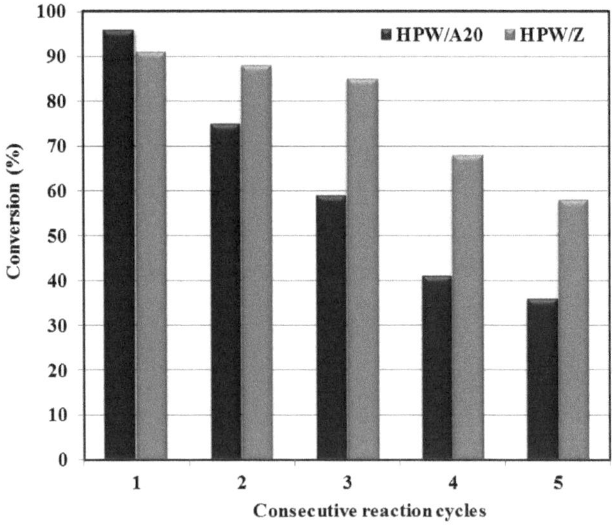

Figure 5.18 Conversion of palmitic acid into methyl palmitate over five consecutive esterification cycles of 6 h at 60 °C using 0.3 g of HPW/Z and HPW/A20 supported catalysts.

5.4.2 HPAs as Redox Catalysts

Fine chemicals constitute a thriving branch within the chemical industry [51, 52]. Most of their synthetic routes involve elaborated protocols pertaining oxidative reactions carried out in homogeneous liquid phase. These reactions usually require catalysts to proceed [34] and classical Lewis acids with remarkable affinity toward O and N atoms (e.g., $AlCl_3$, BF_3, $SnCl_4$, or $TiCl_4$) are among the most widely used of such catalysts [2–4]. The oxidation reactions also require reagents capable of donating O atoms to the catalyst–substrate adduct, which coordinate to the Lewis-acidic metal centers to form oxo- or peroxometallic intermediate species, and then transfer to the substrate to form the oxidized product [36].

One of the most relevant advances undergone by oxidation catalysis is the use of increasingly cheaper and less environmentally harmful oxidants, such as hydrogen peroxide [53], but as they usually lack enough activity because of, for example, low decomposition temperatures or kinetical instability, such oxidants demand the aid of catalysts to raise the reaction rates [54]. Transition-metal-containing molecular species (e.g., Ti, V, Mo, W, Cr, Mn, Fe, and Re) stand out as the most suitable among such catalysts, but they display two major drawbacks: (1) their marked tendency to undergo deactivation through oligomerization, which leads to catalytically inactive, polynuclear oxo-bridged complexes; and (2) the fact that they are usually soluble in the reaction medium, which makes the catalytic processes to run in the homogeneous phase. Under these premises, we analyzed the applicability of insoluble heteropolysalts and HPA-immobilized porous solids as heterogeneous acid catalysts for the green oxidation of organic substrates with H_2O_2 as a clean and eco-friendly oxidant that generates just H_2O as by-product. Our first studies in the field were focused on the oxidation of cyclohexene as a green route for the synthesis of adipic acid.

As mentioned above, H_2O_2 displays the disadvantage of its high thermal instability and many substances can catalyze its decomposition, including most of the transition metals and their compounds. In this regard, we have observed that, despite being extensively employed in oxidation reactions due to the promoting effect of the catalytic activity [55, 56], metal-substituted POMs in which at least one addenda metal center (Mo^{VI} or W^{VI}) is replaced with other transition metals (V^{V} in VPMo, Co^{II} in CoPMo, Ni^{II} in NiPMo, or Fe^{III} in FePMo) are unsuitable when H_2O_2 is used as oxidant because they considerably accelerate decomposition (Fig. 5.19). However, using cesium or ammonium heteropolysalts derived from the plenary HPW or HPMo clusters remarkably reduces the extent to which H_2O_2 decomposes, and this reduction is even more pronounced when the catalysts are previously calcined at 375 °C. It is also worth highlighting that the presence of acetic acid in the reaction medium assists the stabilization of H_2O_2 toward decomposition [56–58].

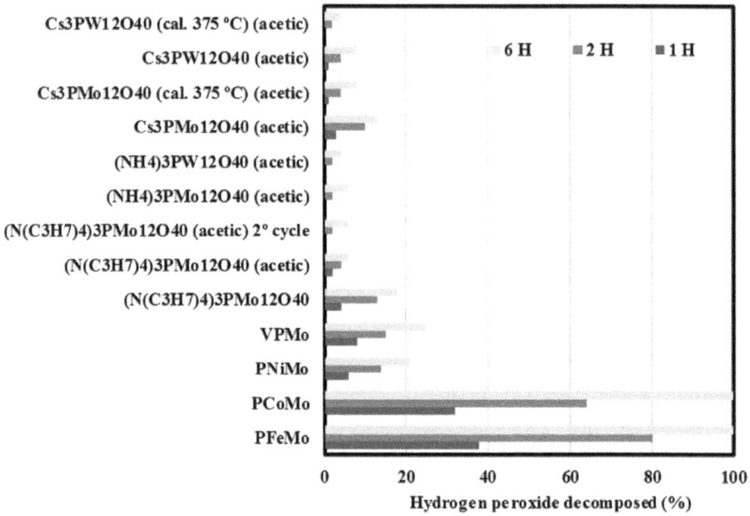

Figure 5.19 Percentage of H_2O_2 decomposed in the presence of different HPAs and heteropolysalts for a period of time from 1 h to 6 h.

Based on the observations above, the insoluble heteropolysalt CsPMo (heterogeneous phase reaction) and its parent heteopoly-acid HPMo (homogeneous phase reaction) were selected as catalysts in the green oxidation of cyclohexene with H_2O_2. As shown in Fig. 5.20, the evolution of the conversion over the time follows the trend of a sigmoid curve, which is typical of catalytic processes that require an induction period. This period might be related to the formation of peroxometallic intermediates on the surface of the HPMo or CsPMo nanoparticles, as observed in general for other similar heteropolysalts [3, 18, 19, 21]. This hypothesis is supported by the observation that the induction period disappears when the catalysts are pre-treated with hydrogen peroxide before running the catalytic cycle. In order to verify the presence of peroxometallic species on the surface of the catalysts, we made use of FT-Raman spectroscopy. The comparison of the Raman spectra of the catalysts before and after running the catalytic cycle indicates that their characteristic Keggin structure remains intact [30, 38], and that the most relevant difference among both spectra is the emergence of a new peak at 853 cm^{-1} for the used catalyst. This new peak is attributed to a υ_s(O–O) mode [59, 60] on the basis of its close location to the stretching O–O vibration characteristic of H_2O_2, which appears at 877 cm^{-1} [61]. These results strongly suggest the existence of peroxo species on the surface of the POM-based catalysts.

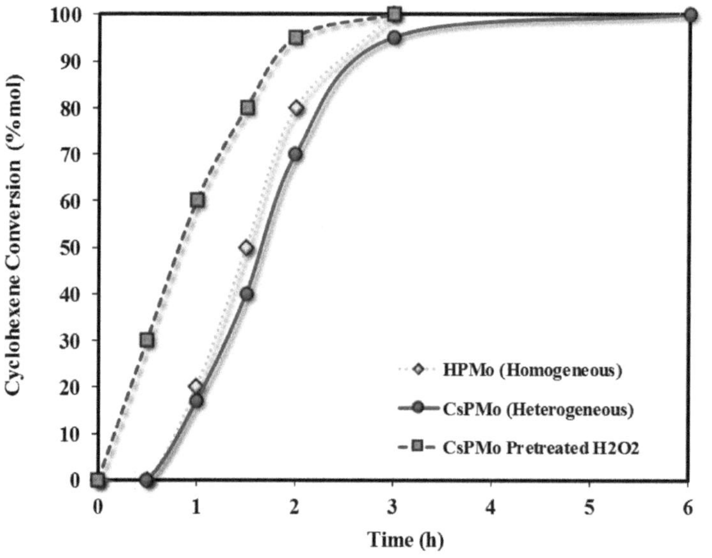

Figure 5.20 Conversions afforded by the catalysts HPMo and its hetero-polysalt CsPMo (fresh and pretreated with H_2O_2) in the oxidation of cyclohexene with H_2O_2 vs reaction time.

In order to analyze the influence of the compositional nature of the Keggin-type POM on the conversion, a large number of PMo and PW-type heteropolysalts were tested under the optimized conditions for the cyclohexene oxidation. The conversion values compiled in Table 5.7 show that the Mo-based compounds afford higher catalytic activity than those based on W regardless of the type of cation accompanying the Keggin-type anion, which is in agreement with the higher oxidant character of $[PMo_{12}O_{40}]^{3-}$ compared to the $[PW_{12}O_{40}]^{3-}$ counterpart [12]. For a given anion, the type of cation influences dramatically the extent to which cyclohexene is converted because the larger the cation, the lower the conversion afforded by the corresponding salt. Thus, the decrease in the conversion observed for the series with quaternary ammonium cations is consistent with the decrease in the porosity determined for such salts, this correlation being particularly relevant in the case of the $(Ph_3PC_3H_7)$PMo salt with almost no porosity (see Fig. 5.12) and negligible catalytic activity. Therefore, the low conversions afforded by the quaternary ammonium or $(Ph_3PC_3H_7)^+$ salts should be attributed to the steric hindrance imposed by the voluminous cations, which hinder the approach of either H_2O_2 or cyclohexene to the POM surface.

The key requirement to be fulfilled by POM-based salts for acting as heterogeneous catalysts is to be insoluble in the reaction medium. In our case, however, certain amounts of catalyst are always solubilized in the course of the reaction as shown in Table 5.7, which compiles the cumulative mass percentage of catalytically active POM species that transfers into the solution for each catalytic cycle. This percentage is highly dependent on the nature of the catalyst: HPAs are completely soluble in the aqueous reaction medium as expected, while ammonium salts display also high solubilities that decrease as the cation becomes larger. In contrast, the lowest solubilities are associated with cesium salts. PMo-based salts are in all cases more soluble than their PW-based analogs and calcination at 375 °C lowers the solubility of a given heteropolysalt substantially. Previous results confirm that the interaction with H_2O_2 is responsible of the high solubilities found for some of these salts [5], and the comparative analysis of the results listed in Table 5.7 reveals certain correlation among the extent of the conversion and the high values of the solubility. All of these results suggest that POM-based nanoparticles transfer into the solution upon formation of peroxometallic species onto their surfaces, thereby pointing out to H_2O_2 being the cause of the solubility observed for these heteropolysalts [5].

Table 5.7 Cyclohexene conversion and cumulative percentage of catalyst mass referred to the initial amount that is transferred to the solution throughout 3 consecutive cycles

	Conversion (%)			Mass Loss (%)		
POM salt	**Cycle 1**	**Cycle 2**	**Cycle 3**	**Cycle 1**	**Cycle 2**	**Cycle 3**
$H_3PMo_{12}O_{40}$	100	–	–	100	–	–
$H_3PW_{12}O_{40}$	100	–	–	100	–	–
$(NH_4)_3PMo_{12}O_{40}$	100	–	–	70	100	–
$(NH_4)_3PW_{12}O_{40}$	75	80	–	30	60	100
$(N(C_2H_5)_4)_3PMo_{12}O_{40}$	80	100	–	40	85	–
$(N(C_3H_7)_4)_3PMo_{12}O_{40}$	70	85	90	20	40	55
$(N(C_4H_9)_4)_3PMo_{12}O_{40}$	30	45	50	10	15	25
$(Ph_3PC_3H_7)_3PMo_{12}O_{40}$	3	4	5	0	2	5
$Cs_3PMo_{12}O_{40}$	95	95	–	25	60	100
$Cs_3PMo_{12}O_{40}$ (375 °C)	100	100	100	12	25	40
$Cs_3PW_{12}O_{40}$	50	60	65	10	15	25
$Cs_3PW_{12}O_{40}$ (375 °C)	75	77	80	3	5	10

5.4.3 Heteropolysalts as Bifunctional Catalysts

In a recent study, we showed that the acidity in heteropolysalt catalysts is not only associated with the presence of protons, but can also originate from cationic metal centers with Lewis acid character (e.g., lanthanides) that are used to assemble POM fragments into large frameworks [13]. Lanthanide-containing species are currently one of the largest and most intriguing groups among the vast family of POMs because combining POM building blocks with rare-earth metal centers constitutes a suitable route for accessing new architectures endowed with additional functionalities besides those inherent to the POM fragments [62]. Thus, incorporation of lanthanide sites allows the resulting POM frameworks to act as acid catalysts [63, 64], and furthermore, it can also lead to bifunctional catalysts suitable for cyanosilylation or oximation reactions owing to the fact that the combination Lewis acidic sites with the Lewis base character of nucleophilic POM surfaces can activate complementary substrates [65, 66].

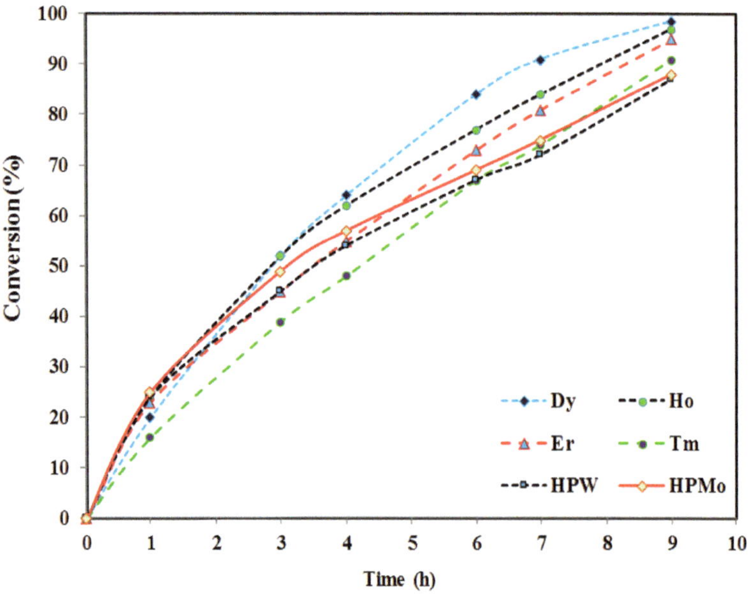

Figure 5.21 Conversions afforded by different **Na-$\beta\beta$-Ln$_4$** catalysts and the two commercial heteropolyacids HPMo and HPW in the oxidation of aniline with H_2O_2 as a function of the reaction time.

We investigated the oxidation reaction of aniline with the eco-friendly H_2O_2 oxidant using a new type of POM-based salts as catalysts acting in heterogeneous phase, hereon referred to as **Na-$\beta\beta$-Ln$_4$**. The POM anion contained in these new heteropolysalts consists in a dimeric assembly of dilacunary Keggin-type fragments linked by four trivalent Ln^{+3} cations. As shown in Fig. 5.21, all **Na-$\beta\beta$-Ln$_4$** salts afford conversions comparable to or even superior to those obtained when commercial HPMo and HPW heteropolyacids catalyze the title reaction in homogeneous phase. It is worth noting that **Na-$\beta\beta$-Ln$_4$** salts demand shorter reaction times than the selected HPAs to achieve similar levels of conversion regardless of acting in heterogeneous phase, and that their catalytic activity depends heavily on the specific type of Ln^{+3} cation because it decreases as the lanthanide atomic number increases.

The characterization of the **Na-$\beta\beta$-Ln$_4$** series was carried out through DRIFT spectroscopy (Fig. 5.22). The bands observed at wavenumbers above 1100 cm^{-1} do not originate from the Keggin-type framework [67], but associate with vibrational modes of the different types of water molecules present in the structures of our highly hydrated heteropolysalts. Thus, the broad band in the 3600–2600 cm^{-1} range is characteristic of υ(O–H) vibrations and should be attributed to interstitial water molecules of crystallization, whereas the peak at ca. 1640 cm^{-1} can be assigned to the H–O–H bending on the basis of the in-plane bending observed for the $H_5O_2^+$ ion [68]. This peak is virtually identical in all four DRIFT spectra, but that located near 1550 cm^{-1} and those observed around 1455 cm^{-1} differ significantly in intensity within the **Na-$\beta\beta$-Ln$_4$** series, in such a way that they become weaker as the Ln atomic number increases. These intensity-variable signals observed in the δ(H–O–H) spectral region might originate from the presence of well-defined water clusters in the vicinity of the Ln^{3+} sites, the existence of which have been recently assessed for porous MOF–POM type composites [69], as well as in the structure of a hybrid compound based on Wells–Dawson POMs and dicopper metal-organic moieties [70]. Considering that Ln_2O_3 oxides become more basic as Ln varies from Dy to Tm [71], a similar trend could be expected for our heteropolysalt series in which the lanthanide ions are confined in the pockets of an oxotungstate framework. Therefore, the decrease in the relative intensities of the above mentioned signals could relate to the Lewis acidity of the Ln sites, with the more acidic Dy atom favoring the clustering of water molecules in its close vicinity over the more basic Tm through stronger hydrogen bonding.

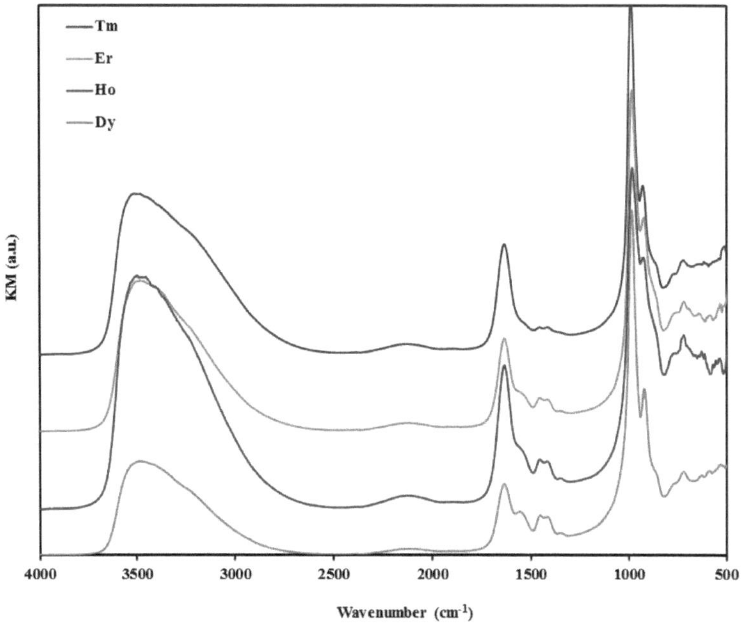

Figure 5.22 DRIFT spectra of the **Na-ββ-Ln$_4$** catalysts (Ln = Dy, Ho, Er, Tm).

Figure 5.23 shows an amplification of the 1700–1300 cm^{-1} region for the DRIFT spectra of the Er-containing heteropolysalt recorded from room temperature up to 300 °C (the full spectra from 600 to 4000 cm^{-1} are given in the inset). The vibration peaks originating from the dilacunary Keggin-type fragments (1000–600 cm^{-1}) remain unchanged during the calcination process, which is consistent with the fact that such POM structural type is thermally stable up to temperatures above 300 °C [72], but close inspection reveals main modifications in the spectral region corresponding to the bonding water molecules, which is in agreement with the analysis of the evolved gases consisting in water exclusively. More specifically, the bands associated with water molecules are located in the 3600–2600 cm^{-1} range and at ca. 1640, 1550, and 1455 cm^{-1}, and they all maintain their original relative intensities up to 168 °C. At higher temperatures, the intensity of the former two signals undergoes remarkable decrease, while variations for those at 1550 and 1455 cm^{-1} are almost negligible. The latter signals start becoming weaker only at temperatures higher than 260 °C, and this behavior indicates that two types of water molecules coexist in the structures of our heteropolysalts: loosely bound water of crystallization that is lost with ease (associated with the 3600–2600 and 1640 cm^{-1} signals), and some other molecules involved in stronger interactions that demand higher temperatures for their release to take place. This second type of water molecules must encompass the aqua ligands coordinated to the Ln sites and those molecules clustered in the close vicinity via strong hydrogen bonding.

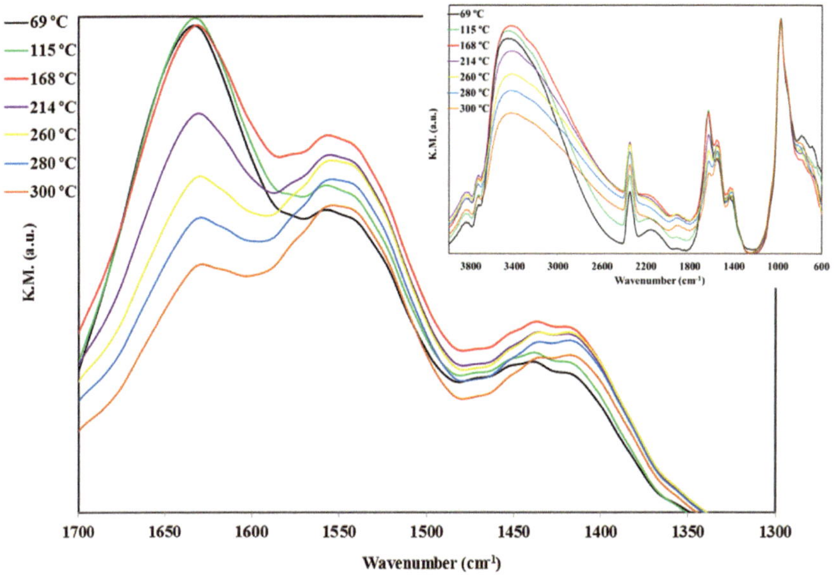

Figure 5.23 Detail of the DRIFT spectra of **Na-$\beta\beta$-Er$_4$** recorded at different temperatures up to 300 °C. The full spectra are given in the inset.

Acknowledgments

This work was financially supported by Generalitat Valenciana through FEDER funds in the framework of the program Prometeo (grants PROMETEO2/2014/010 and PROMETEO/2018/076).

References

1. Hagen, J. (2015). *Industrial Catalysis: A Practical Approach*, 3rd ed, Wiley-VCH, Weinheim, Germany.
2. Parida, K.M., and Mallick, S. (2008). Hydroxylation of phenol over molybdovanadophosphoric acid modified zirconia, *J. Mol. Catal. A Chem.*, **279**, pp. 104–111.
3. Pizzio, L.R., Cáceres, C.V., and Blanco, M.N. (1998). Acid catalysts prepared by impregnation of tungstophosphoric acid solutions on different supports, *Appl. Catal. A Gen.*, **167**, pp. 283–294.
4. Rivera, T.S., Sosa, A., Romanelli, G.P., Blanco, M.N., and Pizzio, L.R. (2012). Tungstophosphoric acid/zirconia composites prepared by the sol–gel method: an efficient and recyclable green catalyst for the one-pot synthesis of 14-aryl-14H-dibenzo[a,j]xanthenes, *Appl. Catal. A Gen.*, **443–444**, pp. 207–213.
5. Alcañiz-Monge, J., Trautwein, G., and Garcia-Garcia, A. (2014). Influence of peroxometallic intermediaries present on polyoxometalates nanoparticles surface on the adipic acid synthesis, *J. Mol. Catal. A Chem.*, **394**, pp. 211–216.
6. Farhadi, S., and Zaidi, M. (2009). Polyoxometalate–zirconia (POM/ZrO_2) nanocomposite prepared by sol–gel process: a green and recyclable photocatalyst for efficient and selective aerobic oxidation of alcohols into aldehydes and ketones, *Appl. Catal. A Gen.*, **354**, pp. 119–126.
7. Leofanti, G., Padovan, M., Tozzola, G., and Venturelli, B. (1998). Surface area and pore texture of catalysts, *Catal. Today*, **41**, pp. 207–219.
8. Tauster, S.J., Fung, S.C., and Garten, R.L. (1978). Strong metal-support interactions. Group 8 noble metals supported on titanium dioxide, *J. Am. Chem. Soc.*, **100**, pp. 170–175.
9. Goodman, D.W. (1995). Model studies in catalysis using surface science probes, *Chem. Rev.*, **95**, pp. 523–536.
10. Campbell, C.T. (2012). Catalyst-support interactions: electronic perturbations, *Nat. Chem.*, **4**, pp. 597–598.
11. (a) Pope, M.T., and Müller, A. (1991). Polyoxometalate chemistry: an old field with new dimensions in several disciplines, *Angew. Chem. Int. Ed. Engl.*, **30**, pp. 34–48; (b) Pope, M.T., and Müller, A., (eds). (1994). *Polyoxometalates: From Platonic Solids to Antiretroviral Activity*, Kluwer Academic Publishers, Dordrecht, The Netherlands; (c) Pope, M.T., and Müller, A., (eds). (2001). *Polyoxometalate Chemistry:*

From Topology via Self-Assembly to Applications, Kluwer Academic Publishers, Dordrecht, The Netherlands; (d) Yamase, T., and Pope, M.T., (eds). (2002). *Polyoxometalate Chemistry for Nano-Composite Design*, Kluwer Academic/Plenum Publishers, New York NY, USA; (e) Sécheresse, F., (ed.) (2013). *Polyoxometalate Chemistry: Some Recent Trends*, World Scientific, Singapore.

12. (a) Hill, C.L., (Guest ed.) (1998). Special monographic issue on polyoxometalates, *Chem. Rev.*, **98**, pp. 1–390; (b) Pope, M.T. (1983). *Heteropoly and Isopoly Oxometalates*, Springer-Verlag, New York, NY, USA; (c) Borrás-Almenar, J.J., Coronado, E., Müller, A., and Pope, M.T., (eds) (2003). *Polyoxometalate Molecular Science*, Kluwer Academic Publishers, Dordrecht, The Netherlands; (d) Moffat, J.B. (2001). *Metal-Oxygen Clusters: The Surface and Catalytic Properties of Heteropoly Oxometalates*, Kluwer Academic/Plenum Publishers, New York NY, USA; (f) Kozhevnikov, I.V. (2002). *Catalysts for Fine Chemical Synthesis: Catalysis by Polyoxometalates*, Vol. 2, John Wiley & Sons, Chichester, UK; (g) van Eldik, R., and Cronin, L., (eds). (2017). *Advances in Inorganic Chemistry, Vol. 69: Polyoxometalate Chemistry*, Academic Press, Elsevier Inc., Cambridge, MA, USA.
13. Trautwein, G., El Bakkali, B., Alcañiz-Monge, J., Artetxe, B., Reinoso, S., and Gutiérrez-Zorrilla, J.M. (2015). Dimeric assemblies of lanthanide-stabilized dilacunary Keggin tunsgstogermanates: a new class of catalysts for the selective oxidation of aniline, *J. Catal.*, **331**, pp. 110–117.
14. (a) McMonagle, J.B., and Moffat, J.B. (1984). Pore structures of the monovalent salts of the heteropoly compounds, 12-tungstophosphoric and 12-molybdophosphoric acid, *J. Colloid Interface Sci.*, **101**, pp. 479–488; (b) Yoshimune, M., Yoshinaga, Y., and Okuhara, T. (2002). Effect of alkaline metal on microporosity of acidic alkaline salts of 12-tungstophosphoric acid, *Microporous Mesoporous Mater.*, **51**, pp. 165–174.; (c) Moffat, J.B. (1989). Implicit and explicit microporosity in heteropoly oxometalates, *J. Mol. Catal.*, **52**, pp. 169–191.; (d) Lapham, D., and Moffat, J.B. (1991). Preparative effects on the surface area and pore structure of microporous heteropoly oxometalates, *Langmuir*, **7**, pp. 2273–2278.
15. Sheldon, R.A., and Downing, R.S. (1999). Heterogeneous catalytic transformations for environmentally friendly production, *Appl. Catal. A Gen.*, **189**, pp. 163–183.
16. (a) Alcañiz-Monge, J., Trautwein, G., Parres-Esclapez, S., and Maciá-Agulló, J.A. (2008). Influence of microporosity of activated carbons as a support of polyoxometalates, *Microporous Mesoporous*

Mater., **115**, pp. 440–446.; (b) Alcañiz-Monge, J., El Bakkali, B., Trautwein, G., and Reinoso, S. (2018). Zirconia-supported tungstophosphoric heteropolyacid as heterogeneous acid catalyst for biodiesel production, *Appl. Catal. B Environ.*, **224**, pp. 194–203.

17. Farneth, W.E., and Gorte, R.J. (1995). Methods for characterizing zeolite acidity, *Chem. Rev.*, **95**, pp. 615–635.
18. Bardin, B.B., Bordawekar, S.V., Neurock, M., and Davis, R.J. (1998). Acidity of Keggin-type heteropolycompounds evaluated by catalytic probe reactions, sorption microcalorimetry, and density functional quantum chemical calculations, *J. Phys. Chem. B*, **102**, pp. 10817–10825.
19. Tropecêlo, A.I., Casimiro, M.H., Fonseca, I.M., Ramos, A.M., Vital, J., and Castanheiro, J.E. (2010). Esterification of free fatty acids to biodiesel over heteropolyacids immobilized on mesoporous silica, *Appl. Catal. A Gen.*, **390**, pp. 183–189.
20. Endalew, A.K., Kiros, Y., and Zanzi, R. (2011). Inorganic heterogeneous catalysts for biodiesel production from vegetable oils, *Biomass Bioenergy*, **35**, pp. 3787–3809.
21. Alcañiz-Monge, J., Trautwein, G., and Marco-Lozar, J.P. (2013). Biodiesel production by acid catalysis with heteropolyacids supported on activated carbon fibers, *Appl. Catal. A Gen.*, **468**, pp. 432–441.
22. Radovic, L.R., and Rodríguez-Reinoso, F. (1997). Carbon materials in catalysis. In: Thrower, P. A., (ed.) *Chemistry and Physics of Carbon*, vol. 25, pp. 243–441, Dekker, New York, NY, USA.
23. Kozhevnikov, I.V., Sinnema, A., Jansen, R.J.J., and van Bekkum, H. (1994). ^{17}O NMR determination of proton sites in solid heteropoly acid $H_3PW_{12}O_{40}$. ^{31}P, ^{29}Si and ^{17}O NMR, FT-IR and XRD study of $H_3PW_{12}O_{40}$ and $H_4SiW_{12}O_{40}$ supported on carbon, *Catal. Lett.*, **27**, pp. 187–197.
24. Izumi, Y., and Urabe, K. (1981). Catalysis of heteropoly acids entrapped in activated carbon, *Chem. Lett.*, **10**, pp. 663–666.
25. Schwegler, M.A., Vinke, P., van der Eijk, M., and van Bekkum, H. (1992). Activated carbon as a support for heteropolyanion catalysts, *Appl. Catal. A Gen.*, **80**, pp. 41–57.
26. Hu, J., Li, K., Li, W., Ma, F., and Guo, Y. (2009). Selective oxidation of styrene to benzaldehyde catalyzed by Schiff base-modified ordered mesoporous silica materials impregnated with the transition metal-monosubstituted Keggin-type polyoxometalates, *Appl. Catal. A Gen.*, **364**, pp. 211–220.

27. Gao, R., Zhu, Q., Dai, W.-L., and Fan, K. (2011). Highly active and green aminopropyl-immobilized phosphotungstic acid on mesocellular silica foam for the O-heterocyclization of cycloocta-1,5-diene with aqueous H_2O_2, *Green Chem.*, **13**, pp. 702–708.

28. (a) Tamiolakis, I., Lykakis, I.N., Katsoulidis, A.P., Stratakis, M., and Armatas, G.S. (2011). Mesoporous Cr_2O_3-phosphomolybdic acid solid solution frameworks with high catalytic activity, *Chem. Mater.*, **23**, pp. 4204–4211; (b) Tamiolakis, I., Lykakis, I.N., Katsoulidis, A.P., Malliakas, C.D., and Armatas, G.S. (2012). Ordered mesoporous Cr_2O_3 frameworks incorporating Keggin-type 12-phosphotungstic acids as efficient catalysts for oxidation of benzyl alcohols, *J. Mater. Chem.*, **22**, pp. 6919–6927.

29. Li, B., Ma, W., Liu, J., Han, C., Zuo, S., and Li, X. (2011). Synthesis of the well-ordered hexagonal mesoporous silicate incorporated with phosphotungstic acid through a novel method and its catalytic performance on the oxidative desulfurization reaction, *Catal. Commun.*, **13**, pp. 101–105.

30. Mestl, G., Ilkenhans, T., Spielbauer, D., Dieterle, M., Timpe, O., Kröhnert, J., Jentoft, F., Knözinger, H., and Schlögl, R. (2001). Thermally and chemically induced structural transformations of Keggin-type heteropoly acid catalysts, *Appl. Catal. A Gen.*, **210**, pp. 13–34.

31. Dimitratos, N., and Védrine, J.C. (2006). Study of Ga modified $Cs_{2.5}H_{1.5}PV_1Mo_{11}O_{40}$ heteropolyoxometallates for propane selective oxidation, *J. Mol. Catal. A Chem.*, **255**, pp. 184–192.

32. Molinari, J.E., Nakka, L., Kim, T., and Wachs, I.E. (2011). Dynamic surface structures and reactivity of vanadium-containing molybdophosphoric acid ($H_{3+x}PMo_{12-x}V_xO_{40}$) Keggin catalysts during methanol oxidation and dehydration, *ACS Catal.*, **1**, pp. 1536–1548.

33. Herzog, B., Wohlers, M., and Schlögl, R. (1997). An in situ DRIFTS study of the active phase of the heteropoly acid catalyst $H_4[PVMo_{11}O_{40}]$ in oxidation reactions, *Mikrochim. Acta*, **14**, pp. 703–704.

34. El Bakkali, B., Trautwein, G., Alcañiz-Monge, J., and Reinoso, S. (2018). Zirconia-supported 11-molybdovanadophosphoric acid catalysts: effect of the preparation method on their catalytic activity and selectivity, *Acta Crystallogr. Sect. C*, **C74**, pp. 1334–1347.

35. Arichi, J., Eternot, M., and Louis, B. (2008). Synthesis of V-containing Keggin polyoxometalates: versatile catalysts for the synthesis of fine chemicals?, *Catal. Today*, **138**, pp. 117–122.

36. Shinachi, S., Matsushita, M., Yamaguchi, K., and Mizuno, N. (2005). Oxidation of adamantane with 1 atm molecular oxygen by vanadium-substituted polyoxometalates, *J. Catal.*, **233**, pp. 81–89.

37. Damyanova, S., and Fierro, J.L.G. (1998). Structural features and thermal stability of titania-supported 12-molybdophosphoric heteropoly compounds, *Chem. Mater.*, **10**, pp. 871–879.

38. Rocchiccioli-Deltcheff, C., and Fournier, M. (1991). Catalysis by polyoxometalates. Part 3.—Influence of vanadium(V) on the thermal stability of 12-metallophosphoric acids from in situ infrared studies, *J. Chem. Soc., Faraday Trans.*, **87**, pp. 3913–3920.

39. Mizuno, N., Suh, D.-J., Han, W., and Kudo, T. (1996). Catalytic performance of $Cs_{2.5}Fe_{0.08}H_{1.26}PVMo_{11}O_{40}$ for direct oxidation of lower alkanes, *J. Mol. Catal. A Chem.*, **114**, pp. 309–317.

40. McGarvey, G.B., and Moffat, J.B. (1991). A study of solution species generated during the formation of 12-heteropoly oxometalate catalysts, *J. Mol. Catal.*, **69**, pp. 137–155

41. Hernández Enríquez, J.M., Cortez Lajas, L.A., García Alamilla, R., Castillo Mares, A., Sandoval Robles, G., and García Serrano, L.A. (2009). Synthesis and characterization of mesoporous and nano-crystalline phosphate zirconium oxides, *J. Alloys Compd.*, **483**, pp. 425–428.

42. Song, I.K., Kitchin, J.R., and Barteau, M.A. (2002). $H_3PW_{12}O_{40}$-functionalized tip for scanning tunneling microscopy, *Proc. Natl. Acad. Sci. U.S.A.*, **99**, pp. 6471–6475.

43. Yang, Y., Wu, Q., Guo, Y., Hu, C., and Wang, E. (2005). Efficient degradation of dye pollutants on nanoporous polyoxotungstate–anatase composite under visible-light irradiation, *J. Mol. Catal. A: Chem.*, **225**, pp. 203–212.

44. Alcañiz-Monge, J., Trautwein, G., and Román-Martínez, M.C. (2011). Effect of counteranion of ammonium salts on the synthesis of porous nanoparticles $(NH_4)_3[PMo_{12}O_{40}]$, *Solid State Sci.*, **13**, pp. 30–37.

45. Alcañiz-Monge, J., Trautwein, G., El Bakkali, B., and Reinoso, S. (2017). A simple approach to develop tailored mesoporosity in nanostructured heteropolysalts, *Chem. - Eur. J.*, **23**, pp. 2387–2395.

46. Alcañiz-Monge, J., Trautwein, G., Pérez-Cadenas, M., and Román-Martínez, M.C. (2009). Effects of compression on the textural properties of porous solids, *Microporous Mesoporous Mater.*, **126**, pp. 291–301.

47. Kozhevnikov, I.V., Kloetstra, K.R., Sinnema, A., Zandbergen, H.W., and van Bekkum, H. (1996). Study of catalysts comprising heteropoly acid $H_3PW_{12}O_{40}$ supported on MCM-41 molecular sieve and amorphous silica, *J. Mol. Catal. A Chem.*, **114**, pp. 287–298.

48. Mallik, S., Dash, S.S., Parida, K.M., and Mohapatra, B.K. (2006). Synthesis, characterization, and catalytic activity of phosphomolybdic

acid supported on hydrous zirconia, *J. Colloid Interface Sci.*, **300**, pp. 237–243.

49. Kulkarni, M.G., Gopinath, R., Charan Meher, L., and Dalai, A.K. (2006). Solid acid catalyzed biodiesel production by simultaneous esterification and transesterification, *Green Chem.*, **8**, pp. 1056–1062.
50. Ertl, G., Knözinger, H., Schüth, F., and Weitkamp, J., (eds). (2008). *Handbook of Heterogeneous Catalysis*, 2nd ed, Wiley/VCH, Weinheim, Germany.
51. Noyori, R., Aoki, M., and Sato, K. (2003). Green oxidation with aqueous hydrogen peroxide, *Chem. Commun.*, pp. 1977–1986.
52. Shaabani, A., and Rezayan, A.H. (2007). Silica sulfuric acid promoted selective oxidation of sulfides to sulfoxides or sulfones in the presence of aqueous H_2O_2, *Catal. Commun.*, **8**, pp. 1112–1116.
53. Bossmann, S.H., Oliveros, E., Göb, S., Siegwart, S., Dahlen, E.P., Payawan, L., Straub, M., Wörner, M., and Braun, A.M. (1998). New evidence against hydroxyl radicals as reactive intermediates in the thermal and photochemically enhanced Fenton reactions, *J. Phys. Chem. A*, **102**, pp. 5542–5550.
54. Babou, F., Coudurier, G., and Vedrine, J.C. (1995). Acidic properties of sulfated zirconia: an infrared spectroscopic study, *J. Catal.*, **152**, pp. 341–349.
55. Sheldon, R.A., and van Santen, R.A. (1995). *Catalytic Oxidation: Principles and Applications,* World Scientific, Singapore.
56. Sheldon, R.A., and Kochi, J.K. (1981). *Metal-Catalyzed Oxidations of Organic Compounds,* Academic Press, New York, NY, USA.
57. Deng, Y., Ma, Z., Wang, K., Hua, L., and Chen, J. (1999). Clean synthesis of adipic acid by direct oxidation of cyclohexene with H_2O_2 over peroxytungstate–organic complex catalysts, *Green Chem.*, **1**, pp. 275–276.
58. Maheswari, P.U., Tang, X., Hage, R., Gamez, P., and Reedijk, J. (2006). The role of carboxylic acids on a Na_2WO_4/H_2WO_4-based biphasic homogeneous alkene epoxidation, using H_2O_2 as oxidant, *J. Mol. Catal. A: Chem.*, **258**, pp. 295–301.
59. Grzywa, M., Łasocha, W., and Rutkowska-Żbik, D. (2009). Structural investigation of tetraperoxo complexes of Mo(VI) and W(VI): X-ray and theoretical studies, *J. Solid State Chem.*, **182**, pp. 973–982.
60. Shi, X.-y., and Wei, J.-f. (2005). Oxidation of alcohols with H_2O_2 catalyzed by bis-quaternary phosphonium peroxotungstates (or peroxomolybdates) under halide- and organic solvent-free condition, *J. Mol. Catal. A*, **229**, pp. 13–17.

61. Dickman, M.H., and Pope, M.T. (1994). Peroxo and superoxo complexes of chromium, molybdenum, and tungsten, *Chem. Rev.*, **94**, pp. 569–584.

62. Miras, H.N., Vilà-Nadal, L., and Cronin, L. (2014). Polyoxometalate based open-frameworks (POM-OFs), *Chem. Soc. Rev.*, **43**, pp. 5679–5699.

63. Boglio, C., Lemière, G., Hasenknopf, B., Thorimbert, S., Lacôte, E., and Malacria, M. (2006). Lanthanide complexes of the monovacant Dawson polyoxotungstate $[\alpha_1\text{-}P_2W_{17}O_{61}]^{10-}$ as selective and recoverable Lewis acid catalysts, *Angew. Chem. Int. Ed.*, **45**, pp. 3324–3327.

64. Kholdeeva, O.A., Timofeeva, M.N., Maksimov, G.M., Maksimovskaya, R.I., Neiwert, W.A., and Hill, C.L. (2005). Aerobic oxidation of formaldehyde mediated by a Ce-containing polyoxometalate under mild conditions, *Inorg. Chem.*, **44**, pp. 666–672.

65. Suzuki, K., Sugawa, M., Kikukawa, Y., Kamata, K., Yamaguchi, K., and Mizuno, N. (2012). Strategic design and refinement of Lewis acid–base catalysis by rare-earth-metal-containing polyoxometalates, *Inorg. Chem.*, **51**, pp. 6953–6961.

66. Zhao, S., Huang, L., and Song, Y.-F. (2013). Highly selective and efficient Lewis acid–base catalysts based on lanthanide-containing polyoxometal ates for oximation of aldehydes and ketones, *Eur. J. Inorg. Chem.*, **2013**, pp. 1659–1663.

67. Bridgeman, A.J. (2003). Density functional study of the vibrational frequencies of α-Keggin heteropolyanions, *Chem. Phys.*, **287**, pp. 55–69.

68. Hashimoto, M., Koyano, G., and Mizuno, N. (2004). In situ IR spectrum of 12-tungstophosphoric acid hexahydrate with planar $H_5O_2^+$, *J. Phys. Chem. B*, **108**, pp. 12368–12374.

69. Duan, C., Wei, M., Guo, D., He, C., and Meng, Q. (2010). Crystal structures and properties of large protonated water clusters encapsulated by metal–organic frameworks, *J. Am. Chem. Soc.*, **132**, pp. 3321–3330.

70. Yang, H., Cao, M., Gao, S., and Cao, R. (2014). A new hybrid constructed from Dawson-like polyoxometalates and dicopper coordination compounds containing a discrete tridecameric water cluster, *J. Mol. Struct.*, **1056–1057**, pp. 141–145.

71. Sato, S., Takahashi, R., Kobune, M., and Gotoh, H. (2009). Basic properties of rare earth oxides, *Appl. Catal. A: Gen.*, **356**, pp. 57–63.

72. Gambaro, L.A., and Briand, L.E. (2004). In situ quantification of the active acid sites of $H_6P_2W_{18}O_{62} \cdot nH_2O$ heteropoly-acid through chemisorption and temperature programmed surface reaction of isopropanol, *Appl. Catal. A Gen.*, **264**, pp. 151–159.

Chapter 6

Transition Metal Oxide–Based Storage Materials

Daniel Malcolm and Laia Vilà-Nadal

WestCHEM, School of Chemistry,
University of Glasgow, Glasgow G12 8QQ, UK

Laia.Vila-Nadal@glasgow.ac.uk

As polyoxometalate (POM) chemistry develops and becomes more varied in the structures that can be formed, the areas of science that POM-based materials can potentially impact also increases. Much of the interest for these molecules has arisen not only due to POMs unique ability toward molecular growth and complex re-arrangements, but also due to their application in diverse areas, such as catalysis, magnetism, molecule-based electronic, and energy storage devices. The focus for POM chemists is slowly switching from obtaining novel structures to the design of functional materials with genuine real-world applications. A review published by Hill in 1998 discussed the potential for the use of POMs in a variety of disciplines. Although this review was published over 20 years ago much of the discussion is still relevant. By far the most active area for applied POM research is in catalysis, energy applications, water oxidation catalysts, and hydrogen production, but as those applications are already highlighted in several excellent reviews,

Polyoxometalates: Advances, Properties, and Applications
Edited by Leire Ruiz Rubio, José Luis Vilas Vilela, Beñat Artetxe, and Juan Manuel Gutiérrez-Zorrilla

ISBN 978-981-4968-14-0 (Hardcover), 978-1-003-27744-6 (eBook)
www.jennystanford.com

this book chapter will focus in other fields where the unique properties exhibited by POMs are still being investigated. The last decade has seen advances in very divergent applications where POMs have played a key role, for instance, in the formation of chemical gardens, as components in flash-memory devices.

6.1 Molecular Metal Oxides in Magnetism and Semiconductors

Nuclear astrophysics is the branch of astrophysics that helps to understand the formation of elements in the periodic table. About 13.8 bn years ago, the most abundant elements in our universe were H (${}^{1}_{1}H$ and ${}^{1}_{2}He$) and He (${}^{4}_{2}He$ and ${}^{3}_{2}He$), with trace quantities of Li and some heavier elements present [1]. It took billions of years of forging within the cores of stars until electrons could join nuclei in atoms and give rise to other elements, including the production of the heavy elements beyond iron. Those remarkable series of events occurred within supernova explosions or neutron-star mergers. Iron, a transition metal (TM) with partially filled d orbitals, is the most prevalent metal on Earth and is the direct consequence of the physics of nucleosynthesis in dying stars [2]. In fact, most of the Earth's crust consists of solid oxides, the result of elements being oxidized by the oxygen in air or water. The prevalence of oxides in Earth is due to the "Great Oxidation Event" or GOE and left fingerprints in the rock record. Over an interval from roughly 2.4 to 2.1 Gyr ago, Earth atmosphere shifted from being fundamentally reducing to oxidizing [3]. Despite decades of intensive investigations, there is no consensus on what exactly caused this oxygen revolution on planet Earth, mostly given the lack of definitive geologic indicators of microbial oxygen production [4]. What matters to us chemists it is not what originated O_2, but its consequences and the fact that we live in an environment in which O_2 is prevalent, making up to 21% of the modern atmosphere. This has conditioned the chemistry we can make and the raw materials we can find. Other metal-oxides are reasonably abundant in the Earth's crust include TMs, such as Mn, Fe, Co, Ni, Cu, and early TMs Ti, V, Cr, Zr, Nb, and W. The d-block TMs have great importance in our lives and key in the modern history and the booming post-industrial digital economy.

For over a hundred years, researchers in physics and chemistry have been intrigued by the magnetic properties of solid materials [5]. Magnetism and technological developments have been connected for centuries—from navigation compasses to motors and generators to modern day magnetic data storage. In the 2020 digital world, we rely on the successful progression of memory in a silicon chip to read and write data. The driving force behind innovations, such as smart phones, voice control, track, and trace apps, driverless cars are reliable memory devices; improving storage capabilities, therefore, is key. Ferromagnetic materials have been the most important materials in the field of information storage from its foundation to the present day [6]. Memories based on ferromagnets, such as magnetic tapes and hard disk drives, were made possible by innovations in the semiconductor technology [7]. The search for reliable memory devices is still underway. One of the candidates are molecular metal oxides, or polyoxometalates (POMs); these are a ubiquitous class of inorganic compounds and span a wide range of structural and electronic features. POMs' discrete molecular structures are composed of metal ions bridged by oxide anions. POMs can also be referred to as molecular metal oxides due to their molecular composition and their position between monomeric entities and bulk oxides. POMs are primarily constituted of early-transition-metal (d-block) elements in their highest oxidation states [8]. Almost all POM structures are anionic and subsequently form salts when in the presence of obligatory cations [9]. This family of compounds is composed of between 6 and 368 metal ions in a single molecule and is frequently assembled under "one-pot" reaction conditions. POMs, especially tungsto- and molybdophospates, and silicates, are highly used in industry for catalysis [10].

In this book chapter, we describe the use of POMs as a nano-molecular computing element. As inorganic oxides POMs are naturally more compatible with current complementary metal–oxide–semiconductor (CMOS) than organic molecules. The electronic properties and structures of POMs are much more compatible with SiO_2 than organic elements (see Fig. 6.1) [11]. In addition, molecular metal oxides or POMs provide greater electronic complementarity as they are highly redox active, can be doped with electronically active heteroatoms, and can shape the semiconducting and metallic states required for low power electronics circuits that are the basis of device components, such as processors, voltage converters and regulators, liquid-crystal display (LCD) displays. Indeed, one of the pillars of the success of silicon nanoelectronics is the achievement of high electrical conductivity and controllable conductivity by doping, but organic molecules have unfortunately not yet shown such attractive properties. Recent results have demonstrated that POMs are ideal CMOS compatible materials since they provide an unrivaled structural diversity of molecules, display a wide range of properties and nuclearities, and are assembled under easy-to-prepare "one-pot" reaction conditions, enabling nanoscale objects to be self-assembled from very small building blocks.

Promising results in flash memory have come to light by exploiting the electronic properties of a POM-based material [12]. This is a tremendous area of interest with immense possibilities, and the knowledge gained by studding POM self-assembly mechanisms could be a true game changer in the field. By combining building blocks, we could design the next material to be used as a component in the next generation of metal-oxide-semiconductors (MOS). POM scientists need to develop new approaches to understand self-assembly that combine a good quantum foundation with increasing scale in terms of the size of the system [13]. New semi-empirical approaches to coping with mixed valence states, dynamic assemblies, kinetic as well as global thermodynamic aspects could provide a good foundation [14].

The real game changer lies in the ability to design proof-of-principle experiments to demonstrate the true potential for the applications of new nanomaterials in sustainable energy devices as the primary objective, and also to be incorporated as components in flash memory RAMs. This book chapter is written with the aim to inspire and motivate a new generation of POM experts. In addition, silicon-based traditional electronic storage devices need to face the evolution that incorporating molecular metal oxides could provide. By combining experimentation and theoretical estimates for new POM materials and their properties, we will gain a deeper understanding faster and more efficiently than with test and experimentation alone. First principles multiscale modeling has allowed a deep understanding of materials by enabling accurate prediction of electronic properties of materials at a low computational cost; for instance, band gaps that are crucial for memory device applications. Preliminary work also shows that we can obtain information of the building blocks formed in the synthetic mixtures, hence, we are closer to the design of new metal-oxide materials with the targeted properties [15]. Ideally, material development and design activities include a close coupling between prediction, synthesis, and characterization of materials. The discovery of materials that can be adapted to the needs of far more demanding technology needs to be accelerated and lowered. The increased use of computing resources, the generation of databases and experimental developments has accelerated these activities substantially [16].

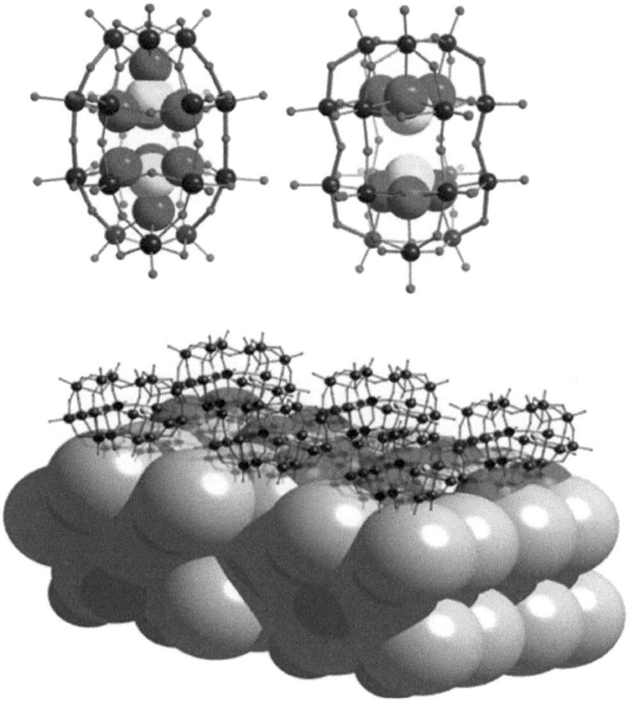

Figure 6.1 Top left: ball-and-stick view on the classic Wells–Dawson structure $[M_{18}O_{54}(SO_4)_2]^{4-}$ M = W and Mo. Top right: ball-and-stick view on the non-classical Wells–Dawson anion $[M_{18}O_{54}(SO_3)_2]^{4-}$ when M = W and Mo. Bottom: idealized vision of a POM layer in a SiO_2.

6.2 Abundance of Raw Materials

Traditionally, chemistry focuses on converting resources into commodities, contributing to the creation of a multitude of synthetic methodologies that produce many molecules and materials currently used in society [17]. However, this linear development model has produced a lot of waste that also reaches the atmosphere, causing significant environmental problems both local and global [18]. This offers a new position for chemistry, which focuses on the creation of new processes of recovery and recycling to facilitate the effective use of resources, as well as on the development of new synthetic methods that use waste as a resource [19]. At the consumer end, the implementation of circular business models (CBMs) could offer an effective way to tackle societal challenges and contribute to sustainable development goals [20].

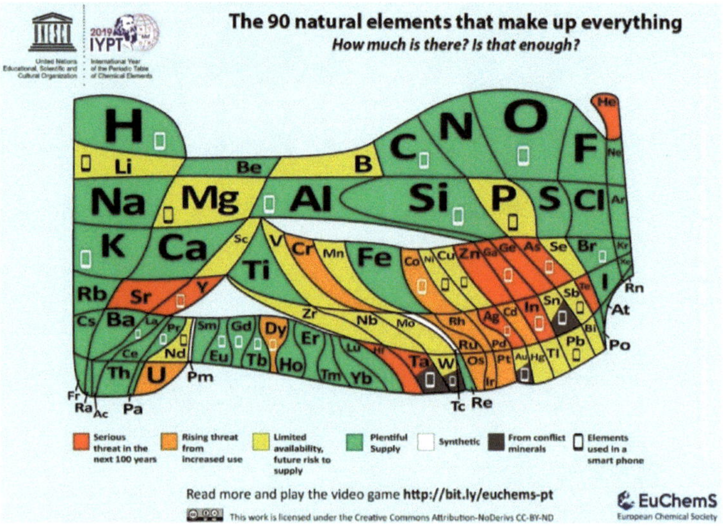

Figure 6.2 Periodic table produced by the European Chemical Society (EuChemS) to highlight element availability and vulnerability as part of the celebrations of the United Nations Educational, Scientific and Cultural Organization (UNESCO) International Year of the Periodic Table [20]. The area occupied by each element is proportional to the amount of that element in the Earth's crust and in the atmosphere. This shape is based on an original idea of J.F. Sheehan [21], as modified by K.A. Carrado [22]. Notice that Sn, Ta, W, and Au are labeled as "from conflict minerals."

Although, unlike their name suggest, rare earths are not that rare in the Earth's crust, their supply is still limited, mainly coming from China [23, 24]. As the old saying goes "if it's not grown it's mined." Metals are critical to modern technology and many are either uncommon, difficult to extract or in unstable regions, see for instance Fig. 6.2, where four elements are partly painted black: tin, tantalum, tungsten, and gold. This is to indicate that the components that come from mines where mineral rights battles are fought over, or where mining profits finance wars [17]. Mainly in the mines of Democratic Republic of the Congo, we must be cautious to ensure that the samples used in production or in our research laboratories do not originate from these mines. Traceability is the mechanism by which ores from the mine through all the processing phases to the final manufacture of a device are tagged and tracked during their lifetime. It is also extended, after the end of the useful life of the unit, to the fate of that feature. It is time to ensure that we do not have elements in our pockets from areas where people have died to remove them.

Another important aspect to consider is the current dependency to certain elements, for instance, there is no scarcity of palladium catalysts available to a chemist planning a Suzuki cross-coupling [25]. The importance of the C–C bond formation in catalyzed cross-coupling reactions was bestowed with the Nobel Prize in 2010 by Heck, Negishi, and Suzuki. Unfortunately, other elements cannot easily provide the same catalytic activity. Although, nickel will continue to be studied as an affordable replacement for palladium, Ni can also supplement to other metals with a variety of inherent properties [26]. In fact, the chemical industry despite benefitting from advances in improved catalyst efficiency produces not only useful supplements, pharmaceuticals, flavors, and pesticides, but also a considerable amount of waste. It is especially the case of pharmaceutical and fine chemical manufacturing, where only a fraction of the volume of waste and unsaleable by-products of synthesis may be the volume of the desired product. The continuous work of researchers in the area for instance, Pérez-Ramírez and his group, is crucial as they have managed a new catalyst that not only cuts the costs of synthesizing fine chemicals, it also reduces the consumption of palladium and decreases the amount of waste [27] (Fig. 6.3).

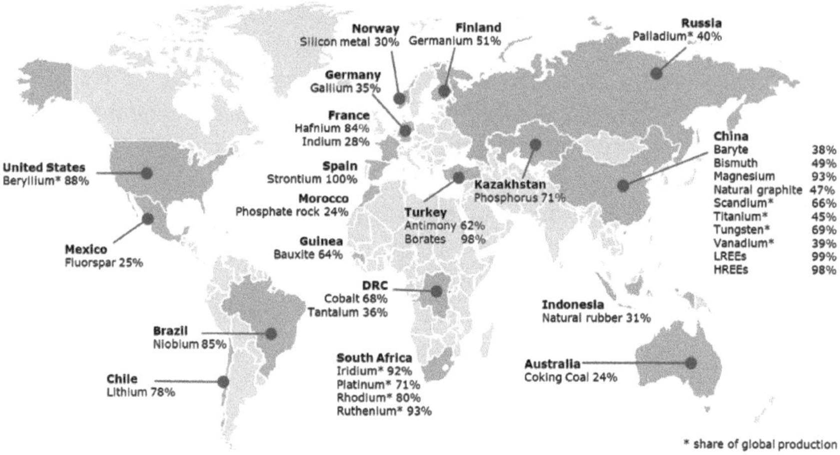

Figure 6.3 Biggest supplier countries of Critical Raw Materials (CRMs) to the EU [28].

We understand that the pandemic of COVID-19 impacts everyone's everyday lives and the global science community in unparalleled ways. But, it is unfortunate that a pandemic is needed to highlight the alert many have been making in Europe since many years: there is a lack of essential mineral resources and processing plants that have been commercially established and a dangerous over-reliance on imports from small overseas sources. The EU list of Critical Raw Materials (CRMs) is updated every three years since first being published in 2011, describes 30 essential raw materials (see Table 6.1) [29]. In fact, this year's table has four additional entries, including lithium. This alkaline metal is important to the transition to electromobility. Access to vital raw materials used in digital and clean technology was labeled as a "strategic security concern" in the European Green Deal [29].

New industrial policies in the coming years will need to address the issue of implementing circular economies for CRMs and ensuring sustainable exploitation of the resources available. Considering that in the next 5–10 years far from reaching a plateau, the demand for critical metals is expected to increase. This increase will be partially driven by the growth of green transport and renewable energy sources, and mostly due to an increased use in electronic devices. As we stated earlier:

phones, tablets, and computers require critical metals to function. Goodenough *et al.* [24] explain that In and Ga are used in touch screens while rare earth elements (REEs), such as Ce are used to polish the screens. LCD screens and light-emitting diode (LEDs) rely on semiconductor technology being made with a variety of metals, such as La, Ce, Eu, Gd, Y, Ga, and are doped with P (n- and p-doping). Circuit boards, hard drives, and permanent magnets use metals, such as Nd, Ta, Ga, and Ge whereas fiber optic internet cables require Eu.

Table 6.1 2020 Critical Raw Materials (new as compared to 2017 in bold) [28]

Antimony	Hafnium	Phosphorus
Baryte	Heavy rare earth elements	Scandium
Beryllium	Light rare earth elements	Silicon metal
Bismuth	Indium	Tantalum
Borate	Magnesium	Tungsten
Cobalt	Natural graphite	Vanadium
Coking coal	Natural rubber	Bauxite
Fluorspar	Niobium	Lithium
Gallium	Platinum group metals	Titanium
Germanium	Phosphate rock	Strontium

As famously stated by former UN Secretary-General Ban Ki-moon, "There can be no Plan B because there is no planet B." The future is not written and possibly if we cannot find enough metals on Earth, we will have to begin searching the rest of our solar system, perhaps starting with asteroids [30]. The technological jump required for space mining to be a reality is still far, and again will be pushed by both greed and curiosity. "Any term that suggest faster-than-light travel remains at the level of speculation" [31], still this reality has been dreamed by many. David S. Mackay a pioneering Moon [32] and Mars [33] scientist. He served as the chief scientist for Astrobiology and Planetary Science and Exploration at the NASA Johnson Space Center (1996–2013) and was the man who taught geology to the first humans that have walked on the Moon during the Apollo mission, said back in 1999:

"Near-future launch costs of about $600 per kilogram for the trip from Earth's surface to orbit, combined with space ferries cycling between high-earth orbits and nearby asteroids (returning, over their lifetimes, 100 tons of materials for each ton of equipment launched from the Earth), suggest future supplies of raw materials in near-earth space at a cost of a few dollars per kilogram. This is comparable to the cost of a house on the Earth." [34]. But before sustainable asteroid mining becomes a reality [35], we need to focus in what is in front of us.

The current pandemic has accelerated the digitalization of the society, work, play and stay connected are a must. In fact, the pandemic has speeded up our technological use and this same technology can accelerate us out from this crisis. We need to engage to promote a societal change, engage as many researchers as possible in the development of a new world of elemental science and share our vision. In this book chapter, we hope to inspire new researchers to work in the area of metal oxides and learn how to recycle better and more, primary raw materials. As we move to a recovery phase, our aim is to map new opportunities for researchers to join in the quest of finding alternative routes to memory storage materials.

6.3 Introduction: Building the Nanoworld

POMs are inorganic metal-oxide clusters with the general formula $[MO_x]_n$ (M = Mo, V, and W, x = 3–7, n = 6–368) which have the unusual ability to self-assemble discrete components into larger, organized structures [36]. They generally contain Groups 5 or 6 TMs in their highest oxidation state, multiple oxides, and often an intrinsic heteroatom anion housed in a tetrahedral cavity [37]. The high oxidation state of their metals also makes them remarkably good reducers, with little to no change observed in their overall structural geometry [38]. POM clusters can further combine to form larger POM-based frameworks. Their ability to carry out "one-pot" synthesis pathways for self-assembly and general reaction purposes is one of, if not the most, useful properties of POM structures as it is vastly superior to complicated multi-step synthesis and is what makes them such an interesting and relevant field of chemistry [39].

When it comes to synthesizing increasingly small molecular machines and structures, there are generally two schools of thought: the bottom up and the top down approaches [39]. Bottom up concerns itself with understanding how to build structures on a very small scale (for POMs this range is generally 0.5–5.5 nm), beginning with influencing individual atoms, and then moving on to the construction of bigger and more complex projects as our knowledge of the fundamentals increases and improves [37]. Top down, on the other hand, is the process of understanding how to scale down current "macro technology" (smaller than 1 μm). It differs from bottom up in that it revolves around learning how to reconcile our current electronics with the inevitable quantum effects they will experience as their size decreases. There currently exists a "no-man's land" where both approaches have yet to venture into, known as the "Nanoworld" [39]. This region of space, ranging from 1 to 1000 nm, is too big for "bottom-up" chemists and too small for "top down" engineers to build complex structures within at this point in time. The Nanoworld is therefore characterized by a "critical size" at which both approaches run into difficulties. For a chemist wanting to build large molecular structures, they will have to consider how to control atom flow into reaction sites due to the possibility of structures developing less rigid, larger pores as a direct result of unwieldy size. This will have a direct effect on the structures ability to regulate access to critical reaction sites. A successful example of self-assembly control and miniaturization technology was in linking two ring-shaped molecules together to form a chain, referred to as a catenane [40]. This has culminated in the 2016 Nobel Prize for Chemistry being awarded to Sauvage, Stoddart, and Feringa for their work in the field of molecular nanomachines [41]. Their research focused on creating ring-shaped molecules that could assemble after binding to specific target molecules, yielding interlocking chains.

It is worth noting that there has been a lot of research in recent years into determining position and effects of critical sizes and those POMs in general have been garnering attention for making strides in breaching the Nanoworld. A specific example involves recent development of a "flexi-crystal" which is a POMzite/large multi-POM network capable of changing size by alteration of thermal conditions without the associated sacrifice in rigidity [42]. POMs, otherwise known as molecular metal oxides, have been known of for centuries but until the last hundred years our understanding of chemistry and engineering simply was not advanced enough to allow for sufficient analysis into their structures [39]. Berzelius and co-workers were the first to describe a POM, the formula of which was determined to be $(NH_4)_3[PMo_{12}O_{40}]$aq. almost a century later, in 1826. There were several attempts to describe the structural bonding of phosphotungsto- and phosphomolybdenum salts from 1900 to 1923 until Keggin finally solved the structure as $H_3[PW_{12}O_{40}]$, which is strikingly similar in formula and appearance to current POM models (see Fig. 6.4). The existence of POM isomers has been known since the 1950's but it was not until the development of single-crystal X-ray diffraction and electrospray ionization-mass spectrometry (ESI-MS) that they could be accurately analyzed and categorized.

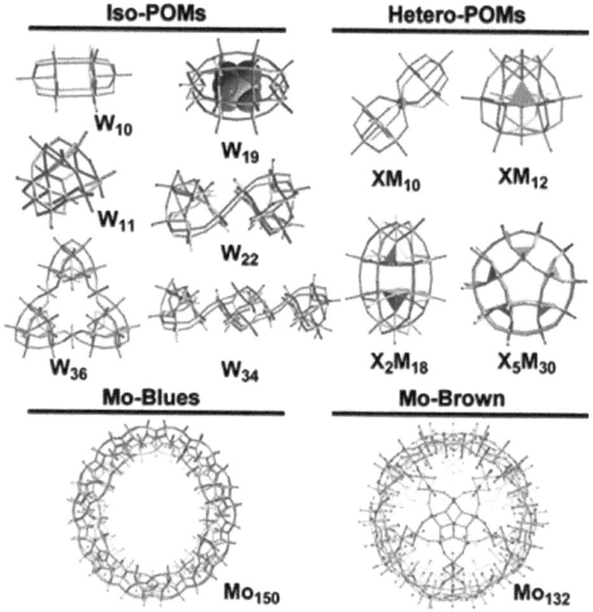

Figure 6.4 Comparison of $H_3[PW_{12}O_{40}]$ structure developed by Keggin and co-workers in 1923 (left) and modern Keggin (α-$[\{XO_4\}_2M_{12}O_{36}]^{n-}$) structure which is treated as most accurate model to date (right) [44]. Similarities include: heteroatom located at center of POM, several distinct triad faces capable of rotation and variation in inter- and intra-triad oxometalate ions.

Since then, models of structures have become more precise and, coupled with advancements in computing, have allowed for not only a greater understanding into POM cages but have also opened the door to tackling new questions, such as precisely how ligands react with POM pores on a molecular level and why different ligands affect POMs in different ways [38]. Many such structures have been identified, many more of these with multiple isomers of varying stabilities and features. Indeed, POM isomerism is one of the most described but least understood aspects of POM chemistry. POMs are set to make great contributions to nanotechnology and have applications that span multiple fields, such as medicine, catalysis, electrochemistry, and electro-optics [43]. More specific examples of their usefulness include: increasing flash drive storage capabilities, providing small-scale cell-like environments for biologists, and trapping harmful gasses, such as CO_2 and CO [37].

6.4 Molecular Metal Oxides: POMs, Synthesis, and Structural Features

POM synthesis generally proceeds via acidification of simple oxo anions, followed up by a condensation reaction with TMs to form the desired M–O–M bonds [37]. Whilst this process of making POMs is generally well accepted, the multiple variables that control isomer selection are overall poorly understood. Self-assembly is very sensitive to reaction conditions, such as temperature, pH, pressure, solvent choice, POM metal identity and concentration, POM heteroatom identity and concentration, presence of (additional) ligands, and presence of a counter ion to name just a few [43, 45]. A slight change in any one variable may yield a completely different and undesirable structure. The process is especially hard to understand from a chemist's point of view, as there are many different fields and variables involved that transcend the already sprawlingly large field of chemistry (by this we are referring to the fact that POM research must consider the fields of physics and computer simulation as well as those of organic and inorganic chemistry).

Figure 6.5 Example oxometalate ion $(XO_4)^{n-}$ fundamental building block with universal oxometalate ion properties. The yellow ball represents the metal atom (X) and the red spheres show the tetrahedral position of the oxygen atoms (O).

The major down-side to one-pot synthesis is that it can produce unintended side-products and have fairly chaotic results [37]. In order to guarantee synthesis of specific species, the reactants must be fined-tuned prior to the reaction. Oxidized TMs are very useful in this regard as their properties are somewhat determined by the materials they are bound to, allowing POMs to be a viable choice for one-pot synthesis as they are a lot easier to tune pre-reaction than many other species. Despite their unique relationship with one-pot synthesis, POM discovery is understandably random. This is evident in Müller's discovery of naturally-occurring molybdenum blue family of compounds [39]. An immediate downside to this method of discovery is that it may be hard to track the effects of all the variables that lead to structure selection, resulting in difficulties with reaction pathway reproduction. Once the pathway has been identified, however, new POMs can be used as a jumping-off point to discover more [37]. Whilst there are many different structures that fall under the banner of POMs, they all follow the same general "blue-prints" for assembling themselves or their fundamental building-blocks.

The archetypical POM building-block usually consists of a central heteroatom (the exception being isopolyoxometalates (isoPOMs)) of the form XO_n (X = heteroatom, n = 3, 4, 6. A common example is $[PO_4]^{3-}$) that serves as the core around which the oxometalates can arrange themselves and form the POM cage. The specific metal element used for the oxometalate units also has a great impact on the nature of the POM (see Fig. 6.5). Mo-based clusters, for example, develop exceptional structural flexibility upon reduction [11]. Mo is also a particularly useful metal for POM synthesis as it is one of the only atoms that displays self-assembly under acidic conditions and Mo reduction is associated with increased electron density on the surface cluster, resulting in protonation [39]. This protonation/pH control is associated with POM growth and is therefore a major factor in controlling POM synthesis. By comparison, W-based clusters are characteristically very strong and reliably yield large multi-POM structures in a way that many other POMs cannot. Pore size and chemical nature is responsible for control over entry of smaller molecules into the cavity. Pore volume and shape are easy to alter, which is essential if accurate control over molecular access to active sites and pores in enzyme-like and cell-like POMs, respectively, is to be implemented [37].

IsoPOMs are the least stable POMs due to the absence of an intrinsic anion [37]. Their usefulness arises from their unique properties, such as a high charge, a very basic oxygen surface, and being able to form the smallest POMs yet observed. Due to their small size and subsequent simpler chemistry, they are routinely used to test out hypotheses concerning the properties of larger structures. The Lindqvist iso anion ($[M_6O_{19}]^{n-}$), composed is of only 6 oxometalate units, being the smallest isopolyanion reported to date [45]. A species of Lindqvist, $[H_xNb_6O_{19}]^{8-x}$, has been utilized in experiments aimed at understanding aqueous reactions involving minerals and other extended structures [37]. IsoPOMs also serve as good building blocks for the formation of larger POM-frameworks, mainly due to their very stable nature, especially compared to larger POMs, such as heteropolyoxometalates (HPOMs).

HPOMs generally feature two redox inactive tetrahedral anions (usually PO_4^{3-} or SO_4^{2-}) symmetrically aligned along the C_3 axis [36, 37, 46]. These POMs can be referred to as classical-POMs as they are the most frequently encountered POM structures. They are associated with a greater degree of stability than IsoPOMs thanks to their anion. Tungsten-based POMs in particular are renowned for their strength and are used to synthesize Keggin- and Wells–Dawson (WD)-like structures with a more than standard number of vacancies [37]. Their intrinsic anion also bestows upon them an overall negative charge, allowing ESI-MS to yield a more accurate structure of the POM.

Although a few HPOMs may share an identical structure, for example, Wells–Dawson, they may differ in their redox capabilities and relative isomer stabilities. These HPOMs can be characterized as classical or non-classical, depending on the redox capabilities of their heteroatom. Using WDs as an example, classical-WDs contain at least one anion of the form XO_4 whilst non-classical WDs contain either a XO_3^{n-} or XO_6^{n-} anion (see Fig. 6.6) [36].

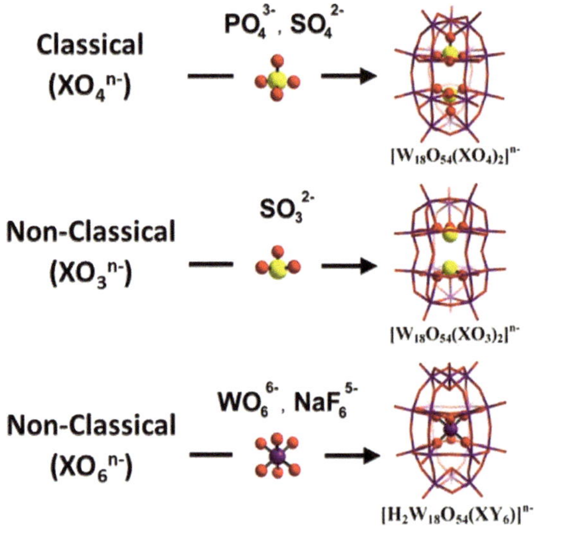

Figure 6.6 Classical and non-classical POMs and their associated heteroatoms. The top graphic (XO_4) represents a classical POM whilst the middle and bottom models (XO_3 and XY_6, Y = O, F), display the two non-classical POM structures. Note that while all POM cages are colored purple for metal ion positions and red for oxygen atoms, the heteroatom for the XY_6 POM is purple, which contrasts with the yellow spheres used to represent the heteroatoms in the other two structures.

Both classical and non-classical WDs are redox active thanks to their oxometalate units, which readily undergo reduction.

The difference is that non-classical WDs feature a redox active heteroatom, whereas classical WD heteroatoms are redox inactive and, therefore, only their oxometalate units undergo reduction. Non-classical heteroatoms are more redox active than the oxometalate units, so will be reduced (I, Te) or oxidized (Se, S) first. One interesting application of non-classical HPOMs is in the field of flash memory [12]. Oxidizable non-classical HPOM heteroatoms form an internal X–X bond, meaning the presence or absence of charge can be used to represent information. This is akin to how binary can be used to communicate data in the form of 1 or 0 or how Morse code uses on or off to transmit communications. The issue with current MOS is that they are hard to scale down past about 10 nm and into the nanoworld. Non-classical HPOMs, specifically those with WD structure, can overcome many of the technical difficulties experienced by alternative solutions as they are small (heights are typically around 1.5 nm), have a large charge range, feature a charge and heteroatom that are easy to alter, and display high thermal stability (storage devices must be able to withstand temperatures of up to 600 K). The internal heteroatom bond described earlier allows the POM to act as a floating gate within the flash drive, allowing it to store and release information in the form of electrons when a voltage is applied. Density functional theory (DFT) calculations and experimental data indicate that $[W_{18}O_{54}(SeO_3)_2]^{4-}$ is the ideal HPOM as W provides the appropriate thermal stability and Se has good general stability, ideal electronic activity, and only achieves an oxidation state of 5+ when the Se–Se bond is formed, allowing for easy detection of oxidation.

6.5 Magnetic Materials

The field of POMs is now taking the next step whereby both the structure and function of the systems can be planned and regulated [47]. However, due to their dynamic existence, with an almost infinite structural diversity, means that the assembly under non-equilibrium conditions of functional nano-molecules and adaptive materials will be possible. This method will be used to access new libraries of building blocks that will contribute to the development of new nano-material structures and functions that are not accessible from near-equilibrium processing techniques and will concentrate on the production of new products, assemblies and devices. In order to drive, direct and trap the self-assembly of molecular metal oxide dependent building blocks, clusters, and materials in solution, we must acknowledge that such processes can be driven by redox reactions, ion exchange, metal unit substitution. By using such non-equilibrium dependent manufacturing, with traditional, static, near equilibrium self-assembly techniques, it would be possible to target at engineering materials with unparalleled structural functionality and adaptive capacity as possible.

Ferromagnetic materials have dominated the field of information storage from its foundation to the present day [48]. At the turn of the 20th century, ferromagnet-based memories, such as magnetic tapes and hard disk drives, allowed by developments in semiconductor technology [48], gave rise to the information era. Spin-based electronics (or spintronics) came to light in this sense in the 1980s [49], offering significant advantages in terms of speed, heat dissipation and energy consumption. The molecular magnetism field seeks to synthesize compounds with magnetic properties that are tailor-made [7]. First, this ambitious objective involves the creation of theoretical models capable of making predictions that are accurate and practical. But the foundations of these models need to explain the magnetic properties of already synthesized compounds before reaching this ideal stage.

In 1998, POMs were described in molecular magnetism as model artifacts or magnetic elements of hybrid molecular materials by Coronado *et al.* [50]. In a 2009 study by Kortz *et al.* [51], their importance to molecular magnetism is also discussed. POMs are ideal candidates as supporting ligands to design magnetic core structures, considering their wide range of structural and electronic properties. POMs give a well-defined rigid environment that restricts the alignment of spin centers and regulates the coupling of intra-molecular exchange. For example, POM ligands induce strong axial magnetic anisotropy, exemplified by $[As_2W_{20}Mn^{II}O_{68}]^{8-}$ [52]. This Mn(II)–POM has the highest axial zero field splitting (ZFS) parameter to date for an isolated Mn^{II} anion(D1/4 + 1.46 cm^{-1})

is a simple example of a POM as a suitable candidate for a single-molecule magnet (SMM)-based molecule. Another example comes from a recent paper where it was demonstrated that a Cr^{III} cation could be "locked" into a high spin state, thus creating a permanent SMM, by enclosing the cation within the tetrahedral cavity of a recently discovered Keggin structure [53]. In a 2020 by Ammari et. al., they created a 1D coordination polymer containing Co cations and $(C_6H_{10}N_2)_2[Co(H_2O)_4P_2Mo_5O_{23}].6H_2O$ anions [54]. In their work, the Co cations are locked in place by the POMs, resulting in long range ferromagnetism. Combined with its ability to act as a semiconductor, polymers of this type would be ideal for creating new spintronic materials. As outlined in a review by Coronado *et al.* [55],three main areas were based on recent developments in magnetic aspects of metal oxide materials:

(1) POMs that encapsulate magnetic clusters (with variable nuclearities and high symmetries) of transition metal ions connected through oxo bridges, enabling the magnetic exchange interactions between the metal (mainly cobalt) ions.

(2) POMs that encapsulate lanthanoid ions are the most successful examples of SMMs. The most remarkable example is the POM $[Ho(W_5O_{18})_2]^{9-}$ (abbreviated, HoW_{10}), which is able to enhance coherence in solid–state molecular spin qubits without resorting to extreme dilution and a relative insensitivity [56].

(3) Mixed-valence POMs, which host several electrons throughout the whole POM framework, usually by hopping.

(4) Heterometallic 3d–4f POMs: an additional advantage of POMs as ligands in which molecular magnetism stems from their diamagnetic structures which can encapsulate paramagnetic metal ions that become well-separated from each other leading to an ideal magnetic insulation. POM-based SMMs based on either 3-d or 4-f metal ions [57, 58].

Research in POM molecular storage devices will generate significant interest in the area of materials design and modeling (see Fig. 6.7). Results will improve synthetic access to metal oxide materials combined with libraries of new building blocks and the proposed way of classifying the resultant structures, silica POM-based memory devices, are emerging as an interesting alternative to pure silicon oxide devices. We believe that the development of POM magnetic materials may enter a new era. The result will be that the familiar and technologically vital metal oxides, designed using topological bottom up approaches to form large-paneled structures, will enter the realm of the digital age. By developing this we can promise that POM-based materials to be competitive in device storage. This ambitious aim involves the creation of theoretical models capable of making accurate and practical predictions. The foundations of the models, therefore, must explain the magnetic properties of synthesized compounds before they reach this ideal stage. POMs provide a rigid, well-defined environment that restricts the alignment of spin centers and regulates the coupling of intramolecular exchange. The use of POMs as nanoscale molecular memory elements offers a fundamentally better electronic complementarity with SiO_2 than organic molecules due to their oxidic nature.

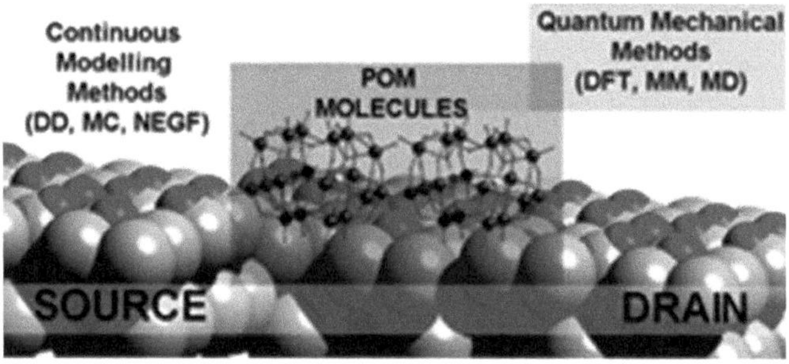

Figure 6.7 Schematic representation of a molecular nanowire sensor and memory cell, combining continuous and quantum modeling methods.

Beyond memory devices, POM magnetism can also be exploited in a diversity of applications, for instance to easily remove POMs from solution in the presence of an external magnet. This method of separation is a lot easier and faster than other methods, such as filtration. The superparamagnetic nature of $CoFe_2O_4$/MIL-88B(Fe)-NH_2/(Py-Ps)PMo, in particular, makes it an ideal magnetic catalyst [59]. Misra *et al.* were able to utilize this technique in a recent paper to purify water samples of impurities, such as heavy metals and microplastics [60]. These magnetic POM-SILPs are good at both attracting and absorbing a wide range of impurities and can be removed from solution via magnetism due to their Fe_2O_3 core. Magnetic water purification, as opposed to filtration, could help to handle large volumes of water and, if removal of particles by simple permanent magnets is possible, can generally be used without a further infrastructure. Magnetic effects can also be developed by maintaining the stability and arrangement of molecules/atoms within a structure. Breitwieser *et al.* demonstrated that inclusion of POMs within a binary superlattice (DDA-$\{Mo_{132}\}$/OA-γ-Fe_2O) can be used to prevent aggregation during annealing at temperatures from 643–723 K and therefore can effectively maintain a degree of structural organization [61]. As the temperature is increased up to the limit of 723 K, the distance between the POMs present in the superlattice progressively decreases, resulting in a greater dipolar magnetic interaction/repulsion between individual POMs.

6.6 Metal Oxides in Memory Devices

In 1973, when Aviram and Ratner speculated about organic molecules as elements in electronic circuits, the idea of molecular electronics was born [62]. Since then, the emerging field of molecular electronics has typically used switchable organic molecules anchored between nano-electrodes, such as rotaxane-based molecular memory arrays, which demonstrate the first molecular-based RAM [40, 63]. However, organic molecular electronics raises questions of high resistance and power, low performance and problematic fabrications from an engineering point of view.

The idea of a POM-based memory device was presented on two ground-breaking concept papers published in 2006 and 2013 by Cronin and co-workers, were they described their unique characteristics and their great potential as elements of molecular memory [64, 65]. This concept was not only very well received, but it also made the community think in a new way about how nanosystems might look in the future [66, 67]. Today, time has come to evolve toward the next generation of nano-electronics devices. As devices shrink and the molecular/atomic level approaches, vast problems with variability and lifetime are becoming more a reality. In 2014 [20], we presented a new approach to the design of flash-RAM incorporating POM clusters within the floating gate to enhance the performance parameters. Our vision was to propose the potential incorporation of POMs as elements into a flash memory cell. Wells–Dawson POMs are a family of metal oxide clusters mainly formed by Mo^{6+} and W^{6+} combined with a main group oxyanion (phosphate, silicate, etc.).

The ideas and results presented in that work show that POMs, as redox-active elements, offer great advantages over the conventional polysilicon memory devices and should engage other researchers in the chemical sciences to develop electronic devices based upon POMs. In the paper, previously published electrochemical data were combined with a theoretical preliminary electronic structural description, by means of DFT, and device modeling in order to evaluate the suitability of POMs as a non-volatile molecular memories (NVMM). Given the multidisciplinary nature, ambition and complexity of the project, this work was a breakthrough in current applications in the field of metal oxides. Projects interested in developing this conceptual research into the next generation of memory devices continue to be carried out by several research groups [61, 67–70].

Although the application of POM memory drives is still in its infancy, there have already been some new projects based on the original "Dawson-like" POM designed by Cronin and colleagues, such as using a Keggin-like structure as part of a larger resistive random-access memory (RRAM) device [71]. One of the downsides of working with quantum machines is that it often presents a whole range of novel problems that need to be considered. A prime example is that of magnetic noise, which can interfere with the POMs' ability to store data [72]. There is no doubt as that as the field progresses there will be more quantum effects that will have to be addressed before quantum processors can be utilized on a wide-scale.

By embedding redox active units via core-shell POM clusters Cronin and co-workers explored that embedded Se, Te, or P as "p-like" oxidative dopants into the cluster core allowing the oxidation of the cluster to a radicals of Se, P or Te with $S = 1/2$. Measuring ca. 3 nm from tip-to-tip, the cross-bar-clusters include different dopant heteroatoms with the formula $[(XYW_{15}O_{54})_4(WO_2)_4]^{32-/36-}$ (X = HP^{III}, Y = Se^{IV}, Te^{IV}, As^{III}) [73]. See Fig. 6.8, in that work DFT theoretical calculations explained the self-selection of dopants and their effect on the electronic properties, while device-models demonstrated transport properties of charge through the clusters that is not only modulated by the heteroatoms, but also by the orientation of the cluster between the electrodes. Through the study of single molecules and their electron and energy states, we will investigate whether electron spin or multi-electron state charging manipulation is feasible for memory and logic applications. Our previous research has shown that four nanoscale molecular metal oxide clusters with a cross-bar architecture can be self-assembled. These preliminary results describe a roadmap toward a new type of electronic component that combines cross-bar-cluster architectures of CMOS-compatible molecules with semiconductors and the dimensions of the molecules are comparable to the smallest lithographic dimensions manufactured, potentially driving single binding molecules [74].

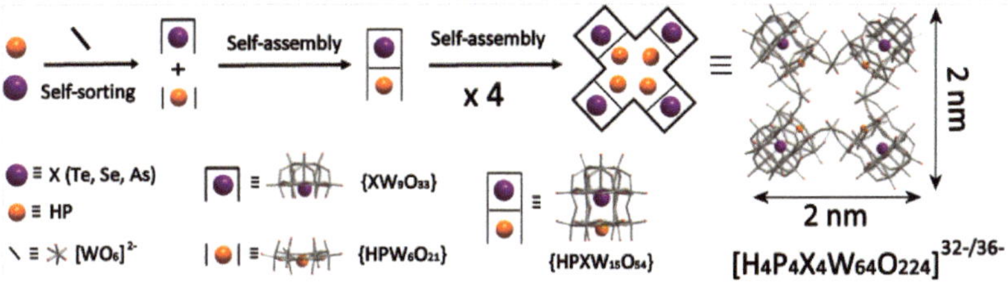

Figure 6.8 Molecular structure of the nano-metal oxide crosses with embedded hetero-atoms showing self-sorting and self-assembly of heteroatom dopants templated nano-scale cross-bar polyoxoanions $[H_4P_4X_4W_{64}O_{224}]^{32-/36-}$ (X = Te, Se, As, P) [73] (Reproduced with permission from [73]. Copyright 2018 American Chemical Society).

6.7 Future Challenges and Research

The design of the interfaces is crucial to stablish POM-electronics and whereby controls the operation, hence top-down control of architectures at the nanoscale is essential. However, the bottom-up self-assembly of molecules with the right properties has the potential for devices to be made. In addition to having the correct properties, molecular metal oxide clusters, designed by main-group dopant heteroatoms, have been shown to be consistent with CMOS architectures. Although their structures are limited to elliptical or spherical forms [12, 68, 69]. Through the development of a new sensor technology aimed at the electronic characterization of single and few molecules, the approach will enable direct testing of the electron states of each molecule. This technology will be used to demonstrate multivalued flash memory cells, an area of major industrial interest, to increase the density and decrease flash memory power consumption, and to explore the potential of specific molecules for spintronic and multi-electron states. In the coming years, we should address the following issues to push forward the field of POM memory devices:

Immobilizing the POM in the surface by developing binding groups on the molecules to allow selective placement of single molecules into defined positions in devices defined by Si technology (CMOS) processes [74]. The molecules need to be sufficiently large, by means of controlling their geometries and/or charged states to repeal other molecules and aid single molecular placement.

Test the quantum states on single or a few POM molecules. This will allow for single POM molecule flash memories with single and multi-value memory states. We will need to develop techniques that will allow the precise placement of single molecules in Si technology to allow single molecular devices to advance in this area.

Deposition of single molecule POMs on the substrate could be achieved by a variety of approaches (see Fig. 6.9). For instance, non-covalent deposition of the POMs onto surfaces by absorption from solution [47], Langmuir–Blodgett deposition [75], and electrostatic layer-by-layer self-assembly (ELSA) [76]. Given the inherent anionic nature of POMs, they are ideally appropriate for the ELSA technique. By means of deposition with positively charged species it is possible to create alternative films easily layer-by-layer.

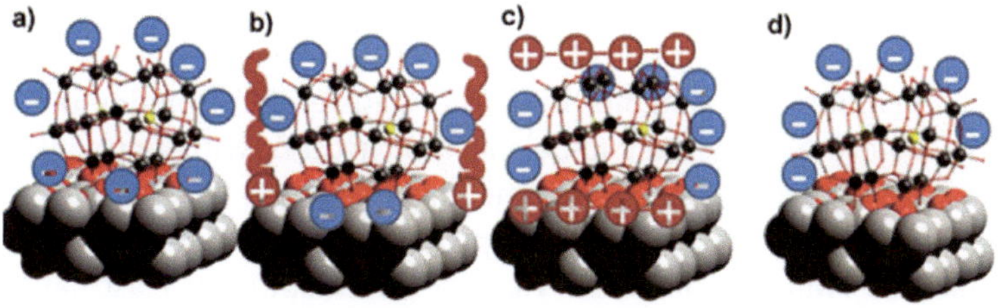

Figure 6.9 Examples of controlled (a) adsorption of POMs into a SiO_2 surface: (a) adsorption; (b) Langmuir–Blodgett deposition; (c) electrostatic layer-by-layer self-assembly (ELSA); and (d) covalent bonding of POM to SiO_2 surface.

6.8 Conclusion

The discoveries that cluster structures are centered in a hierarchy of templates and templating subunits have yet to be explored, are part of the main aspects of the latest discoveries at the boundaries of metal oxide cluster research. It has only just emerged, more and more, that POM -based reaction networks can be treated as complex chemical structures that contain inter-dependent self-assembly, self-template building blocks. The archetypal model for the exploration of inorganic chemical networks can also be complex interactive "systems" as describe with POM building blocks. POMs now enter a new process in which the structure and function of the systems can be planned and regulated. However, their dynamic existence, with an almost infinite structural diversity, means that the assembly under non-equilibrium conditions of functional nano-molecules and adaptive materials will be created.

In addition to the importance of creating catalytically active or multifunctional materials that have multiple functions encoded into a single material, previous studies have shown. A combination of a cluster capable of redox activity, a cluster capable of binding to a small molecule, and a cluster capable of photochemical activation, for example, could create a material with all these characteristics. The challenge is to choose the ideal POM and metal-ion connector, as well as to design their assembly in an effective and high-yielding way. We trust that the structural explosion of POMs will now lead to an explosion of functions that benefits from the transferable building-blocks, the ability to construct unbalanced structures with unparalleled characteristics and emergent functions.

We need to be aware of proposing new memory devices that use either metal oxides that are abundant or at least have in place recycling mechanisms. We could be developing the best single molecule device, but its cost could be so prohibitive that the scaling up and incorporation into real life nanoelectronics will never be possible. Academic research often aims for proof of principle techniques that could potentially improve the technology of the future. Memory modules include priceless metals and other elements which need to be disposed of in controlled ways. Perhaps it is time to develop better and more transparent recycling strategies for memory devices.

Acknowledgments

The authors would like to thank the University of Glasgow, WestCHEM, the EPSRC, the Royal Society for financial support as well as members of the Complex Chemistry Section in the School of Chemistry at the University of Glasgow, past and present.

References

1. Schwerdtfeger, P., Smits, O.R. and Pyykkö, P. (2020). The periodic table and the physics that drives it, *Nat. Rev. Chem.*, **4**, pp. 359–380.
2. Frey, P.A. and Reed, G.H. (2012). The ubiquity of iron, *ACS Chem. Biol.* **7**, pp. 1477–1481.
3. Lyons, T.W., Reinhard, C.T. and Planavsky, N.J. (2014). The rise of oxygen in Earth's early ocean and atmosphere, *Nature*, **506**, pp. 307–315.
4. Brocks, J.J., Logan, G.A., Buick, R. and Summons, R.E. (1999). Archean molecular fossils and the early rise of eukaryotes. *Science*, **285**, pp. 1033–1036.
5. Železný, J., Wadley, P., Olejník, K., Hoffmann, A. and Ohno, H. (2018). Spin transport and spin torque in antiferromagnetic devices. *Nat. Phys.*, **14**, pp. 220–228.

6. Coey, J.M.D. (2019) Magnetism in d^0 oxides, *Nat. Mater.*, **18**, pp. 652–656.
7. Cinchetti, M., Dediu, V.A. and Hueso, L.E. (2017). Activating the molecular spinterface. *Nat. Mater.*, **16**, pp. 507–515.
8. Izarova, N.V., Pope, M.T. and Kortz, U. (2012). Noble metals in polyoxometalates, *Angew. Chemie – Int. Ed.*, **51**, pp. 9492–9510.
9. Misra, A., Kozma, K., Streb, C. and Nyman, M. (2020). Beyond charge balance: counter-cations in polyoxometalate chemistry, *Angew. Chemie – Int. Ed.*, **59**, pp. 596–612.
10. Paille, G., Gomez-Mingot, M., Roch-Marchal, C., Lassalle-Kaiser, B., Mialane, P., Fontecave, M., Mellot-Draznieks, C. and Dolbecq A. (2018). A fully noble metal-free photosystem based on cobalt-polyoxometalates immobilized in a porphyrinic metal–organic framework for water oxidation, *J. Am. Chem. Soc.*, **140**, pp. 3613–3618.
11. Vilà-Nadal, L., Sarasa, J., Rodríguez-Fortea, A., Igual, J., Kazansky, L. and Poblet, J. (2010). Towards the accurate calculation of 183W NMR chemical shifts in polyoxometalates: the relevance of the structure, *Chem. – An Asian J.*, **5**, pp. 97–104.
12. Busche, C., Vilà-Nadal, L., Yan, J., Miras, H.N., Long, D.-L.,. Georgiev, V.P., Asenov, A., Pedersen, R.H., Gadegaard, N., Mirza, M.M., Paul, D.J., Poblet, J.M. and Cronin, L. (2014). Design and fabrication of memory devices based on nanoscale polyoxometalate clusters, *Nature*, **515**, pp. 545–549.
13. Weinstock, I.A., Schreiber, R.E. and Neumann, R. (2018). Dioxygen in polyoxometalate mediated reactions, *Chem. Rev.*, **118**, pp. 2680–2717.
14. Anyushin, A.V., Kondinski, A. and Parac-Vogt, T.N. (2020). Hybrid polyoxometalates as post-functionalization platforms: from fundamentals to emerging applications, *Chem. Soc. Rev.*, **49**, pp. 382–432.
15. Zheng, Q., Vilà-Nadal, L., Lang, Z., Chen, J.J., Long, D.-L., Mathieson, J.S., Poblet, J.M. and Cronin, L. (2018). Self-sorting of heteroanions in the assembly of cross-shaped polyoxometalate clusters, *J. Am. Chem. Soc.*, **140**, pp. 2595–2601.
16. Bo, C., Maseras, F. and López, N. (2018). The role of computational results databases in accelerating the discovery of catalysts, *Nat. Catal.*, **1**, pp. 809–810.

17. Cole-Hamilton, D.J. (2019). Elements of scarcity. *Chem. Int.*, **41**, pp. 23–28.

18. Editorial (2011). Elements in short supply. *Nat. Mater.* **10**, p. 157.

19. Slootweg, J.C. (2020). Using waste as resource to realize a circular economy: circular use of C, N and P. *Curr. Opin. Green Sustain. Chem.*, **23**, pp. 61–66.

20. Keijer, T., Bakker, V. and Slootweg, J.C. (2019). Circular chemistry to enable a circular economy, *Nat. Chem.*, **11**, pp. 190–195.

21. Sheehan, W. F. (1976). *Chemistry*, **49**, 17–18.

22. Carrado, K.A. (1993). Chemistry for kids – presenting the fun side of the periodic table, *J. Chem. Educ.*, **70**, pp. 658–659.

23. Nakamura, E. and Sato, K. (2011). Managing the scarcity of chemical elements, *Nat. Mater.*, **10**, pp. 158–161.

24. Goodenough, K.M. Schilling, J. Jonsson, E., Kalvig, P., Charles, N., Tuduri, J., Deady, E.A., Sadeghi, M., Schiellerup, H., Müller, A., Bertrand, G., Arvanitidis, N., Eliopoulos, D.G., Shaw, R.A., Thrane, K. and Keulen, N. (2016). Europe's rare earth element resource potential: an overview of REE metallogenetic provinces and their geodynamic setting, *Ore Geol. Rev.*, **72**, pp. 838–856.

25. Miyaura, N., Yamada, K. and Suzuki, A. (1979). Our continuous discovered. *Tetrahedron Lett.*, **20**, pp. 3437–3440.

26. Tasker, S.Z., Standley, E.A. and Jamison, T.F. (2014). Recent advances in homogeneous nickel catalysis, *Nature*, **509**, pp. 299–309.

27. Chen, Z., Vorobyeva, E., Mitchell, S., Fako, E., Ortuño, M.A., López, N., Collins, S.M., Midgley, P.A., Richard, S., Vilé, G. and Pérez-Ramírez, J. (2018). A heterogeneous single-atom palladium catalyst surpassing homogeneous systems for suzuki coupling, *Nat. Nanotechnol.*, **13**, pp. 702–707.

28. European Commission, (2011). Report on critical raw materials in the circular economy. Luxembourg: Publications Office of the European Union, 2011.

29. European Commission, (2020). Report on critical raw materials in the circular economy. Luxembourg: Publications Office of the European Union, 2020.

30. Elvis, M. (2012). Let's mine asteroids—for science and profit, *Nature*, **485**, p. 549.

31. Ford, L., and Roman, T. (2000). Negative energy, wormholes and warp drive, *Sci. Am.*, **282**, pp. 46–53.

32. Pieters, C.M., Taylor, L.A., Noble, Keller, L.P., Hapke, B., Morris, R.V., Allen, C.C., Mckay, D.S. and Wentworth, S. (2000). Space weathering on airless bodies: resolving a mystery with lunar samples, *Meteorit. Planet. Sci.*, **35**, pp. 1101–1107.

33. McKay, D.S., Gibson Jr., E.G., Thomas-Keprta, K.L., Vali, H., Romanek, C.S., Clemett, S.J., Chillier, X.D.F., Maechling, D.R., and Zare, R.N. (1996). Search for past life on Mars: possible relic biogenic activity in martian meteorite ALH84001, *Science*, **273**, pp. 924–930.

34. Interview David S. McKay of the NASA Johnson Space Center–October 21, 1990. *Scientific American*, [Last accesed: Novmber 6, 2020], available at: https://www.scientificamerican.com/article/ive-read-references-in-bo/.

35. Editorial. (2019). Sustainable space mining. *Nat. Astron.*, **3**, p. 465.

36. Vilà-Nadal, L. Mitchell, S.G., Long, D.-L., Rodríguez-Forte, A., López, X., Poblet, J.M. and Cronin L. (2012). Exploring the rotational isomerism in non-classical Wells–Dawson anions {W18X}: a combined theoretical and mass spectrometry study, *Dalt. Trans.*, **41**, pp. 2264–2271.

37. Vilà-Nadal, L. and Cronin, L. (2017). Design and synthesis of polyoxometalate-framework materials from cluster precursors, *Nat. Rev. Mater.*, **2**, pp. 1–15.

38. López, X., Carbó, J.J., Bo, C. and Poblet, J.M. (2012). Structure, properties and reactivity of polyoxometalates: a theoretical perspective, *Chem. Soc. Rev.*, **41**, pp. 7537–7571.

39. Gouzerh, P. and Che, M. (2006). From Scheele and Berzelius to Müller Polyoxometalates (POMs) revisited and the "missing link" between the bottom up and top down approaches, *L'acqtualité quimie*, **298**, pp. 9–22.

40. Mena-Hernando, S. and Pérez, E.M. (2019). Mechanically interlocked materials. Rotaxanes and catenanes beyond the small molecule, *Chem. Soc. Rev.*, **48**, pp. 5016–5032.

41. Press release: The Nobel Prize in Chemistry 2016. NobelPrize.org. Nobel Prize Outreach AB 2022. Fri. 21 Jan 2022. https://www.nobelprize.org/prizes/chemistry/2016/press-release/.

42. Zhan, C., Cameron, J.M., Gabb, D., Boyd, T., Winter, R.S., Vilà-Nadal, L., Mitchell, S.G., Glatzel, S., Breternitz, J., Gregory, D.H. Long, D.-L., Macdonell, A. and Cronin, L. (2017). A metamorphic inorganic framework that can be switched between eight single-crystalline states, *Nat. Commun.*, **8**, p. 14185.

43. Janusson, E., De Kler, N., Vilà-Nadal, L., Long, D.L. and Cronin, L. (2019). Synthesis of polyoxometalate clusters using carbohydrates as reducing agents leads to isomer-selection, *Chem. Commun.*, **55**, pp. 5797–5800.

44. Pope, M. and Kortz, U. (2012). Polyoxometalates. In. *Encyclopedia of Inorganic and Bioinorganic Chemistry*, Wiley and Sons, USA.

45. Vilà-Nadal, L. and Cronin, L. (2017). *Handbook of Solid State Chemistry*, Part 4. Nano and Hybrid Materials, Chapther 1. Self-Assembly of Molecular Metal Oxide Nanoclusters, Wiley-VCH, Germany, pp. 1–20.

46. Vilà-Nadal, L., Romo, S., López, X. and Poblet, J.M. (2012). Structural and electronic features of Wells–Dawson polyoxometalates, *NATO Sci. Peace Secur. Ser. B Phys. Biophys.*, pp. 171–183.

47. Song, Y.-F. (2018). *Polyoxometalate-Based Assemblies and Functional Materials.*, Spinger, Germany, p. **176**.

48. Dietl, T., Ohno, H., Matsukura, F., Cibert, J. and Ferrand, D. (2000). Zener model description of ferromagnetism in zinc-blende magnetic semiconductors, *Science*, **287**, pp. 1019–1022.

49. Žutić, I., Fabian, J. and Sarma, S.D. (2004).Spintronics: fundamentals and applications, *Rev. Mod. Phys.*, **76**, pp. 323–410.

50. Clemente-Juan, J.M. and Coronado, E. (1999). Magnetic clusters from polyoxometalate complexes, *Coord. Chem. Rev.*, **193–195**, pp. 361–394.

51. Kortz, U., Müller, A., van Slageren, J., Schnack, J., Dalal, N.S. and Dressel, M. (2009). Polyoxometalates: fascinating structures, unique magnetic properties, *Coord. Chem. Rev.*, **253**, pp. 2315–2327.

52. Pichon, C., Mialane, P., Rivière, E., Blain, G., Dolbecq, A., Marrot, J., Sécheresse, F. and Dubocet, C. (2007). The highest D value for a MnII ion: investigation of a manganese(II) polyoxometalate complex by high-field electron paramagnetic resonance, *Inorg. Chem.*, **46**, pp. 7710–7712.

53. Gumerova, N.I., Roller, A., Giester, G., Krzystek, J., Cano, J. and Rompel, A. (2020). Incorporation of CrIII into a Keggin polyoxometalate as a chemical strategy to stabilize a labile $\{Cr^{III}O_4\}$ tetrahedral conformation and promote unattended single-ion magnet properties, *J. Am. Chem. Soc.*, **142**, pp. 3336–3339.

54. Ammari, Y., Baaalla, N., Hlil, E.K. and Abid, S. (2020). Structure, optical and magnetic properties of a novel homometallic coordination polymers: experimental and computational studies, *Sci. Rep.*, **10**, pp. 1–12.

55. Gaita-Ariño, A., Luis, F., Hill, S. and Coronado, E. (2019). Molecular spins for quantum computation, *Nat. Chem.*, **11**, pp. 301–309.

56. Shiddiq, M., Komijani, D., Duan, Y., Gaita-Ariño, A., Coronado, E., and Hill, S. (2016). Enhancing coherence in molecular spin qubits via atomic clock transitions, *Nature*, **531**, pp. 348–351.

57. Reinoso, S. (2011). Heterometallic 3d–4f polyoxometalates: still an incipient field, *Dalt. Trans.*, **40**, pp. 6610–6615.

58. Reinoso, S. and Galán-Mascarós, J.R. (2010). Heterometallic 3d–4f polyoxometalate derived from the weakley-type dimeric structure, *Inorg. Chem.*, **49**, pp. 377–379.

59. Xie, W. and Wang, H. (2020). Synthesis of heterogenized polyoxometalate-based ionic liquids with Brönsted–Lewis acid sites: a magnetically recyclable catalyst for biodiesel production from low-quality oils, *J. Ind. Eng. Chem.*, **87**, pp. 162–172.

60. Misra, A., Zambrzycki, C., Kloker, G., Kotyrba, A., Anjass, M.H., Franco Castillo, I. and Mitchell, S.G. (2020). Water purification and microplastics removal using magnetic polyoxometalate-supported ionic liquid phases (magPOM–SILPs). *Angew. Chemie – Int. Ed.,* **59**, pp. 1601–1605.

61. Breitwieser, R., Garnier, A., Auvray, T., Ngo, A.-T., Baptiste, B., Menguy, N., Proust, A., Petit, C., Volatron, F. and Salzemann, C. (2019). Protective effect of polyoxometalates in {Mo132}/maghemite binary superlattices under annealing, *Front. Chem.*, **7**, pp. 1–8.

62. Joachim, C., Gimzewski, J.K., and Aviram, A.A. (2000). Electronics using hybrid-molecular and mono-molecular devices, *Nature*, **408**, pp. 541–548.

63. Fernandez, A. Ferrando-Soria, J., Pineda, E.M., Tuna, F., Vitorica-Yrezabal, I.J., Knappke, C., Ujma, J., Muryn, C.A., Timco, G.A., arran, P.E., Ardavan, A. and Winpenny, R.E.P. (2016) Making hybrid [n]-rotaxanes as supramolecular arrays of molecular electron spin qubits, *Nat. Commun.*, **7**, pp. 1–6.

64. Long, D.L. and Cronin, L. (2006). Towards polyoxometalate-integrated nanosystems, *Chem. – A Eur. J.*, **12**, pp. 3698–3706.

65. Vilà-Nadal, L., Mitchell, S.G., Markov, S., Busche, C., Georgiev, V., Asenov, A., and Cronin, L. (2013). Towards polyoxometalate-cluster-based nano-electronics, *Chem. – A Eur. J.*, **19**, pp. 16502–16511.

66. Joo, N., Renaudineau, S., Delapierre, G., Bidan, G., Chamoreau, L.-M., Thouvenot, R., Gouzerh, P. and Proust A. (2010). Organosilyl/-germyl polyoxotungstate hybrids for covalent grafting onto silicon surfaces: towards molecular memories, *Chem. – A Eur. J.*, **16**, pp. 5043–5051.

67. Douvas, A.M., Makarona, E., Glezos, N., Argitis, P., Mielczarski, J. A., and Mielczarski, E. (2008). Polyoxometalate-based layered structures for charge transport control in molecular devices, *ACS Nano*, **2**, pp. 733–742.

68. Laurans, M., Trinh, K., Francesca, K.D., Izzet, G., Alves, S., Derat, E., Humblot, V., Pluchery, O., Vuillaume, D., Lenfant, S., Volatron, F. and Proustm A. (2020). Covalent grafting of polyoxometalate hybrids onto flat silicon/silicon oxide: insights from poms layers on oxides, *ACS Appl. Mater. Interfaces*, **42**, pp. 48109–48123.

69. Balliou, A., Papadimitropoulos, G., Skoulatakis, G., Kennou, S., Davazoglou, D., Gardelis, S. and Glezos, N. (2016). Low-dimensional polyoxometalate molecules/tantalum oxide hybrids for non-volatile capacitive memories. *ACS Appl. Mater. Interfaces*, **8**, pp. 7212–7220.

70. Dalla Francesca, K., Lenfant, S., Laurans, M., Volatron, F., Izzet, G., Humblot, V., Methivier, C., Guerin, D., Proust, A. and Vuillaume, D. (2019). Charge transport through redox active $[H_7P_8W_{48}O_{184}]^{33-}$ polyoxometalates self-assembled onto gold surfaces and gold nanodots, *Nanoscale*, **11**, pp. 1863–1878.

71. Chen, X., Huang, P., Zhu, X., Zhuang, S., Zhu, H., Fu, J., Nissimagoudar, A. S., Li,W., Zhang, X., Zhou, L., Wang, Y., Lv, Z., Zhou, Y. and Han, S.-T. (2019). Keggin-type polyoxometalate cluster as an active component for redox-based nonvolatile memory. *Nanoscale Horiz.*, **4**, pp. 697–704.

72. Escalera-Moreno, L. and Baldoví, J.J. (2019). Unveiling the effect of magnetic noise in the coherence of single-molecule quantum processors. *Front. Chem.*, **7**, pp. 1–7.

73. Zheng, Q., Vilà-Nadal, L., Lang, Z., Chen, J.-J., Orcid, Long, D.-L., Mathieson, J.S., Poblet, J.M. and Cronin, L. (2018). Self-sorting of heteroanions in the assembly of cross-shaped polyoxometalate clusters, *J. Am. Chem. Soc.*, **140**, pp. 2595–2601.

74. Roy, T., Tosun, M., Kang, J.S., Sachid, A.B., Desai, S.B., Hettick, M., Hu, C.C., and Javey, A. (2014). Field-effect transistors built from all two-dimensional material components, *ACS Nano*, **8**, pp. 6259–6264.

75. Xu, L., Tetreault, A.R., Khaligh, H.H., Goldthorpe, I.A., Wettig, S.D., and Pope, M.A. (2019). Continuous Langmuir–Blodgett deposition and transfer by controlled edge-to-edge assembly of floating 2D materials, *Langmuir*, **35**, pp. 51–59.

76. Liu, S., Volkmer, D. and Kurth, D.G. (2003). Functional polyoxometalate thin films via electrostatic layer-by-layer self-assembly. *J. Clust. Sci.*, **14**, pp. 405–419.

Chapter 7

Polyoxometalate-Based Redox Flow Batteries

Ángela Barros, Unai Eletxigerra, Estibaliz Aranzabe, and Marta Hernaiz

Surface Chemistry and Nanotechnology Unit,
Tekniker, C/Iñaki Goenaga, 5, 20600 Eibar, Gipuzkoa, Spain

angela.barros@tekniker.es, unai.eletxigerra@tekniker.es

This chapter holds the state of the art of an emerging kind of redox flow batteries (RFBs) based on polyoxometalates (POMs). Energy storage systems (ESS) play a key role in the energetic transition and RFBs represent a potential solution in the field thanks to their high energy density and efficiency. Specifically, POMs are attracting a great deal of interest as active species in this sort of device due to their rich electrochemical properties.

7.1 Introduction

The world is currently experimenting an energy transition from fossil fuels to renewable energy sources. Environmental concerns about the perils of fossil fuels, together with the long-term environmental damages of greenhouse gas emissions, have given

Polyoxometalates: Advances, Properties, and Applications
Edited by Leire Ruiz Rubio, José Luis Vilas Vilela, Beñat Artetxe, and Juan Manuel Gutiérrez-Zorrilla

ISBN 978-981-4968-14-0 (Hardcover), 978-1-003-27744-6 (eBook)
www.jennystanford.com

rise to an urgency feeling toward the development of an efficient, safe, and affordable energy system [1]. In this sense, politicians have committed different international treaties on climate change, such as the Paris Agreement (2015) or the Green Deal (2019). Their objectives are reaching a climate neutral world by mid-century [2]. Renewable energies are the most promising solution, where wind and solar power are the most deployed ones [3]. The Earth receives just in 1h enough solar radiation to meet worldwide energy requirements for a year. Moreover, capturing a small percentage of potential wind energy could also contribute significantly to meeting the world's electrical energy requirements. However, both of them are intermittent and often unpredictable, and therefore create significant challenges for the electric grid operators that need to introduce different strategies to avoid frequency instabilities. In this scenario, electrical energy storage systems (ESS) are highly valuable tools to compensate the fluctuation of renewable energy sources acting as a reserve that is filled during production peaks and emptied out during periods of demand. What it is more, electrical ESS has been considered as a key enabler of the smart grid or future grid, which is expected to integrate a significant amount of renewable energy sources [4].

From the beginning of the development of renewable energies, human being has fought against the inability to obtain an ES technology which allows the employment of renewable sources at large-scale. Generally speaking, there are two main ways in which energy is stored nowadays: physically and chemically [5]. Table 7.1 holds additional information about these sorts of ESS.

Table 7.1 ESS classification depending on whether the energy is physically or chemically stored

Physical	Chemical
Potential	Flow battery
Hydro pump	Sodium–sulfur battery
Compressed air	Lithium-ion battery
Kinetic	Lead–acid battery
Fly wheel	
Electromagnetic	
Superconductor	
Supercapacitor	

It must be noticed that physical ES englobes by far the most widely employed technology: water-pumping collects more than 90% of the total power volume in ESS worldwide [6], as it allows the storage of a huge volume of energy, long service at a high efficiency, and low cost. Nevertheless, it is highly restrained geographically since it may have a negative impact on the environment and ecology. Thereby, in recent years, electrochemical ES technology is gaining more attention to achieve better regulation and even distribution of the electrical energy. This kind of ESS offers additional advantages, such as a standalone modular design, no geographical requirement, high efficiency, and fast response, which are essential features in all kinds of load leveling applications. In spite of the existence of different kinds of batteries, such as the mentioned in Table 7.1, lithium-ion batteries (LIBs) have been the most commercialized ones, dominating the ES market [7]. The LIB has attractive advantages, such as high performance and being useful as mobile ESS, but its poor safety and short life cycle make it a bit challenging for large-scale ES.

Trying to overcome these limitations, other stationary electrochemical storage systems are being paid special attention recently thanks to its availability to store higher amounts of energy. Safety, durability, and reliability are the main requirements for a stationary ESS [4]. In this sense, redox flow batteries (RFBs) have currently attracted a great deal of interest in the field. They are affordable as stationary systems, where volume and weight are not a problem, or in other words, energy density (amount of energy stored in a given system per unit volume or weight) is important, but not critic for their future application. All of it is based on their availability to decouple power and energy due to their unique architecture, because the energy-bearing complexes are stored externally rather than in the electrode compartment. On the one hand, power is determined by the size of the stack and on the other hand, the amount of electrolyte stored in the tank defines the total energy of the battery [8]. Consequently, the system results in advantages, such as flexible modular design and operation, excellent scalability, moderate maintenance costs, and long-life cycling [9]. These properties may enable RFBs to play a key role in the future electrical grid ES [10].

7.2 RFBs

In general terms, a RFB is based on two ES tanks which contain the electrolytes (the anolyte or negative electrolyte, in which the oxidation half-reaction takes place and the catholyte or positive electrolyte, which holds the reduction half-reaction), these electrolytes are formed by the redox active species solved in an appropriate media (supporting electrolyte). A flow system carries the solutions to each tank. The electrochemical cell, shown in Fig. 7.1, contains two electrodes and an ion-selective membrane which allows the proton flux in order to keep the charge equilibrium. It is important to note that a whole RFB is usually formed by a stack, a group of electrochemical cells connected in series through a bipolar plate [11].

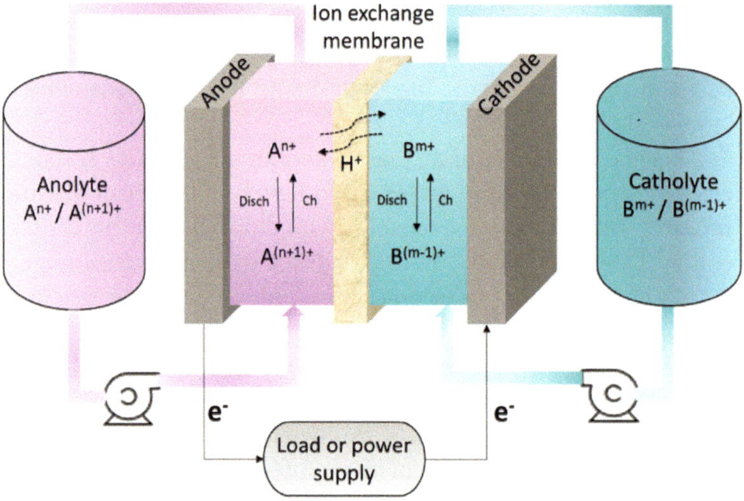

Figure 7.1 Scheme of a RFB electrochemical cell with two external tanks containing electrolytes A and B.

The whole electrochemical process, in the discharge mode, consists in an anolyte solution flowing through an electrode where it reacts generating electrons. These electrons flow through the external circuit in order to produce the reduction reaction on the catholyte. These processes are described by the following reactions:

$$A^{n+} \rightarrow A^{(n+1)+} + e^- \quad \text{[Anolyte]}$$

$$B^{m+} + e^- \rightarrow B^{(m-1)+} \quad \text{[Catholyte]}$$

The redox active species determines the overall performance of the RFB. The energy density of the system is limited by the redox potential and solubility of the species due to the cell architecture. In addition, the cell voltage is determined by the equilibrium potentials of active species in the cathodic and anodic half-cells, whereas the capacity (maximum amount of energy provided by the battery) is dependent on the effective concentration, which is the solubility multiplied by the number of electrons transferred in the redox reaction [12]. Another highlighted yield parameter is the efficiency, understood as the ratio between the capacity of charge and discharge of the battery. Efficiency is described by the Coulombic efficiency (CE), the voltage efficiency (VE), and the energy efficiency (EE). CE is measured in terms of current and indicates the crossover, the electrolyte osmotic rate and secondary reactions. The choice of electrode materials, the manufacturing process, and the distribution of the solution will greatly influence the VE. In other words, VE coincides with the polarization properties of the battery. EE or round trip efficiency informs about the global battery performance and indicates the capacity conversion of the system.

The emerging RFBs importance can be appreciated in the graphic presented in Fig. 7.2, due to the number of published articles have increased dramatically since 2011.

Nonetheless, this type of system was developed in the 1970s and includes a wide diversity of batteries [13]. Although several research lines are active nowadays, there are some classic technologies which may be mentioned. The first RFB dates from 1970. It was the iron/chromium (Fe/Cr) technology, developed by NASA for photovoltaic applications [14]. Nevertheless, this project was not fully successful because it displayed low output voltage and efficiency, so as a high crossover between iron and chromium [15]. Followed by this prototype, during 1980s, emerged the all-vanadium redox flow battery (VRFB) which is actually the most widespread. This system was especially important because thanks to having a unique redox active species, the crossover contamination was avoided [16]. Being the most developed technology in RFBs field, a deep insight into this system is given in the next section.

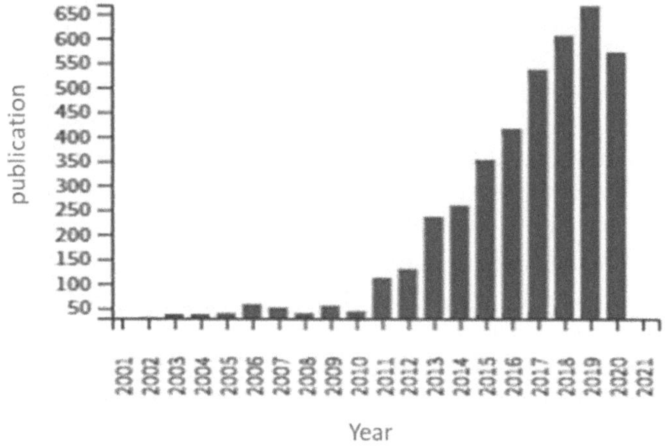

Figure 7.2 Development of the research activity in the field of RFBs in the last years, data collected from Web of Science (Clarivate Analytics).

In addition, despite the zinc-bromine first appeared in the nineteenth century, the whole RFB system (ZBRFBs) was not completely developed until 2000s. It is the second most industrialized RFB, following VRFB and is considered a hybrid RFB; in other words, they do not store all the energy in liquid or gas phase, but the zinc is in solid state during the charge process and liquid during the discharge, while the bromine is always solved in the electrolyte [17]. This fact is one of the main drawbacks in the cell functionality because it might trigger dendrite formation and consequently, results in an electrical shorting. The cell exhibits a high cell voltage, so as a high theoretical specific energy of 60–85 Wh kg^{-1} and low cost [18]. Their performance is comparable to that for VRFBs, having no cycle-life limitations with lifetimes between 11 and 14 years and 80% of EE [19]. However, the low working current density results in the necessity of bigger cells for a given power output [20].

It is noticeable that the main components are usually common to all RFBs, varying the underlying redox chemistry, being this one of the most popular open research lines in material science nowadays. In this way, RFBs could be classified paying attention to the supporting electrolyte (aqueous or non-aqueous), the redox active species or the architecture of the electrochemical cells (symmetric or asymmetric). Moreover, it could be distinguished between pure and hybrid RFBs, whose differences are the physical state of the electroactive material; while in pure-RFBs the electrolytes are both liquids (anolyte and catholyte), hybrid-RFBs count with one solid electroactive species [10].

7.2.1 All-VRFBs

As it was above mentioned, VRFBs are widely commercialized (the Vanitec website cites 26 companies which produce this kind of technology). One of the main advantages of VRFBs is that they are symmetric, with a unique species in solution (vanadium ions at different oxidation states: V^{2+}, V^{3+}, VO^{2+}, VO_2^+), avoiding the severe cross-contamination issue of other electrochemical cells. Although the currently achieved energy density is lower than for LIBs (25–35 Wh L^{-1} vs 250 Wh L^{-1}), the charge/discharge cycles are considerably higher than the obtained for other batteries (15 000–20 000 vs 5000 cycles), which is one of the most remarkable advantages of this kind of system [21], especially for stationary ES. In addition, the fast electrochemical kinetics allows an immediate response to surge demand from the electrical grid [22].

Even if VRFBs are worldwide distributed, they present some drawbacks. On the one hand, although the environmental impact for the vanadium electrolyte is lower than for other acid-based batteries, vanadium is considered a critical raw material (CRM), which significantly increases the cost of the resultant battery [23]. Additionally, the supporting electrolyte is sulfuric acid, which implies the generation of an acidic media environmentally dangerous and highly corrosive [9]. On the other hand, the electrolyte performance is limited by the poor specific energy density (20–35 Wh L^{-1}), associated with the moderate solubility of redox active species (1.5–2 M) in the potential window. In regard to the temperature stability, it is reduced to a range (10–40°C), which implies that this technology requires cooling devices. It is also noticeable the limited chemical stability of the electrolyte, caused by side reactions, which requires corrective procedures to prevent capacity fading over longer periods of time. In addition, VRFBs are intrinsically safe due to the fact that the electrolyte is water-based and therefore not flammable. Nowadays, electrolyte spillage is the main risk associated with VRFBs, because it is corrosive and requires careful handling and preventive measures.

Despite their limitations, VRFBs have already been commercialized all over the world in demand of the growing need of cost effective and reliable ESS. One of the largest RFBs is the located in Minami Hayakita Substation in Japan, built in 2015, rated 15 MW and 60 MWh. In Europe, there is one in the ES station at Fraunhofer ICT in Pfinztal, Germany, rated 2 MW and 20 MWh was commissioned in 2019. Moreover, UniEnergy Technologies, WA, USA has installed a number of systems rated 2 MW and 8 MWh. However, the largest project so far is the 200 MW and 800 MWh storage station designed by Rongke Power of China, which will be by far the largest electrochemical ES plant in the world [10].

7.2.2 Alternatives to all-VRFBs

Even though VRFBs is head of other technologies, the community has not yet decided on a universal battery chemistry. In order to overcome the limitations of vanadium, different RFB systems are being developed, including new chemistries and cell configurations.

Considering the advantages of vanadium technology, some developments are focused on increasing the energy density through the improvement of the vanadium solubility. In the 2000s, the positive electrode was substituted by an air electrode, so as a bromine solution in the vanadium-air (VORFB) [24] and vanadium-bromine (VBRFB) [25] flow systems, respectively. Nevertheless, the limited impact of these strategies in the overall battery performance has pointed out the need of investigating completely new electrolytes.

Since 2010s, the use of organic molecules as redox active materials was widely accepted taking into account that they are sustainable molecules based on Earth abundant elements

(C, H, O, N, and S). Moreover, they are easily structurally tuneable and present wide cell-voltage, high solubility, fast electrode reactions, so as chemical and electrochemical stability. Numerous aqueous organic systems have been explored, some examples are the carbonyl-based electrolytes [26], and other ones with more complex structures, such as quinone-based [27], quinoxaline-based [28], polymer-based [29], or even viologen-based electrolytes [30]. The employment of organics is foreseen as an option to assess the high demand for ES with the utilization of materials which allow a large-scale production with reduced costs. However, current investigations are working on the improvement of the degradation rate, one of the main drawbacks of this type of electrolytes.

In this sense, inorganic structures have also been recently studied as active species for RFBs. Among them, polyoxometalates (POMs) have been especially selected thanks to their interesting electrochemical properties, so as wide compositional versatility, which enables an easy tunability by the modification of the synthetic procedures to obtain the desired properties in the final material. Nevertheless, there have been relatively few investigations related to POM–RFBs until recent years. What it is more, first published work in the area was developed by Pratt *et al.* between 2013 and 2014 [31].

Additionally, trying to overcome the limitations given by the stable electrochemical window of the supporting electrolyte media, organic compounds have been studied in recent years. In aqueous media, if a higher voltage than 1.5 V is applied, then water splitting occurs, producing hydrogen and oxygen, limiting the operating voltage window. Contrarily, acetonitrile, whose stability is in a 5 V window, is widely used for this purpose [32]. As result, all-organic RFBs were developed in 2012. Nonetheless, it must be taken into account the environmental damages that may cause these volatile solvents in contrast to water.

Another interesting research line is the one which relies on more complex cell architectures, including membraneless systems by laminar flux [33] or biphasic batteries based on immiscible electrolytes [34]. The elimination of the physical barrier between electrolytes may imply a significant decrease in the final battery cost, although this kind of RFBs are still at a low maturity level.

Besides this new kind of cell, other proposals tried to improve the dependence of the energy density with the concentration of the active species. The selected option to fulfill this purpose was the development of insoluble solid active species. For example, the semi-solid RFBs, system based on an electrolyte formed by a suspension of solid particles that form a slurry, also called a semi-solid electrode [35]. There are also some emerging RFBs based on more complex technologies, such as solar RFBs which incorporate photoelectron-conversion electrodes [36] or metal–CO_2 RFBs, presented as an eco-friendly and economical solution that reduce the amount of CO_2 while storing energy [37].

The most relevant developments in RFBs are outlined in the timeline in Fig. 7.3.

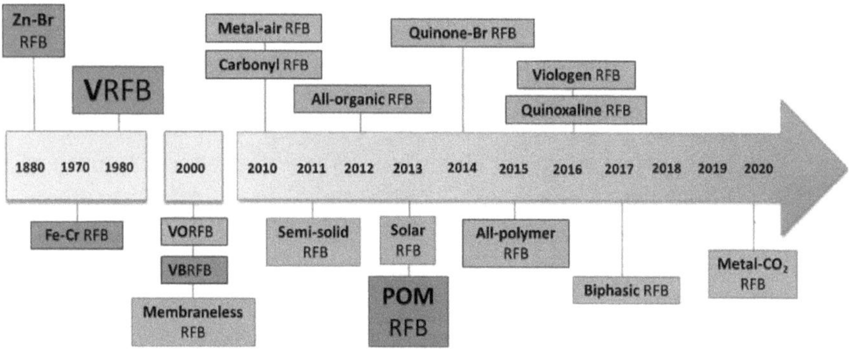

Figure 7.3 Timeline of the development of RFBs, a brief summary of stationary EES history.

7.3 POM-Based RFBs

Thanks to its rich structural diversity, so as rich electrochemistry, POMs have been used in RFBs in recent years. Specific characteristics of interest in POMs as active species in RFBs can be summarized as follows:

1. POMs as electron reservoirs: high current densities are enabled [38] due to the fast electron transfer provoked by the delocalization of the reduction electrons added to POMs. They usually exhibit high redox kinetic constants: $k_0 \sim 10^{-2}$ cm s^{-1} [39].

2. Low permeability or crossover: POMs do not permeate through cation exchange membranes used in RFBs because of its large size and anionic nature [40].
3. High stability: the net charge of the POMs does not change in oxidation or reduction reactions [41] because the electron transfer is often associated to protonation or cationic association processes, which result in an increased stability.
4. High energy density: the great majority of the POMs are highly soluble, with the maximum concentration determined by the electrolyte and the presence of counterions. Concentrations of 2.7 M [31] could be reached, much higher than those for vanadium (1.5 M) in water [42]. This fact, together with the availability of the multi-electron transfers per molecule, gives rise to higher energy densities [43].
5. Tunable redox properties: the redox potential of POM units can be modified because they are sensitive to the concentration of species and the pH of the electrolyte.
6. Thermal stability of inorganic structures allows to work under safety conditions up to 80°C.
7. Scalability of synthetic procedures: POMs are prepared by simple synthetic procedures which enables easy scalability.
8. Low degradability: in contrast to organic compounds, inorganic POM clusters are highly stable in aqueous solution at specific pH ranges.
9. Cost: since the main starting materials in POMs synthesis are metal oxides, they are easily commercially available.

7.3.1 Fundamental Redox Mechanisms of POMs at Electrodes and ESS

Although there are several archetypical structures and systematic compositional variations on them that can lead to an enormous catalog of metal-oxoanions, electrochemical studies reveal that POMs can be classified according to their electronic properties into two main groups: class I polyanions are easily and reversibly reduced (e.g., Keggin and Wells–Dawson type structures), whereas the reduction of class II clusters is more difficult and shows irreversible nature (e.g., Anderson–Evans structures) [44].

Structurally, this fact can be inferred from the polarization of pM–O bond of the MO_6 subunits toward one (class I) or two (class II) terminal oxygen atoms [45].

The well-known Keggin and Wells–Dawson structures belong to the first group and represent the most explored species in electrochemical POM-based ESS. The α-Keggin-type $[XM_{12}O_{40}]^{n-}$ heteropolyanion exhibits ideal T_d symmetry and it is constituted by four $\{M_3O_{13}\}$ trimers formed by three edge-sharing $\{MO_6\}$ octahedra, which are linked to each other and to the central XO_4 tetrahedron (X = heteroatom) by corner sharing. Rotation on 60° of one or more trimers leads to different Baker–Figgis isomers (α, β, γ, δ, and ε) [46]. The Keggin structure is highly susceptible to compositional modifications, especially to those related to the replacement of addenda metals (M = Mo, W) or heteroatoms (X = B, Si, Ge, P, As, S, Co, etc.). However, when it comes to vanadates only the bicapped $[PV_{14}O_{42}]^{9-}$ derivative have been isolaten to date [47].

Additionally, some defect structures can be prepared by the removal of one or more metal atoms from the parent plenary cluster. These lacunary derivatives can act as multidentate oxygen donor ligands toward additional 3d metal ions (Mn, Fe, Co, etc.) to regenerate the parent cluster, which results in metal substituted species. Mono-, di- or trilacunary Keggin type species can be synthesized in aqueous solution. The trivacant anions show two different isomers named as *A* or *B* depending on whether the three $\{MO_6\}$ octahedra eliminated belong to a $\{M_3O_{13}\}$ edge-sharing trimer or to a $\{M_3O_{15}\}$ corner-sharing triad, respectively. The strongly related family of Wells–Dawson $[X_2M_{18}O_{62}]^{n-}$ anions, is constituted by two $[A\text{-}\alpha\text{-}XM_9O_{34}]$ trilacunary units sharing corners through the belt $\{MO_6\}$ octahedra [48].

Related to electrochemical properties of POMs, several studies have been developed. For example, the redox potentials of molybdenum POMs (heteropolymolybdates) under neutral conditions are more positive than those of the corresponding tungsten POMs (heteropolytungstates): $[PMo_{12}O_{40}]^{3-} > [PW_{12}O_{40}]^{3-}$, $[SiMo_{12}O_{40}]^{4-} > [SiW_{12}O_{40}]^{4-}$. In addition, the first redox potentials of POMs with the same framework are linearly related to the anionic charge of the POMs in any solvent: $[SM_{12}O_{40}]^{2-} > [PM_{12}O_{40}]^{3-} > [SiM_{12}O_{40}]^{4-}$ (M = Mo, W). The redox potentials of the Keggin-type POMs $[SiW_{12}O_{40}]^{4-}$, $[PW_{12}O_{40}]^{3-}$, $[PMo_{12}O_{40}]^{3-}$ and $[GeMo_{12}O_{40}]^{4-}$ and the Wells–Dawson-type POM $[P_2W_{18}O_{62}]^{6-}$ are related to the acidity and basicity of the solvent. Keggin-type POMs have several isomers that are generated from the 60° rotation of each $\{M_3O_{13}\}$ unit. Redox waves of the β-form appear at more positive potentials than those of the corresponding α-form, e.g., β-$[PM_{12}O_{40}]^{3-} > \alpha$-$[PM_{12}O_{40}]^{3-}$ (M = Mo, W). Moreover, other interesting research line is based on metal-substituted POMs. Electrochemical studies reveal that mostly, the substitution of an addenda metal by other transition metal implies a displacement to more positive tungsten redox potential. Among these transition metals, Mn, Co, Fe, or V can be found [49].

The vast majority of electrochemical studies for POMs take place in aqueous media. Then, it is important to notice that POMs have a high pH dependence, so as usually present a pH domain in which the molecule is stable [44]. In this sense, it is simpler to compare redox potentials in non-aqueous media or in stable pH conditions because protons can act as a counter cation and proton-coupling constants strongly depend on the anionic charge and structure of POMs. Each reduction step increases the negative charge density on the POMs and hence, their basicity. In contrast, in non-aqueous media there are other factors due to the absence of protons. In this case, cations are highly important in order to maintain the whole charge of the species between oxidation and reduction steps, usually, sodium or lithium cations. What it is more, the addition of small salts also generates an increasing in the number of electrons [50]. It is often common to combine non-aqueous with aqueous solvents, in which case the addition of acidic species results in a higher number of electrons in the process.

As it has been already mentioned, POMs have special electrochemical properties, such as multielectron transfer, undergoing through fast and reversible redox reactions or the combination of different oxidation states in one molecule. This has allowed alternative uses in ESS, such as electrode material and specifically as cathode in LIBs. For this task, Keggin molybdenum-based POMs have been mainly used, however, new studies by Sonoyama, N. *et al.* showed that vanadium-based POMs have exhibited better capacity (up to 200 Ah kg^{-1}) and higher cycling stability than Mo-based POMs [51]. On the other hand, Hartung *et al.* have revealed the efficacy of using POMs, such as $Na_6[V_{10}O_{28}]\cdot16H_2O$ as anode in sodium-ion batteries (NIBs). The results showed a reversible capacity of approximately 176 Ah kg^{-1} with an average discharge potential of 0.4 V vs Na^+/Na, as well as a high cycling stability [52].

Nonetheless, POMs have been mainly employed as electroactive species of the electrolyte. Considering the emerging importance of the RFBs, many POMs have been tested in order to develop a promising system able to replace all-vanadium technology.

7.3.2 State of the Art in POM–RFBs

Since Pratt *et al.* presented the first POM–RFB in 2013 [31], this research line has been gradually growing. This type of battery emerges from the drawbacks of VRFBs, which POMs mitigate thanks to their high stability in water and their capacity to undergo several reversible and stable electron transfer processes. The vast majority of studies related to POM–RFBs are based on the selection of a POM as charge carrier and suitable media conditions. The supporting electrolyte heavily influences the stability of POMs as well as imposes limitations to the potential window. In this sense, POM–RFBs can be classified in aqueous or non-aqueous depending on the nature of the supporting electrolyte. There is a recent trend in studying non-aqueous electrolytes in order to avoid the water splitting problem, increasing the voltage operational range. However, it is important to note the better efficiency obtained in the aqueous media and the more environmentally friendly systems in contrast to the organic solvents employed for the non-aqueous ones. What it is more, POMs exhibit lower solubility in organic solvents than in water, potentially limiting the energy density of the cell. This solubility may be improved if POMs are organically functionalized.

Additionally, the developed systems are organized depending on the sort of architecture chosen for the cell: symmetric (holds a unique species as electroactive material) or asymmetric (formed by two kinds of active species). Recent studies tend to favor symmetric cells as they avoid the cross-contamination issue as in the case of all-vanadium systems. However, in the attempt to obtain higher efficiencies and cell potentials, researchers have also studied asymmetric configurations. Other strategy in terms of increasing the cell potential is to modify the POM structure, especially by substituting some addenda metals by different transition metals. In this way, redox processes are separated in two potential ranges, because the substitution usually implies the inclusion of a more electropositive atom, at the same time, the addenda metal potential is displaced to more positive values. All of it is translated in a higher energy density, since the delivered energy and power are proportional to the working voltage.

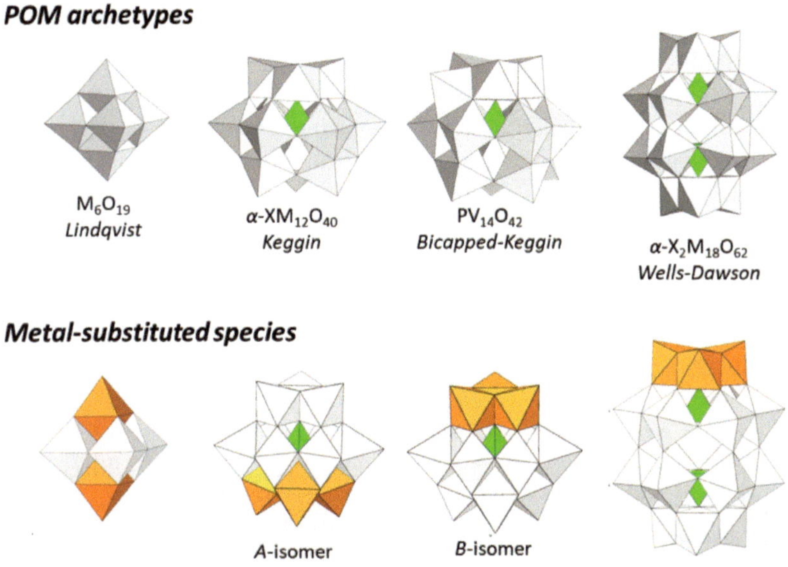

Figure 7.4 Structure of the most employed POMs in RFBs. Color code: gray octahedra: addenda metal; orange octahedra: substituted unit; green tetrahedra: heteroatom.

Finally, POM–RFBs may also be classified according to the family of the chosen POM. Although this classification might be complicated due to the research line is still in a low maturity level, as it was above-mentioned, Keggin and Wells–Dawson families are the most employed ones for this kind of device in base of their electrochemical properties, especially their redox reversibility. Therefore, in order to modify the electrochemical properties of the POM, different strategies and compositions have been tested. For instance, most reported POM–RFBs are tungsten or molybdenum Keggin-type POMs from the plenary species to the trisubstituted or even bicapped derivative. Wells–Dawson-type is also commonly selected: tungsten-based in plenary shape, trisubstituted and hexavanadates, better known as Lindqvist species. All these structures are shown in Fig. 7.4.

A careful analysis of the POM–RFBs published articles in the last ten years has been carried out. Table 7.2 holds the classification of the POM–RFBs developed for the moment. The information includes, on the one hand, the electrolyte characteristics, and on the other hand, the most representative electrochemical parameters. Moreover, in the following lines all of them are explained in detail.

First POM–RFB was developed by Pratt *et al.* [31], comparing the effect of the supporting electrolyte on the RFB performance with the Keggin-type vanadium-trisubstituted POM $[HSiV_3W_{12}O_{40}]^{6-}$ as redox active species, utilizing this two, 3-electron redox couples: ($[SiV^V_3W^{VI}_9O_{40}]^{7-}/[SiV^{IV}_3W^{VI}_9O_{40}]^{10-}$ and $[SiV^{IV}_3W^{VI}_9O_{40}]^{10-}/[SiV^{IV}_3W^V_3W^{VI}_6O_{40}]^{13-}$). On the one hand, the aqueous (0.5 M H_2SO_4) battery demonstrated CE greater than 95% with relatively low capacity fading over 100 cycles. In addition, the system recovered from 48h of reversed polarity after only one cycle. The EY (defined here as the observed capacity during charge or discharge divided by the theoretical capacity) of the flow cell decreased from 90 to 80% during the first 100 cycles. On the other hand, the non-aqueous (PC 0.5 M TBAOTf) system had an initial CE of 87%, an open-circuit voltage improved from the aqueous analog in 0.3 V and a higher operating voltage (1.1 V). Nevertheless, the cycling rate and EY were 65 and 15% lower than the aqueous flow systems. What it is more, after 10 cycles, the EY of the system dropped by half. Thus, regardless the improved voltage, the efficiency of the cell is limited in the non-aqueous system. It is important to notice that depending on the media, the counter cation was inorganic (K^+) or organic ($((CH_3CH_2CH_2CH_2)_4N)^+$) for improving POM solubility in aqueous and non-aqueous electrolytes, respectively. In addition, these results were compared to the acidic form of the plenary Keggin-type $H_4SiW_{12}O_{40}$ in order to appreciate the influence of the vanadium-substitution. Figure 7.5 shows the cyclic voltammetry (CV).

Table 7.2 Classification of the state of the art of POM–RFBs

RFBs	POM	Conc. (mM)	SE	Efficiency	V_{cell} (V)	J (mA/cm^2)	En. Dens. (Wh/L)	Refs
Non-aqueous	***A*-α-TBA$_6$ [HSiV$_3$W$_9$O$_{40}$]**	20	PC 0.5 M TBAOTf	CE: 87% EE: 80% EY: 15%	1.1	0.5 (0.05–2.5V)	0.48	[31]
	[V$_6$O$_7$(OR)$_{12}$]	10	100 mM [NBu$_4$][PF$_6$] in ACN	CE: 97%	1.7	–	–	[56]
	[TiV$_5$O$_7$(OCH$_3$)$_{13}$]$^-$	5	100 mM [NBu$_4$][PF$_6$] in ACN	CE: 91%	2.30	–	–	[57]
	[Ti$_2$V$_4$O$_7$(OCH$_3$)$_{14}$]	5	100 mM [NBu$_4$][PF$_6$] in ACN	CE: 89%	2.74	–	–	[58]
	Li$_6$[P$_2$W$_{18}$O$_{62}$] // Li$_3$[PMo$_{12}$O$_{40}$]	10	DMF	CE: 70%	1.3	–	–	[61]
	Li$_3$[PMo$_{12}$O$_{40}$]	10	0.1 M LiTf in ACN	CE: 68% EE: 40%	0.36	0.2	–	[55]
		10	DMF	CE: 93% VE: 50% EE: 50%	0.7 60% SoC	(0–05–0.8 V) 50% SoC		[61]
		100	ACN	CE: 90% EE: 40%	0.35 80% SoC	2		
Aqueous	***A*-α-K$_6$[HSiV$_3$W$_9$O$_{40}$]**	20	H$_2$SO$_4$ 0.5 M	CE: 95% EE: 90–80% EY: 91%	0.6	2 (0.05–1.4V)	0.48	[31]

(*Continued*)

Table 7.2 *(Continued)*

RFBs	POM	Conc. (mM)	SE	Efficiency	V_{cell} (V)	J (mA/cm^2)	En. Dens. (Wh/L)	Refs
	***A*-α-$K_6[PV_3W_9O_{40}]$**	20	H_2SO_4 1 M	CE: 85% EE: 50% EY: 50–40%	0.6	0.5	0.036	[53]
	***B*-α-$K_6[PV_3W_9O_{40}]$**	20	H_2SO_4 1 M	CE: 70–50% EY: 90–60%	–	–	–	[53]
	$K_8H[P_2V_3W_{15}O_{62}]$	20	H_2SO_4 1 M	CE: 95%	–	–	–	[53]
	$K_4Na_7[SiFe_3W_9(OH)_3O_{34}]_2(OH)_3$	10	HAc/Ac^- 0.5 M	CE: 83% EE: 50% EY: 77–55%	0.4	2.5 (0.05–1.5V)	–	[54]
	cis-$Li_4[V_2W_4O_{19}]$	500	$HCit/Cit^-$ 0.5 M	CE: 45–38% EY: 20–16%	–	–	–	[54]
	$H_6[CoW_{12}O_{40}]$	200–800	H_2SO_4 0.5 M	CE: 99% EE: 86%	1.2–1.4	25–100	15.4	[59]
	$H_4[SiW_{12}O_{40}]$// $Na_5[H_4PV_{14}O_{42}]$	1	1 M LiCl Prerreduced hydrazine	CE: 94% VE: 65%	1.05	–	–	[62]
	$Li_6[P_2W_{18}O_{62}]$//HBr	2	Li_2SO_4 H_2SO_4 1 M	CE: 97.3%	–	50 (0.4–1.5 V) 100% SoC	225	[43]

Notes: CE: Coulombic efficiency, VE: voltage efficiency, EE: energy efficiency, EY: electrochemical yield, //: asymmetric cell or nothing if symmetric. HAc/Ac^-: acetic acid/acetate buffer, $HCit/Cit^-$: citric acid/citrate buffer, PC: polypropylene carbonate, TBAOTf: $(CH_3CH_2CH_2CH_2)_4N(CF_3SO_3)$), LiTf: lithium trifluoromethanesulfonate, ACN: acetonitrile, DMF: dimethylformamide.

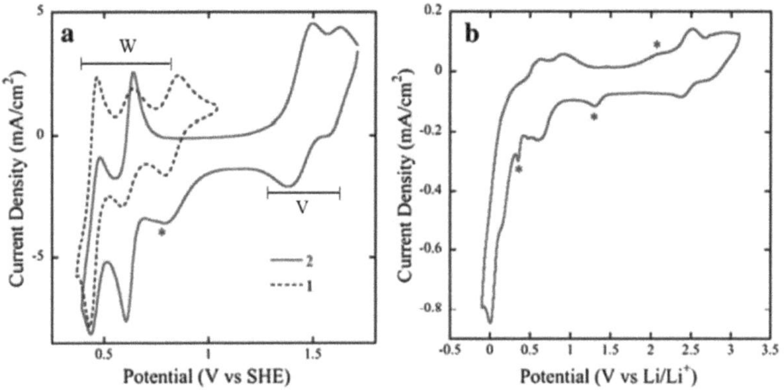

Figure 7.5 (a) CV of $K_6[HSiV_3W_9O_{40}]$ (continuous line, 2) in comparison with its parent compound $H_4SiW_{12}O_{40}$ (dashed line, 1) in 0.5 M H_2SO_4 as supporting electrolyte using a glassy carbon working electrode with a scan rate of 50 mV s^{-1}; (b) CV of $((CH_3CH_2CH_2CH_2)_4N)_4$ $[H_3SiV_3W_9O_{40}]$ in 0.5 M TBAOTf in propylene carbonate using a glassy carbon working electrode with a scan rate of 50 mV s^{-1}. The peaks marked with an asterisk are attributable to the supporting electrolyte. © Copyright license provided by Elsevier and Copyright Clearance Center.

CV (a) shows the influence of the vanadium substitution on the electrochemical properties of the compound, while all redox processes for $H_4SiW_{12}O_{40}$ take part in the same voltage range (0.4–0.7 V), for the vanadium substituted species, there are two voltage ranges separated in 1 V: two redox pairs for tungsten (0.4–0.6 V), and also two processes for vanadium (1.3–1.7 V), all of them being reversible reactions. The comparison between (a, b) CVs shows the easier reducibility of $K_6[HSiV_3W_9O_{40}]$ in aqueous media compared to non-aqueous media, which may have influenced on the efficiency of the cell. Finally, it is also worth mentioning that no POM decomposition processes were observed by ultraviolet–visible (UV–vis), Fourier transform infrared spectroscopy (FTIR), powder X-ray diffraction (PXRD) or nuclear magnetic resonance (^{29}Si-NMR and ^{51}V-NMR) after cyclability studies in a flow cell, which makes this POM into a suitable electroactive species to take part into a RFB.

Following their previous work, Pratt *et al.* tested two Keggin-type POMs (*A*-α-$[PV_3W_9O_{40}]^{6-}$ and *B*-α-$[PV_3W_9O_{40}]^{6-}$) [53] and one Wells–Dawson cluster ($[P_2V_3W_{15}O_{62}]^{9-}$), comparing the influence of different charge densities, bonding subunits and the heteroatom in H_2SO_4 (1 M). First of all, CV and bulk electrolysis experiments on the Keggin compounds (*A*-*α*-$[PV_3W_9O_{40}]^{6-}$ and *B*-*α*-$[PV_3W_9O_{40}]^{6-}$) established that the vanadium centers of these compounds could be used as the positive electrolyte ($[PV_3^{IV}W^{VI}_9O_{40}]^{9-}$/$[PV_3^{V}W^{VI}_9O_{40}]^{6-}$), and the tungsten centers could be used as the negative electrolyte ($[PV^{IV}_3W^{VI}_9O_{40}]^{9-}$/ $[P^{VI}V_3W^{V}_3W^{VI}_6O_{40}]^{12-}$) since these electrochemical processes are separated by about 1 V (due to the vanadium substitution). The results showed that *A*-*α*-$[PV_3W_9O_{40}{}^{6-}]$ had CE above 80%, while for *B*-*α*-$[PV_3W_9O_{40}]^{6-}$, fluctuated between 50 and 70% during cycling. These results may be caused by the higher rigidity of the *B*-bonding sharing edges, which implies a lower reversibility of the processes. The EY was between 40 and 50% for *A*-*α*-$[PV_3W_9O_{40}]^{6-}$, and ^{31}P-NMR showed small amounts of $[PV_2W_{10}O_{40}]^{5-}$ and $[PVW_{11}O_{40}]^{4-}$ formed with cycling. The EY for *B*-*α*-$[PV_3W_9O_{40}]^{6-}$ decreased from 90 to around 60% due to precipitation of the compound on the electrode, but there were no decomposition products detected in the solution by ^{31}P-NMR, and infrared data on the electrode suggested that the cluster remained intact. In overall, the $[P_2V_3W_{15}O_{62}]^{9-}$ (Wells–Dawson structure) suggested higher charge density clusters were not as suitable as the Keggin structures for a RFB due to the poor stability and inaccessibility of the highly reduced materials.

Pratt *et al.* continued with their research line in the aqueous media (0.5 M acetic acid/acetate buffer), selecting an iron-substituted Keggin-type POM: $K_4Na_7[SiFe_3W_9(OH)_3O_{34}]_2(OH)_3$ to contrast with the vanadium-substituted Lindqvist species: cis-$[V_2W_4O_{19}]^{4-}$ [54]. On the one hand, the iron-containing dimer, $[(SiFe_3W_9(OH)_3O_{34})_2(OH)_3]^{11-}$, cycled between $[(SiFe_3W_9(OH)_3O_{34})_2(OH)_3]^{11-}$/$[(SiFe_3W_9(OH)_3O_{34})_2(OH)_3]^{14-}$ and $[(SiFe_3W_9(OH)_3O_{34})_2(OH)_3]^{17-}$/$[(SiFe_3W_9(OH)_3O_{34})_2(OH)_3]^{14-}$ for the positive and negative electrolytes, respectively. This compound demonstrated a CE of 83% after 20 cycles with an EY of 55%. These results, significantly worse than those obtained for the vanadium-substituted compound [31], may be due to the iron is less electropositive and therefore the difference between potentials lower. On the other hand, the CV of the Lindqvist ion, cis-$[V_2W_4O_{19}]^{4-}$, showed non reversible redox processes. In a flow cell configuration, cis-$[V_2W_4O_{19}]^{4-}$ had a CE of 45% (for a

2-electron process) and an EY of 16% after 20 cycles. The poor performance of cis-$[V_2W_4O_{19}]^{4-}$ was attributed primarily to its higher charge density. To sum up, these results showed that POM size and charge density are both important parameters to consider for improving battery performance.

The work of Pratt *et al.* shows that the kind of heteroatom influences directly in the efficiency of the battery because of the different bonding of the centered tetrahedron. In this sense, studies comparing Si or P as heteroatom (as in [31] and [53] with the $[HSiV_3W_9O_{40}]^{6-}$ and $[PV_3W_9O_{40}]^{6-}$) reveal a better performance for the silicotungsten. This fact has been explained by the shorter Si–O bond in the central tetrahedra than P–O, which implies a shorter and therefore stronger W–O bond. Consequently, the silicon tetrahedron is binding stronger to the polyoxoanion compared to the phosphate anion, increasing the stability of the compound. Additionally, Pratt *et al.* demonstrated that the structure of the polyanion is critical in terms of reversibility. Sharing corners is more reversible rather than sharing edges, where the bond is stronger, as it has been confirmed by other studies [53] with *A*-α-$K_6[PV_3W_9O_{40}]$ and *B*-α-$K_6[PV_3W_9O_{40}]$. Moreover, this work also points out the importance of the POM solubility and how strongly influences on its stability. Finally, charge density of the polyanion is also important as it directly affects the reduced form.

Taking into account the importance of the POM structure in the battery performance, other authors, such as Chen *et al.* [55], selected a simpler structure, easier to synthesize: $Li_3PMo_{12}O_{40}$ in a non-aqueous media (0.1 M LiTf in acetonitrile, ACN). This system showed high CE (90%) and high cycling stability (>500h). Related to the POM electrochemical characterization, a 1-electron redox couple was accessed on each side. The charge/discharge reaction was $PMo_{12}O_{40}{}^{4-}/PMo_{12}O_{40}{}^{3-}$ on the catholyte side and $PMo_{12}O_{40}{}^{4-}/PMo_{12}O_{40}{}^{5-}$ on the anolyte side, as it is shown in Fig. 7.6.

In contrast to Fig. 7.5, whilst this molybdenum Keggin-type shows two redox reaction in the same potential range, the vanadium-substituted $[HSiV_3W_9O_{40}]^{6-}$ exhibits three electron-transfer processes in two different potential ranges due to the vanadium and tungsten redox reactions are separated. Moreover, the employment of an organic solvent for the supporting electrolyte does not imply worse electrochemical performance, whereas for the former POM it is deterrent.

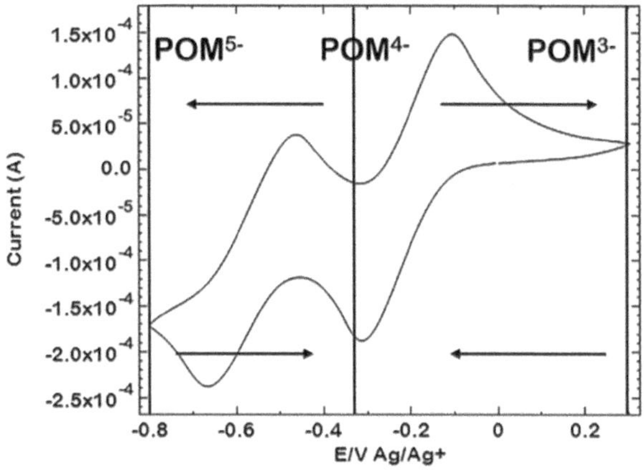

Figure 7.6 Cyclic voltammogram for 0.01 M $Li_3PMo_{12}O_{40}$ with 0.1 M Lithium trifluoromethanesulfonate (LiTf) in acetonitrile, showing the 1-electron transfers accessed in the charge/discharge experiments. © Copyright license provided by Elsevier and Copyright Clearance Center.

Additionally, taking advantage of the size and charge density of POMs, different separators were tested during this work. Both Li-exchanged Nafion-117 and aramid nanofiber (ANF)-based membranes allowed stable charge/discharge cycles. However, the ANF membrane, which separation is based on size, permitted operation at higher current densities than Nafion-117 ion-exchange membrane. In this sense, this research shows the importance of the right selection of the cell materials. POMs allow the employment of ANF membranes in substitution to the expensive Nafion-117, which implies a significant reduction on the final battery cost.

Other important parameter to take into account in the POM–RFB setup is the solubility of POMs in the media. VanGelder *et al.* [56, 57, 58] paid special attention to it in non-aqueous media (100 mM $[NBu_4][PF_6]$ in ACN). The success of this kind of system requires the increasing compatibility of these inorganic compounds with organic solvents. In order to carry out their research, ether-functionalized polyoxovanadate-alkoxide cluster

Lindqvist-type POMs were selected: $[V_6O_7(OR)_{12}]$ (R = CH_3, C_2H_5) and $[V_6O_7(OR)_9(OCH_2)_3CR']$ (R = CH_3, C_2H_5 and R′ = CH_3, CH_2OCH_3, $CH_2OC_2H_4OCH_3$). Ether-functionalization yields a 12-fold increase in solubility, which directly dictates the energy density of the RFB. These clusters exhibit four redox processes, spanning nearly a 2 V window and demonstrate rapid electron-transfer kinetics. Electrochemically, the metal-oxide core remains intact upon deep charge-discharge cycling, enabling extremely high CE (~97%) with minimal overpotential losses (~0.3 V). In addition, VanGelder *et al.* have developed a new family of compounds based on the substitution of vanadium centers by titanium, resulting in the following species: $[TiV_5O_7(OCH_3)_{13}]^-$ and $[Ti_2V_4O_7(OCH_3)_{14}]$. This structural modification implies a 740 and 210% increase in energy density, respectively, together with a significant increase in cell voltage, from 1.60 V in $[V_6O_7(OR)_{12}]$ to 2.30 V and 2.74 V for the titanium-substituted compounds.

Liu, Y. *et al.* [59] developed an aqueous (0.5 M H_2SO_4) symmetric battery using $H_6[CoW_{12}O_{40}]$ as active species with cobalt as heteroatom, susceptible to undergo several redox processes. The all-$H_6[CoW_{12}O_{40}]$ RFB exhibited a volumetric specific energy of 15.4 Wh L^{-1} and a high CE (99%) and EE (86%). The $[CoW_{12}O_{40}]^{6-}$ anion undergoes a two-step 2-electron redox reaction (for a total of four electrons) at the negative electrode. The voltammogram revealed three pairs of redox reaction peaks between −0.36 and 1.44 V vs SHE (the standard hydrogen electrode). Following previous studies about the POM electrochemistry [60], the reduction and oxidation peaks between 0.74 and 1.44 V correspond to the $[Co^{II}W_{12}O_{40}]^{6-}/[Co^{III}W_{12}O_{40}]^{5-}$ redox reaction, whereas the two pairs of redox peaks between −0.36 and 0.10 V correspond to the $[Co^{II}W_{12}O_{40}]^{6-}/H_2[Co^{II}W_{12}O_{40}]^{6-}$ reduction and $H_2[Co^{II}W_{12}O_{40}]^{6-}/H_4[Co^{II}W_{12}O_{40}]^{6-}$ oxidation. This process is a multi-electron transfer, in the case of tungsten, the reaction is 2-electron and for cobalt a single electron one. Moreover, the potential difference between Co^{II}/Co^{III} and W^{V}/W^{VI} in the heteropolyacid anion (1.2–1.4 V) is an appropriate voltage for the RFBs.

In order to show the importance of the supporting electrolyte Cao *et al.* [61] compared two different POM-based cells, both in non-aqueous media. They studied the effect of the

POM concentration between 10 and 100 mM, so as the organic solvent (N,N-dimethylformamide (DMF) and acetonitrile (ACN)). On the one hand, both the voltage range and number of redox cycles available for $Li_3PMo_{12}O_{40}$ were increased by changing the organic solvent from ACN to DMF. While both solvents have high polarity and low viscosity, giving rise to high conductivity compared with other organic solvents, this POM in DMF can access an extra quasi-reversible, stable and 2-electron redox couple. The redox flow cell voltage obtained for 10 mM $Li_3PMo_{12}O_{40}$ in DMF was 0.7 V for the 2-electron transfers (SoC of 60%) for 30 cycles, compared to 0.35 V for 1-electron transfer (SoC of 80%) for 30 cycles in ACN. On the other hand, in an attempt to boost the energy density by operating the RFB at higher concentration of 100 mM $Li_3PMo_{12}O_{40}$, both solvents demonstrated good performance at faster charging speeds; this represents the highest concentration reported to date for a non-aqueous POM–RFB. The CE was nearly 90% in all the cases with over 30 cycles. Although this POM had already been tested by Chen *et al.* in a non-aqueous media, these results show the importance of the optimization of the supporting electrolyte. In comparison, the symmetric cell based only on $Li_3[PMo_{12}O_{40}]$ showed worse results than the asymmetric one combining the Keggin-type $Li_3[PMo_{12}O_{40}]$ (catholyte) with the Wells–Dawson $Li_6[P_2Mo_{18}O_{62}]$ (anolyte), since the asymmetric cell allows an increase in the cell voltage. $Li_6[P_2W_{18}O_{62}]$ exhibited three 2-electron transfer reactions in the potential range between −2.05 and −0.5 V vs Ag/Ag^+ and these can be coupled with $Li_3PMo_{12}O_{40}$ to utilize a total of 4-electron redox couples between −2.0 V and 0.3 V in an asymmetric cell. For these asymmetric RFBs, both solvents showed promising results with strong stability and higher energy densities. However, DMF exhibited better performance because of the larger number of electron transfers and wider cell voltage. The obtained cell voltage is 1.3 V (SoC of 50%), the CE nearly 100% and the EE 70% over the course of 20 cycles.

Friedl, J. *et al.* [62] also developed an asymmetric POM–RFB in 2018 in an attempt to obtain higher cell voltages. The selected POMs were $[SiW_{12}O_{40}]^{4-}$ as anolyte and $[PV_{14}O_{42}]^{9-}$ as catholyte in aqueous media. During CVs studies (results shown in Fig. 7.7), both polyanions exhibit multi-electron redox reactions in the range of 0.2–0.7 V vs SHE. On the one hand, $[SiW_{12}O_{40}]^{4-}$ experiments three redox processes, two 1-electron transfers and one 2-electron. However, this last one is not able to take part in the RFB, as it occurs at −0.37 V vs SHE and the POM irreversibly modifies the electrode which leads to its decomposition and catalysis of the hydrogen evolution reaction close to this voltage [63]. On the other hand, $[PV_{14}O_{42}]^{9-}$ gives redox overlapped multi-electron transitions of many redox reactions, difficulting the determination of the exact standard potentials and peak separations.

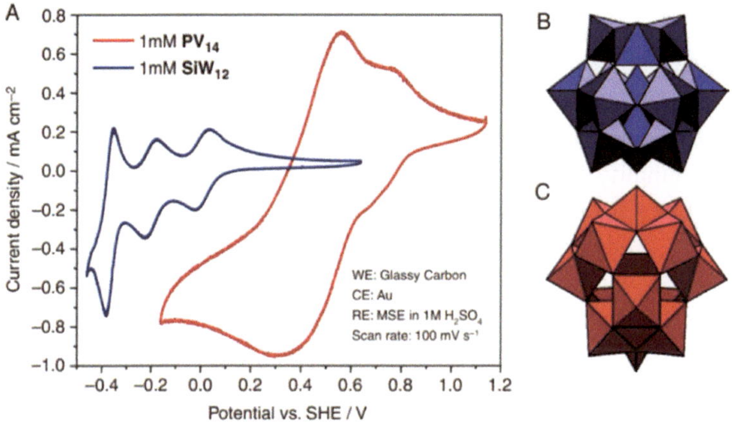

Figure 7.7 CVs and structure of the used POMs. (A) CVs of 1 mM $[SiW_{12}O_{40}]^{4-}$ and 1 mM $[PV_{14}O_{42}]^{9-}$ at a scan rate of 100 mV s^{-1}. (B) Polyhedral representation of $[SiW_{12}O_{40}]^{4-}$. (C) Polyhedral representation of $[PV_{14}O_{42}]^{9-}$. © Copyright license provided by Elsevier and Copyright Clearance Center.

The CE of this asymmetric RFB is 94% during cycling and the VE 65%. Whilst in the symmetric cell the selectivity among species is not as critical, in the asymmetric system, the membrane must be more selective and possibly more expensive in order to avoid cross-contamination. In the Friedl *et al.* system, the crossover of $[SiW_{12}O_{40}]^{4-}$ was monitored through the commercial membranes Nafion-117 (thickness 183 μm) and Fumatech's FUMASEP F-1040 (thickness 40 μm). In contrast with the Fe^{+} cation (selected for the studies as reference), both membranes avoided the permeation of the large polyanion and what it is more, no vanadium was found in the half-cell of $[SiW_{12}O_{40}]^{4-}$ after 14 days of RFB activity.

Although until this moment all the experiments had been carried out in a laboratory scale, Friedl *et al.* [64] developed an upscale setup of their asymmetric cell in two sizes: 25 and 1400 cm^2. On the one hand, the smaller one exhibited a CE of 96%, an EE of 64%, and a 90% of its theoretical capacity was reached (192 mA h). On the other hand, the larger one exhibited higher CE (99.13%) and EE (86.13%). This fact is explained in base of the superior atmospheric containment of the 1400 cm^2 cell, so that the amount of oxygen in the tanks is prevented and the oxidation of $[PV_{14}O_{42}]^{9-}$ avoided. Therefore, the voltage is higher in the larger cell because whilst the CE is mainly given by the redox electrochemistry and the electrode materials, cell construction and its materials determine the VE.

Finally, it is interesting to comment on the studies developed by Chen *et al.* [65], who have reached the highest energy capacity for a POM–RFB (225 Wh L^{-1}) for the moment. Chen attributes the lower reached values until his study to have not yet lived up to POMs potential in terms of storing a large number of electrons in a reversible manner. Thus, they were able to show that the polyoxoanion $[P_2W_{18}O_{62}]^{6-}$ displays notable proton-coupled electron redox activity, which allows this molecule to reversibly accept up to 18 protons and electrons in aqueous solution. This was proved by CV measurement in aqueous solution at different pH conditions and concentrations, which revealed new waves at higher concentrations and a remarkable pH dependence in the amount of electron-transfers.

Based on these results, they developed a three-electrode electrochemical flow cell, which showed that an increase in the concentration of the cluster in aqueous solution at low pH dramatically increases the amount of charge reversibly store on $Li_6[P_2W_{18}O_{62}]$. In each case, the charge passed equated to that which would be expected to bring about an 18-electron reduction of the POM. At low POM concentrations, it is apparent that most of the charge passed during the reduction process cannot be extracted during reoxidation of the POMs solution. Analysis of the headspace of the POMs holding tank indicates that the charge that is not recovered by electrochemical reoxidation is instead liberated as hydrogen.

Once the availability of the Wells–Dawson-type POM had been proved, they decided to explore the properties of the highly reduced $Li_6[P_2W_{18}O_{62}]$, as an electrolyte in a RFB. Therefore, they constructed a system with the cluster as the negative redox couple, using HBr/Br_2 as the positive electrolyte. A discharge capacity density of 42.6 Ah L^{-1} can be achieved at a concentration of 0.1 M with a CE of 96%, as well as an energy density of 43.2 Wh L^{-1}. On increasing the concentration of $Li_6[P_2W_{18}O_{62}]$ to 0.3 and 0.5 M, higher capacities of 131 and 230 Ah L^{-1} can be achieved at 20°C, corresponding to practical energy densities of 130 and 225 Wh L^{-1}, respectively. Meanwhile, the EE at all of the tested concentrations is 76%.

In conclusion, the studies developed by Chen *et al.* revealed the influence on the efficiency and capacity of the final RFB depends on the POM and its properties: the higher the POM concentration and the lower the pH are, the most electron storage is allowed. What it is more, higher efficiencies are potentially expected if using POMs as redox active species in a RFB.

7.4 Conclusion

RFBs have been boosted in recent years as stationary ESS. Even if the all-vanadium RFBs are widespread commercialized, alternative redox active species have been proposed in order to overcome some of its drawbacks. Apart from organic nature compounds, POMs are being deeply studied as electroactive redox species in RFBs field. In this sense, taking into account the state of the art of the POM–RFBs, it may be concluded that POMs are a suitable redox active species for these kinds of device. Nevertheless, there are many aspects to consider in the selection of the proper POM, as not all of them owns the same electrochemical properties. For instance, the supporting electrolyte media may vary the redox reaction which take place, so as the electron-transfers might not be that efficient in both aqueous and non-aqueous media. Additionally, the POM selection must be careful in terms of structure, so as composition, because both charge density and compositional atoms (addenda metal and heteroatoms) are directly responsible of the battery performance. For example, some addenda metal substitution might imply large energy densities, or an organic functionalization might increase the solubility on the electrolyte and as consequence, also the energy density. Moreover, the cell configuration must be taken into account because although some cross-contamination might occur, an asymmetric cell usually implies having a larger operational cell voltage.

To sum up, POM–RFBs are a promising ESS susceptible to be implied in the energy transition to the future electrical grid ES. However, this research field is yet in a low maturity level and open to reach better battery performance.

Acknowledgments

This work has been developed within the framework of Almagrid project, CER-20191006. Call for proposals: accreditation and granting of aid for technological centers of excellence "Cervera".

References

1. Markard, J. (2018). The next phase of the energy transition and its implications for research and policy, *Nat. Energy*, **3**, pp. 628–633.
2. The Paris Agreement, UNFCCC (2015). https://unfccc.int/sites/default/files/english_paris_agreement.pdf (accessed December 20, 2021).
3. Ginley, D., Green, M.A. and Collins, R. (2008). Solar energy conversion toward terawatt, *MRS Bulletin*, **33**, pp. 355–364.
4. Yang, Z., Zhang, J., Kintner-Meyer, M.C.W., Lu, X., Choi, D., Lemmon, J.P. and Liu, J. (2011). Electrochemical energy storage for green grid, *Chem. Rev.*, **111**, pp. 3577–3613.
5. Zhang, H., Li, X. and Zhang, J. (2018) *Redox Flow Batteries: Fundamentals and Applications*, Zhang, H., Li, X., Zhang, J., (eds) Chapter 1 "Large-scale energy storage," CRC Press, New York, pp. 1–39.
6. Fatih-Biro, L. (2010). World Energy Outlook 2010, IEA, Paris https://www.iea.org/reports/world-energy-outlook-2010 (accessed December 20, 2021).

7. Wei, Z., Meng, S., Xiong, B., Ji, D. and Tseng, K.J. (2016). Enhanced online model identification and state of charge estimation for lithium-ion battery with a FBCRLS based observer, *Appl. Energy*, **181**, pp. 332–341.
8. Dunn, B., Kamath, H. and Tarascon, J.M. (2011). Electrical energy storage for the grid: a battery of choices, *Science*, **334**, pp. 928–935.
9. Wang, W., Luo, Q., Li, B., Wei, X., Li, L. and Yang, Z. (2012). Recent progress in redox flow battery research and development, *Adv. Funct. Mater.*, **23**, pp. 970–986.
10. Sánchez-Díez, E., Ventosa, E., Guarnieri, M., Trovò, A., Flox, C., Marcilla, R., Soavi, F., Mazur, P., Aranzabe, E. and Ferret, R. (2021). Redox flow batteries: status and perspective towards sustainable stationary energy storage, *J. Power Sources*, **481**, pp. 228804–228827.
11. Weber, A.C., Mench, M.M., Meyers, J.P., Ross, P.N., Gostick, J.T. and Liu, Q. (2011). Redox flow batteries: a review, *J. Appl. Electrochem.*, **41**, pp. 1137–1164.
12. Pan, F. and Wang, Q. (2015). Redox species of redox flow batteries: a review, *Molecules*, **20**, pp. 20499–20517.
13. Bartolozzi, M., (1989). Development of redox flow battery. A historical bibliography, *J. Power Sources*, **27**, pp. 219–234.
14. Thaller, L.H. (1974). Electrically rechargeable redox flow cells. *Proceedings of 9th Intersociety Energy Conversion Engineering Conference*, NASA TM X-71540, San Francisco, CA, pp. 26–30.
15. Weber, A., Mench, M., Meyers, J., Ross, P., Gostick, J. and Liu, Q. (2011). Redox flow batteries: a review, *J. Appl. Electrochem.*, **41**, pp. 1137–116.
16. Skyllas-Kazacos, M., Rychcik, M., Robins, R.G., Fane, A.G. and Green, M.A. (1986). New all-vanadium redox flow cell, *J. Electrochem. Soc.* **133**, pp. 1057–1058.
17. Mears, L.D., Gotschall, H.L., Key, T. and Kamath, H. (2003). *EPRI-DOE Handbook of Energy Storage for Transmission & Distribution Applications*, EPRI, Palo Alto, CA and the US Department of Energy, Washington, DC.
18. Rajarthnam, G.P. (2016). *The Zinc/Bromine Flow Battery: Fundamentals and Novel Materials for Technology Advancement*, University of Sydney, Faculty of Engineering & Information Technologies, School of Chemical & Biomolecular Engineering, Darlington.
19. Wang, C., Li, X., Xi, X., Zhou, W., Lai, Q. and Zhang, H. (2016). Bimodal highly ordered mesostructure carbon with high activity

for $Br^2/Br-$redox couple in bromine-based batteries, *Nano Energy*, **21**, pp. 217–227.

20. Zhang, L., Zhang, H., Lai, Q., Li, X. and Cheng, Y. (2013). Development of carbon coated membrane for zinc/bromine flow battery with high power density, *J. Power Sources*, **227**, pp. 41–473.
21. Knehr, K. and Kumbur, E.C. (2011). Open circuit voltage of vanadium redox flow batteries: discrepancy between models and experiments, *Electrochem. Commun.*, **13**, pp. 342–345.
22. Lucas, A. and Chondrogiannis, S. (2016). Smart grid energy storage controller for frequency regulation and peak shaving, using a vanadium redox flow battery, *Int. J. Elec. Power*, **80**, pp. 26–36.
23. Viswanathan, V., Crawford, A., Stephenson, D., Kim, S., Wang, W. and Li, B. (2014). Cost and performance model for redox flow batteries, *J. Power Sources*, **247**, pp. 1040–1051.
24. Kaneko, H., Negishi, A., Nozaki, K., Sato, K. and Nakajima, M. (1994). Redox battery. U.S. Patent 5,318,865.
25. Skyllas-Kazacos, M. (2003). Novel vanadium chloride/polyhalide redox flow battery, *J. Power Sources*, **124**, pp. 299–302.
26. Huskinson, B., Marshak, M.P., Suh, C., Er, S., Gerhardt, M.R., Galvin, C.J., Chen, X., Aspuru-Guzik, A., Gordon, R.G., and Aziz, M.J. (2014). A metal-free organic-inorganic aqueous flow battery, *Nature*, **505**, pp. 195–198.
27. Huskinson, B., Marshak, M., Gerhardt, M. and Aziz, M.J. (2014). Cycling of a quinone-bromide flow battery for large-scale electrochemical energy storage, *ECS Trans.*, **61**, pp. 27–30.
28. Lin, K., Gomez-Bombarelli, R., Beh, E.S., Tong, L., Chen, Q., Valle A., AspuruGuzik, A., Aziz, M.J. and Gordon, R.G. (2016). A redox-flow battery with an alloxazine-based organic electrolyte, *Nat. Energy*, **1**, pp. 16102–16109.
29. Janoschka, T., Martin, N., Martin, U., Friebe, C., Morgenstern, S., Hiller, H., Hager, M.D. and Schubert, U.S. (2015). An aqueous, polymer-based redox-flow battery using non-corrosive, safe, and low-cost materials, *Nature*, **527**, pp. 78–81.
30. Liu, T., Wei, X., Nie, Z., Sprenkle, V. and Wang, W. (2016). A total organic aqueous redox flow battery employing a low cost and sustainable methyl viologen anolyte and 4-HOTEMPO catholyte, *Adv. Energy Mater*, **6**, pp. 1501449–1501457.
31. Pratt, H.D., Hudak, N.S., Fang X. and Anderson, T.M. (2013). A polyoxometalate flow battery, *J. Power Sources*, **236**, pp. 259–264.

32. Lide, D.R. (2007) *CRC Handbook of Chemistry and Physics*, 2006–2007 ed, CRC Press, Boca Raton, FL.

33. Ferrigno, R., Stroock, A.D., Clark, T.D., Mayer, M. and Whitesides, G.M. (2002). Membraneless vanadium redox fuel cell using laminar flow, *J. Am. Chem. Soc.*, **124**, pp. 12930–12931.

34. Navalpotro, P., Palma, J., Anderson, M. and Marcilla, R. (2017). A membrane-free redox flow battery with two immiscible redox electrolytes, *Angew. Chem. Int. Ed. Engl.*, **56**, pp. 12460–12465.

35. Duduta, M., Ho, B., Wood, V.C., Limthongkul, P., Brunini, V.E., Carter, W.C. and Chiang, Y.M. (2011). Semi-Solid lithium rechargeable flow battery, *Adv. Energy Mater.*, **1**, pp. 511–516.

36. Liu, P., Cao, Y., Li, G.R., Gao, X.P., Ai, X.P. and Yang, H.X. (2013). A solar rechargeable flow battery based on photoregeneration of two soluble redox couples, *ChemSusChem.*, **6**, pp. 802–806.

37. Wang, K., Wu, Y., Cao, X., Gu, L. and Hu, J. (2020). A Zn–CO_2 flow battery generating electricity and methane, *Adv. Funct. Mater.*, pp. 1908965–1908973.

38. Bominaar, E.L., Achim, C., Borshch, S.A., Girerd, J.J. and Münck, E. (1997). Analysis of exchange interaction and electron delocalization as intramolecular determinants of intermolecular electron-transfer kinetics, *Inorg. Chem.*, **36**, pp. 3689–3701.

39. Mariusz-Kozik, M., Casan-pastor, N., Hammer, C.F. and Baker, L.C.W. (1988). Ring currents in wholly inorganic heteropoly blue complexes. Evaluation by a modification of Evans' susceptibility method, *J. Am. Chem. Soc.*, **110**, pp. 7697–7701.

40. Song, I.K., Kaba, M.S., Nomiya, K., Finke R.G. and Barteau, M.A. (2007). Scanning tunneling microscopy (STM) and tunneling spectroscopy (TS) studies of polyoxometalates (POMs) of the Wells–Dawson class, *J. Mol. Catal. A Chem.*, **262**, pp. 216–226.

41. Friedl, J., Al-Oweini, R., Herpich, M., Keita, B., Kortz U. and Stimming, U. (2014). Electrochemical studies of tri-manganese substituted Keggin Polyoxoanions, *Electrochim. Acta*, **141**, pp. 357–366.

42. Li, L., Kim, S., Wang, W., Vijayakumar, M., Nie, Z., Chen, J., Zhang, J., Xia, G., Hu, J., Craff, G., Liu, J. and Yang, Z. (2011). A stable vanadium redox flow battery with high energy density for large-scale energy storage, *Adv. Energy Mater.*, **1**, 394–400.

43. Chen, J.J., Symes, M.D. and Cronin, L. (2018). Highly reduced and protonated aqueous solutions of $[P_2W_{18}O_{62}]^{6-}$ for on-demand hydrogen generation and energy storage, *Nat. Chem.*, **10**, pp. 1042–1047.

44. Keita, B. and Nadjo, L. (2007) *Encyclopedia of electrochemistry*, Bard, A.J., Stratmann, M., (eds) Chapter 22 "Electrochemistry of isopoly and heteropoly oxometalates," Wiley-VCH, London, pp. 607–695.

45. López, X., Carbó, J.J., Bo, C. and Poblet, J.M. (2012). Structure, properties and reactivity of polyoxometalates: a theoretical perspective, *Chem. Soc. Rev.*, **41**, pp. 7537–7571.

46. Baker, L.C.W. and Figgis, J.S. (1970). New fundamental type of inorganic complex: hybrid between heteropoly and conventional coordination complexes. Possibilities for geometrical isomerisms in 11-, 12-, 17-, and 18-heteropoly derivatives, *J. Am. Chem. Soc.*, **92**, pp. 3794–3797.

47. Selling, A., Andersson, I., Pettersson, L., Schramm, C.M., Downey, S.L. and Grate, J.H. (1994). Multicomponent polyanions. 47: the aqueous vanadophosphate system, *Inorg. Chem.*, **33**, pp. 3141–3150.

48. Dawson, B. (1953). The structure of the 9(18)-heteropoly anion in potassium 9(18)-tungstophosphate, $K_6(P_2W_{18}O_{62})\cdot 14H_2O$, *Acta Crystallogr.*, **6**, pp. 113–126.

49. Ueda, T. (2018). Electrochemistry of polyoxometalates: from fundamental aspects to applications, *ChemElectroChem*, **5**, pp. 823–838.

50. Keita, B., Lucs, T. and Nadjo, L. (1986). New aspects of the electrochemistry of heteropolyacids: reduction currents as a probe of solvent-electrolyte interactions, *J. Electroanal. Chem.*, **208**, pp. 343–356.

51. Uematsu, S., Quan, Z., Suganuma, Y. and Sonoyama, N. (2012). Reversible lithium charge-discharge property of bi-capped Keggin-type polyoxovanadates, *J. Power Sources*, **217**, pp. 13–20.

52. Hartung, S., Bucher, N., Chen, H.Y., Al-Oweini, R., Sreejith, S., Borah, P., Yanli, Z., Kortz, U., Stimming, U., Hoster, H.E. and Srinivasan, M. (2015). Vanadium-based polyoxometalate as new material for sodium-ion battery anodes, *J. Power Sources*, **288**, pp. 270–277.

53. Pratt, H.D. and Anderson, T.M. (2013). Mixed addenda polyoxometalate "solutions" for stationary energy storage, *Dalton Trans.*, **42**, pp. 15650–15655.

54. Pratt, H.D., Pratt, W.R., Fang, X., Hudak, N.S. and Anderson, T.M. (2014). Mixed-metal, structural and substitution effects of polyoxometalates on electrochemical behaviour in a redox flow battery, *Electrochim. Acta*, **138**, pp. 210–214.

55. Chen, J.J.J. and Barteau, M.A. (2017). Molybdenum polyoxometalates as active species for energy storage in non-aqueous media, *J. Energy Storage*, **13**, pp. 255–261.

56. VanGelder, L.E. and Matson, E.M. (2018). Heterometal functionalization yields improved energy density for charge carriers in non-aqueous redox flow batteries, *J Mater. Chem. A*, **6**, pp. 13874–13882.

57. VanGelder, L.E., Petel, B.E., Nachtigall, O., Martinez, G., Brennessel, W.W. and Matson, E.M. (2018). Organic functionalization of polyoxovanadate-alkoxide clusters: improving the solubility of multimetallic charge carriers for nonaqueous redox flow batteries, *Chem. Sus. Chem.*, **11**, 23, pp. 4139–4149.

58. VanGelder, L.E., Kosswattaarachchi, A.M., Forrestel, P.L., Cook, T.R. and Matson, E.M. (2018). Polyoxovanadate-alkoxide clusters as multi-electron charge carriers for symmetric non-aqueous redox flow batteries, *Chem. Sci.*, **9**, pp. 1692–1699.

59. Liu, Y., Lu, S., Wang, H., Yang, C., Su, X. and Xiang, Y. (2017). An aqueous redox flow battery with a tungsten-cobalt heteropolyacid as the electrolyte for both the anode and cathode, *Adv. Energy Mater.*, **7**, pp. 1601224–1601230.

60. Prados, R.A. and Pope, M.T. (1976). Low-temperature electron spin resonance spectra of heteropoly blues derived from some 1:12 and 2:18 molybdates and tungstates, *Inorg. Chem.*, **15**, pp. 2547–2553.

61. Cao, Y., Chen, J.J.J. and Barteau, M.A. (2020). Systematic approaches to improving the performance of polyoxometalates in non-aqueous redox flow batteries, *J. Energy Chem.*, **50**, pp. 115–124.

62. Friedl, J., Holland-Cunz, M.V., Cording, F., Pfanschiling, F., Wills, C., McFarlane, W., Schricker, B., Fleck, R., Wolfschmidt, H. and Stimming, U. (2018). Asymmetric polyoxometalate electrolytes for advanced redox flow batteries, *Energy Environ. Sci.*, **11**, 3010–3018.

63. Keita, B. and Nadjo, L. (1987). New aspects of the electrochemistry of heteropolyacids: Part IV. Acidity dependent cyclic voltammetric behaviour of phosphotungstic and silicotungstic heteropolyanions in water and N,N-dimethylformamide, *J. Electroanal. Chem. Inter. Electrochem.*, **227**, pp. 77–98.

64. Friedl, J., Pfanschilling, F.L., Holland-Cunz, M.V., Fleck, R., Schricker, B., Wolfschmidt, H. and Stimming, U. (2019). A polyoxometalate redox flow battery: functionality and upscale, *Clean Energy*, **3**, pp. 278–287.

Chapter 8

Polyoxometalates with Anticancer, Antibacterial and Antiviral Activities

Manuel Aureliano,[a,b] Dorinda Marques-da-Silva,[c,d] Ana Serrano,[a,b] João Martins,[a,b] Leonor Faleiro,[a,e] Custódia Fonseca,[a,b] Gil Fraqueza,[b,f] and Ricardo Lagoa[c,d]

[a]*FCT, Universidade do Algarve, 8005-139 Faro, Portugal*
[b]*CCMar, Universidade do Algarve, Portugal*
[c]*ESTG, Polytechnic Institute of Leiria, 2411-901 Leiria, Portugal*
[d]*UCIBIO, Faculty of Science and Technology, University NOVA of Lisbon, 1099-085 Lisbon, Portugal*
[e]*ABC-RI, Universidade do Algarve, Portugal*
[f]*ISE, Universidade do Algarve, 8005-139 Faro, Portugal*

maalves@ualg.pt

Polyoxometalates (POMs) are versatile clusters used in a large scale of fields, and have become of a great interest in biomedicinal studies, where many compounds have proved to have anticancer, antibacterial, and antiviral activities. The present chapter covers recent advances of decavanadate (V_{10}) and other POMs as potential candidates for anticancer, antibacterial, and antiviral effects with a particular interest in focus on their mechanisms of action. Decavanadate as well as others several types of POMs demonstrated anticancer activities and inhibition of bacterial proliferation at different concentration values ranging from nM to μM. Moreover, in most cases, when compared to previously

Polyoxometalates: Advances, Properties, and Applications
Edited by Leire Ruiz Rubio, José Luis Vilas Vilela, Beñat Artetxe, and Juan Manuel Gutiérrez-Zorrilla

ISBN 978-981-4968-14-0 (Hardcover), 978-1-003-27744-6 (eBook)
www.jennystanford.com

approved drugs, POMs showed to be more efficient. However, even if the antibacterial, antiviral, and anticancer activities of decavanadate and POMs are well-documented and put forward, their mechanism of action is still not completely understood. The decavanadate inhibition of cancer cell proliferation could be associated with the inhibition of P-type ATPases and cytoskeleton dynamics, whereas the Dawson-type POMs antibacterial activity is suggested to be associated with the strong inhibition of sialyl and sulfotransferases along with phosphatases. Furthermore, several POMs have revealed antiviral activity by targeting viral proteins, thus, preventing the first stage of viral attachment and inhibiting fusion of virions with host cells. Putting it all together, POMs, including decavanadate, present anticancer, antibacterial, and antiviral activities. It seems that some POMs are better suited for antibacterial, anticancer, and antiviral activities than others, although the mechanism of action and the biomolecular targets associated with the biological activities are not clearly recognized. In any case, POMs are promising inorganic drugs and further biomedical, microbial, and biochemical *in vivo* and *in vitro* studies must be conducted to deduce which POM is tuned against a particular cancer cell, bacteria, or viral particle.

8.1 Introduction

Polyoxometalates (POMs) are metal oxo anions of transition metals, such as Mo, W, and V, amenable to a variety of structural transformations. Besides that, POM structures can either include other elements or have one of the major metal oxo anions (e.g., WO_6, VO_6) missing and/or substituted by other metals, such as Co or Mn. Examples of some of those structures can be observed in Fig. 8.1. The diversity of these POMs structures was previously summarized elsewhere [1, 2]. Thus, due to all these diverse factors, POMs can present outstanding physicochemical properties for which biochemical and biomedical applications remain yet to be determined. Nevertheless these inorganic clusters have vigorously been studied in environmental, chemical, and industrial fields, having applications in catalysis, prevention of corrosion, and macromolecular crystallography, among others [3–6]. After a research in the Web of Knowledge, it was found

10347 articles, and that the number of articles about "POMs," in the last 10 years (2010–2019) more than triplicate in comparison to the previous period (2000–2009) (Fig. 8.2A).

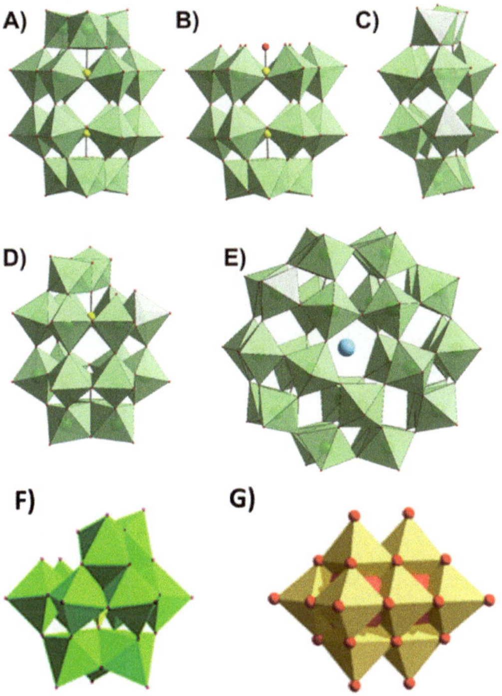

Figure 8.1 Polyhedral representation of polyoxotungstate structures (A–E), polyoxovanadate (F), and decavanadate (G) structures. (A) Wells–Dawson anion, P_2W_{18}; (B) tri-lacunary Wells–Dawson anion, P_2W_{15}; (C) hexa-lacunary Wells–Dawson anion, P_2W_{12}; (D) mono-lacunary Wells–Dawson anion, P_2W_{17}; (E) Preyssler anion, P_5W_{30}; (F) mono-lacunary Keggin, MnV_{11}, (G) isopolyvanadate, V_{10}. Color code (A–E): $\{WO_6\}$, mint; P, yellow; O, red; Na, blue; (F) $\{VO_6\}$, green; Mn, yellow; O, red; (G) $\{VO_6\}$, yellow; O, red.

Regarding to the biological applications of POMs, a fundamental review in 1998 referred that the main activities of POMs in medicine are represented by antiviral and antitumor ones [7]. In the last 20 years, several papers described POMs as potential inhibitors of bacteria, viruses, tumor proliferation, as well as potential drugs for the treatment of several diseases, such as Alzheimer's and diabetes [1, 2, 8–25]. The total number of papers in these three areas is only about 220 but it also increased in the last years since 70% are from the last 10 years. In particularly, we found 104 articles for anticancer studies, 84 for antibacterial studies, and 42 for antiviral studies, being the later area the less representative (Fig. 8.2B).

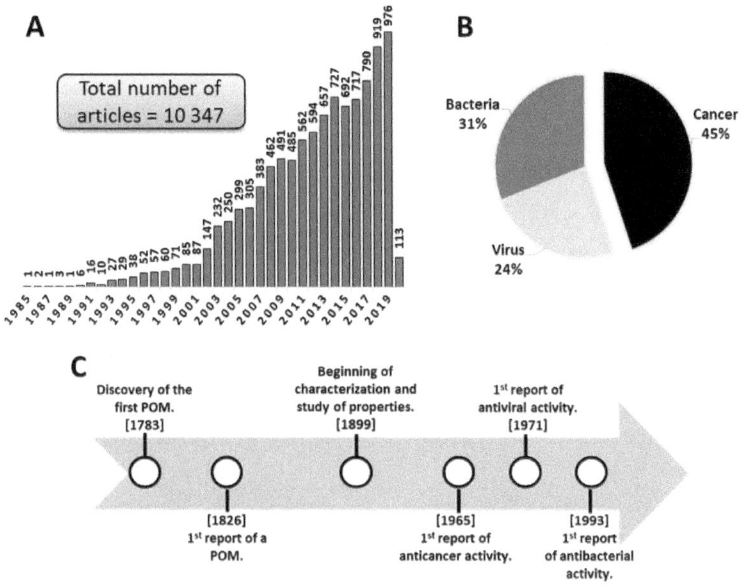

Figure 8.2 Publication of studies involving POMs. (A) Number of publication along the last 35 years; (B) relative contribution of POMs in anticancer, antibacterial, and antiviral studies; (C) chronological timeline representing the discovery of POMs, their first report as well as their first studies related to characterization and anticancer, antiviral, and antibacterial activities.

In fact, two centuries have passed since the first discovery of a POM $[(NH_4)_3[PMo_{12}O_{40}]]$ all the way in 1783 by the D'Elhuyar brothers [26]. Only 43 years later that same POM was firstly reported by Berzelius [27]. The beginning of characterization and study of their properties started in 1899 by Rosenheim [28] and, finally, the first reports of anticancer, antiviral, and antibacterial activities of POMs emerged in 1965, 1971, and 1993, respectively [29–31] (Fig. 8.2C).

The increasing resistance effect of cancer cells and the high toxicity of chemotherapeutic agents, together with growing cancer incidence all around the world, beg for a new therapeutic drug. In this aspect, POMs have been selected by some researchers as alternative antitumor substances with promising results in tumor growth suppression [8–12, 14–17, 22, 25, 32–39]. Similarly, antibiotic resistance represents a real threat to the world public health seen as a high proportion of bacteria are resistant to antibiotics. These multi-resistant bacteria can cause several type of infections such as urinary and blood but also pneumonia when spread through the lungs. In late December 2019, it was reported clusters of patients with atypical pneumonia [40]. Further investigation leads to the identification of the source of pneumonia clusters, a novel coronavirus, SARS-CoV-2 [40]. Thus, in this Chapter 8, we will describe several studies within these three fields of research, compare modes of action of decavanadate and other POMs, and ultimately combine these considerations with the available evidence of protein–decavanadate and protein–POMs interactions in a way to better understand the biochemical mechanisms and processes of anticancer, antibacterial and antiviral activities of POMs.

8.2 Antitumor Activity of POMs

The number of papers that have been published regarding POMs with antitumor activities clearly increased in the past 10 years, representing about 85% of all the papers in this field (Fig. 8.3).

Whereas the first study described for V_{10} and polyoxovanadates (POVs) in cancer was in 2009 [35], polyoxotungstates (POTs) and polyoxomolybdates (POMos) were first reported in 1965 [29]. Note that the first article to report anticancer activity of a POM, it was also the first one to test it *in vivo* [29]. However, *in vivo* studies remain scarce up now. More recently (2011), the anticancer activity of a POM-nanoparticle complex [33] was evaluated (Fig. 8.3).

After a brief research in Web of Science about POMs and cancer, 107 papers were found. Among them, studies with decavanadate, POVs, POTs, POMos, and other types of POMs, such as polyoxoniobates, can be found. For the analysis of each type of POM, their state was considered, for example, if they are pure, hybrid, encapsulated, nanoparticle, or other POM-based compound. For cancer studies, the contribution of the several POMs showed to be different being POTs and POMos the ones with a higher percentage of studies (Fig. 8.4).

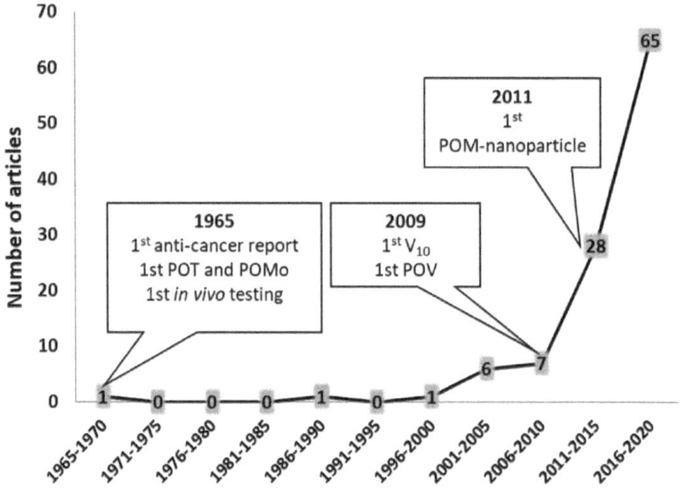

Figure 8.3 Number of papers about cancer and POMs since 1965. The first studies described for POTs, POMos, POVs, and V_{10} as well as for POM-nanoparticle are highlighted.

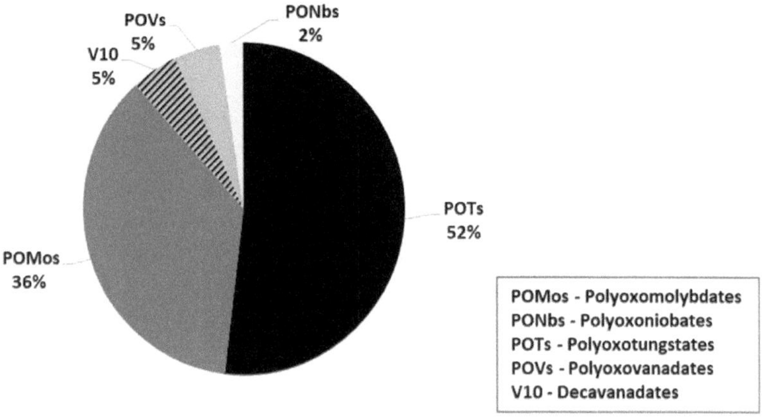

Figure 8.4 Studies for anticancer activities of different POMs. Studies found after a research in the portal Web of Science using the keywords "cancer" and "polyoxometalates."

8.2.1 Decavanadate and POVs

Recently, in 2010, Li *et al.* synthesized two decavanadate compounds $(H_2tmem)_3[V_{10}O_{28}]$ and $(H_2en)_3[V_{10}O_{28}]$, and tested their antitumor activity *in vitro* against human lung carcinoma cells (A549) and murine leukemia cells (P388) [12]. Both compounds exhibited lower inhibition than cisplatin, but the one with a higher lipophilic effect, showed higher inhibitory activity. Such results were explained by the presence of higher methyl substitutes in the first compound that confers a higher lipophilic effect, thus enhancing its uptake through the lipid bilayer of the cell membrane [12]. The cytotoxicity of both POVs was tested on human normal hepatocytes (LO2) and resulted in an IC_{50} of 6.5 μM for $(H_2tmem)_3[V_{10}O_{28}]$ and 7.2 μM for $(H_2en)_3[V_{10}O_{28}]$. More recently, a decavanadate-based compound was synthesized with carnitine resulting in a V_{10}-carnitine complex that proved to exhibit antitumor activity against various cancer cell lines, including specific toxicity against human cancer cells, whereas normal human cells were not affected even for high concentrations of the complex [39]. This complex presented IC_{50} values of 0.72 and 1.8 μM against human lung adenocarcinoma cell line (A549) and human breast adenocarcinoma cell line (MDA-MB-231), respectively. When V_{10}-carnitine was compared to the antitumor drug cisplatin for its cytotoxicity, it exhibited lower cytotoxicity against A549 than cisplatin, while its IC_{50} for MDA-MB-231 cells was 1.7 μM against 700 μM of cisplatin meaning that it is 400 times more effective.

Zhai *et al.* have synthesized a cobalt-based decavanadate (CoV_{10}) and tested its biological activity against human liver (SMMC-7721) and ovary (SKOV-3) cancer cell lines [35]. Results showed that CoV_{10} exhibited antitumor activity as it could inhibit cell proliferation by 95% and 90% of SMMC-7721 and SKOV-3, respectively, at 6.25 μg/mL with an IC_{50} value estimated lower than 0.26 μg/mL for both cancer cell lines. The potency of CoV_{10} as an antitumor drug was compared with the already approved drug fluorouracil (5-FU) and verified to be more efficient. In fact, at 1.5 μg/mL, the activity of CoV_{10} against SMMC-7721 was similar to 5-FU at 25 μg/mL [35], that is, 15 times more efficient.

Besides the studies described above, Table 8.1 summarizes the studies published with V_{10} and others POVs in cancer in the last five years.

Table 8.1 Studies published during the past five years on the anticancer activity of polyoxovanadates (V_{10} and others POVs)

POVs	Cell line	Effects	Refs
V_{10}	HeLa	Induced apoptosis	[41]
	Hep-2		[41]
	HepG2		[41]
	MDA-MB-231		[41]
Other POVs	MCF-7	DNA, BSA, and HSA binding	[42]
	MDA-MB-231	Induced apoptosis G2/M phase cell cycle arrest	[42]
	U-87	n.d.	[43]
	SMMC-7721	Cell cycle arrest, DNA damage, induced apoptosis	[44]

8.2.2 POMos and POTs

Besides POVs, several papers have been published regarding other types of POMs, such as POMos and POTs in antitumor studies [8–12, 14–17, 22, 25, 33, 39, 45]. Indeed, a POMo, $[NH_3Pr]_6[Mo_7O_{24}]$ (named as PM-8), was proven to enhance antitumor activity against various human cancer cell lines and human cancer xenografts, with higher inhibitory rates than already approved drugs, such as cisplatin and 5-FU [9, 25, 36]. Another POMo from the same research group, $[Me_3NH]_6[H_2Mo_{12}O_{28}(OH)_{12}(MoO_3)_4]2H_2O$ (named as PM-17), also proved to enhance antitumor effect against pancreatic cancer cells (AsPC-1) transplanted in nude mice and against human gastric adenocarcinoma cells (MKN45) *in vitro* [36]. This POMo demonstrated interferences with the solid tumor growth with inhibition rates of 33.5 and 68.3% at a concentration of 125 and 500 μg/body/day in a dose-dependent manner. The *in vitro* inhibition of PM-17 showed an IC_{50} of 175 and 40 μg/mL against AsPC-1 and MKN45 cells, respectively [36]. Moreover, recent findings describe that the degradability of an organic POMo, based in Mo_6O_{18}, is the key to inhibit human malignant glioma cells (U251), besides having the capacity to cross the blood brain barrier. This new synthesized complex points out a new type of anticancer agents [46].

A cobalt-based POT, $(Himi)_2[Bi_2W_{20}O_{66}(OH)_4\ Co_2(H_2O)_6Na_4]$ (BWCN) was studied by Wang *et al.* against human cancer cell lines [11]. The inhibitory rates of BWCN, at 160 μmol/L, were approximately 85, 78, and 57% for cell lines of human gastric cancer (SGC-7901), colon cancer (HT-29), and liver cancer (HepG2), respectively. Comparative results between BWCN, 5-FU, and oxaliplatin showed that the cobalt-based POT exhibited greater inhibitory action with rates of 78, 25, and 42%, respectively, on HT-29 cells [11]. Raza *et al.* [34] have evaluated the cytotoxicity of seven POTs against lung carcinoma cell lines (H157) and human corneal epithelial cells (HCEC). Once more, all compounds showed high antitumor activity in a dose-dependent manner. The most efficient POT, $Na_{16}[(O_3POPO_3)_4W_{12}O_{36}]$ presented 65% of cytotoxicity at 100 μM for against H157.

As referred above, the number of biological studies with POMs, more focused in their anticancer activities, has been growing [2, 10, 11, 38, 39, 42, 46, 47]. In what it refers to POTs and POMos, we can see in Fig. 8.5 the types of cancer studied in the last five years using these POMs. We found 18 papers for POTs and 21 for POMos on cancer POMos were more studied in breast cancer whereas POTs in cervix cancer. Among them, only three POT studies [14, 46, 48] and only one POMos study [49] were performed *in vivo*.

A copper heteropolyoxotungstate, (abbreviated PW_9O_{34}), has been studied against human osteosarcoma derived cell line (MG-63) [38] revealing a good antitumor activity with an IC_{50} value against MG-63 of 22 μM. Furthermore, its potency as an antitumor agent was tested against the reference drug cisplatin: at 25 μM, this POT induced a decrease of 60% in cell viability, whereas cisplatin only exhibited 15%, that is, four times less. Keggin type POTs, $CoW_{11}CpZr$, $CoW_{11}CpTi$, and $CoW_{11}CpFe$, were tested against human cancer cells HEK-293 and MCF-7 for their antitumor activity [15]. $CoW_{11}CpFe$ presents higher antitumor effect and lower IC_{50} compared to the other two POMs tested. Cell survival of MCF-7 and HEK-293 after 24h of incubation with 40 μM of this POT was approximately 42 and 45% with IC_{50} values of 6.3 and 8.7 μM, respectively. It has been suggested that this higher activity could be related to the smaller size of $CoW_{11}CpFe$, thus being more effective in its penetration into the cell. Besides their biological activity, Dianat *et al.* demonstrated that these three POMs had ctDNA-binding properties. Indeed, $CoW_{11}CpZr$, $CoW_{11}CpTi$, and $CoW_{11}CpFe$ could bind, through electrostatic interaction, the groove or the outside stacking of ctDNA, producing a change in the redox potential of POMs, rendering them more negative [15].

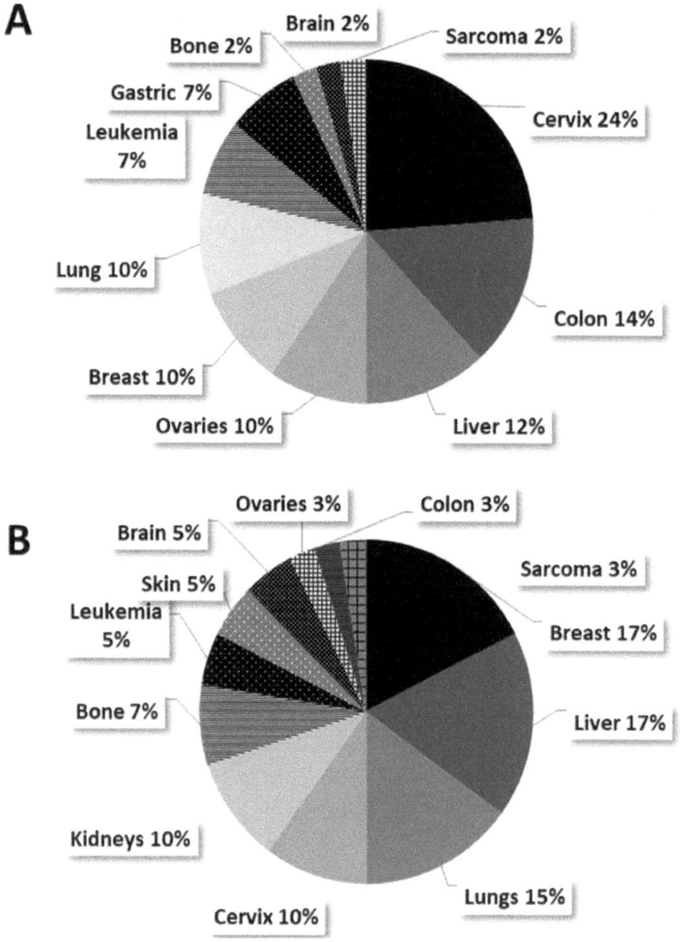

Figure 8.5 Percentages of types of cancer studied in the last five years with POMs. (A) Studies using in POTs; (B) studies using POMos. The pattern covers the cancers more represented to the ones less represented.

A recent study demonstrated the antitumor activity of an iron hepta-tungsten phosphate POM complex, $Na_{12}H[Fe(HPW_7O_{28})_2]$ (IHTPO), against large cell lung cancer (NCi-H460), human hepatoma (HepG2), leukemia (K-562), and lung carcinoma (A549) *in vitro,* and against S180 sarcoma transplanted in mice *in vivo* [14]. IHTPO showed high inhibitory rate only at high concentration (500 μM) with a high IC_{50} against all cell lines tested *in vitro*. In fact, when compared to cisplatin, IHTPO had IC_{50} values of 150.5, 141.5, 63.8, and 114.5 μM for A549, H460, HepG2, and K-562, respectively, whereas cisplatin showed IC_{50} values of 12.1, 1.82, 1.33, and 2.73 μM [14], representing 5–12 times less. Even though cytotoxic effects were only seen at a higher concentration with IC_{50} values superior than 60 μM, IHPTO proved to be more efficient against S180 sarcoma transplanted mice. Apparently, this POT complex could promote splenocytes proliferation and increase IgG, IgG2a/b antibody levels in tumor-bearing mice. The inhibition of tumor growth was observed at 40, 60, and 80 mg/kg with inhibitory rate of 38.9, 42.5, and 45.7%, respectively. Nevertheless, the inhibitory rate of the chemotherapeutic drug cytoxan (CTX) was 72.1% at 50 mg/kg. Even if this POT exhibited lower antitumor activity than the already approved chemotherapeutic drugs, such as cisplatin and CTX, the interesting part is that IHTPO activity might be correlated to an immunomodulatory activity [14].

Fu *et al.* have synthesized an amphiphilic organic–inorganic hybrid POT, $[(C_{16}H_{33})_2NCONH(CH_2)_3SiNaP_5W_{29}O_{110}]$ (Na-lipidP_5W_{29}), to improve biocompatibility, bioactivity, and biospecificity [16]. Basically, a long chain organoalkoxysilane lipid was grafted into a lacunary Preyssler-type, $K_{14}[NaP_5W_{29}O_{107}]$ (P_5W_{29}) in order to produce the desired complex. The hybrid POT, Na-lipidP_5W_{29}, was tested for its antitumor activity against human colorectal cancer cells (HT29) and the results were compared to the parent POT, P_5W_{29}, and to 5-FU. At all concentrations tested Na-lipidP_5W_{29} exhibited higher inhibitory rates than its parent POT and 5-FU. In fact, after 36h of incubation, Na-lipid P_5W_{29} (10 μM) had an inhibitory rate of 83% with an IC_{50} of 2.1 μM whereas P_5W_{29} (10 μM) presented an inhibition ratio of 75% (IC_{50} = 3.6 μM) and 5-FU (100 μM) an inhibition of 27%. The cytotoxic effect of the studied POT was also tested against human umbilical vein endothelial cells (HUVECs) at a concentration lower than 100 μM, and the inhibitory rates were below 20%. Finally, it was suggested that the higher antitumor effect of Na-lipidP_5W_{29} was due to its higher capacity to penetrate the cell, since it can spontaneously assemble into a vesicle [16]. An *in vivo* study was performed with a Keggin-type POT, $PW_{11}O_{39}$ (PW_{11}) against colorectal cancer [50]. To improve bioactivity and decrease the toxicity effect of this POT, an organometallic derivative of PW_{11} was synthesized and encapsulated to form nanoparticles of Pt^{IV}-PW11-DSPE-PEG2000 (NPs). Results showed that these NPs were more efficient in inhibiting the growth of WT20 cancer cells and treating human colorectal cancer in mice, than the classic cisplatin.

8.3 Mechanisms of Action of POMs as Anticancer Agents

The mechanisms of action of POMs as antitumor agents are not yet fully understood. Several possible mechanisms leading to apoptosis have been evoked but nothing can concretely be concluded [10, 11, 25, 39, 45]. Recently, the putative mechanisms of POMs as anticancer agents were reviewed [2]. In this section, we will resume some of them.

8.3.1 Effects in Mitochondria, Oxidative Stress, and Mechanisms of Cell Death

Zhao *et al.* studied the expression of Bax, Bim (two apoptotic promoting family members of Bcl-2) and Bcl-2 in HepG2 cells treated with P_2W_{18} [10]. It was observed that Bax and Bim expression was affected by P_2W_{18} whereas Bcl-2 expression was not affected. On the other hand, Ogata *et al.* have evaluated that PM-17 induced caspase-3 activation on both MKN45 and AsPC-1 cells [9]. Wang *et al.* also measured the cleaved-caspase-3 levels on HT-29 cells treated with the cobalt-based polyoxometalate BWCN. Results showed an increased expression of procaspase-3 and cleaved-caspase-3 in a dose-dependent manner [11]. Others studies suggested that some POMs, such as PM-8, entered into the mitochondrion leading to the inhibition of adenosine triphosphate (ATP) synthesis and induced apoptosis [25]. When the POMo enters in the mitochondrion it can assemble with a Flavin mononucleotide (FMN) and yield a 1:1 POM–FMN complex [25]. FMN, also known as riboflavin-5′-phosphate, functions as a prosthetic group serving as an electron carrier by being alternately oxidized (FMN) and reduced (FMNH2). FMN is found in mitochondria and participates in steps of the electron transport chain. The electrochemical proton gradient resulting from these transfers drives ATP synthesis. Therefore, by forming a complex with FMN, PM-8 inhibits the production of ATP in the cell, resulting in cell degeneration and apoptosis. Other study has also shown cell cycle arrest and apoptosis as a possible mechanism (this time for a POT, PW_9Cu), as well as, the induction of oxidative stress in the cell [38]. Recently, the cell cycle arrest at G2/M phase in association with an increase of p21 and decrease of cyclin A and cyclin B1, and induction of apoptosis mediated through mitochondria apoptotic pathway in association with the increase of BH3-only proteins Noxa and Hrk by POM-HDACi PAC-320 were described [47].

Regarding studies with POVs, it was referred that *in vivo* intraperitoneal and intravenous administration of decavanadate (V_{10}) induce different changes in several stress markers, such as reduced GSH content, overall rate of reactive oxygen species (ROS) production, lipid peroxidation, and antioxidant enzyme activities, when compared to the ones observed in vanadate studies [51–53]. Moreover, it was suggested that the mitochondrial fraction tends to accumulate more vanadium upon the administration of decavanadate than the administration of vanadate, in addition to inducing different changes in antioxidant mitochondrial enzymes activities [53]. By targeting mitochondria, V_{10} might induce, directly or indirectly, cell death processes [54]. It was also described that the inhibition of the mitochondrial respiratory chain by decavanadate can lead to sustained mitochondrial depolarization which demands a high supply of metabolic energy and is lethal for the cell [55, 56]. In fact, *in vitro* studies with hepatic and cardiac mitochondria have shown that mitochondrial depolarization (IC_{50}, 40 nM) and oxygen consumption (IC_{50}, 99 nM), which cause reduction of cytochrome b (complex III), are strongly affected by decavanadate [56]. Mitochondria are a well-known organelle responsible for many features and processes of cell death, such as apoptosis and necrosis, and calcium homeostasis. It was shown in cardiomyocytes that the decavanadate species induces necrotic cell death, whereas no significant caspase-3 increase was observed [54]. This suggests that caspase-3 activation, one of the most established markers of apoptosis, is not affected in these cells. The anti-proliferation activity of another POV, V_{18}, mediated by the arrest of MCF-7 cells in the G2/M phase and induction of apoptosis and necrosis was also described [42].

Tumor cells need redox homeostasis adaptations to cope with the higher ROS production associated to rapid growth and metastasis, so disturbance of antioxidant systems is a plausible anticancer strategy. León *et al.* (2014) found that PW_9Cu concentrations that induced osteosarcoma cells death *in vitro* also increased ROS and decreased the GSH/GSSG ratio in the cells [38]. Moreover, the cytotoxicity of the compound was alleviated with addition of GSH (1 mM) or mixture of vitamin C and E (50 μM each) to cell media, although only partially, suggesting that oxidative stress is a mechanism of POMs to induce cancer cell death.

8.3.2 Autophagy and POMs

The autophagy is characterized by a degradation process where obsolete parts of the cell are digested inside lysosomes. This process seems to begin with the confinement of an organelle through unknown membranes, originating an autophagosome which then merges with a lysosome [57]. Ogata *et al.* have observed the presence of vesicles containing cytosolic components and organelles, described as figurative autophagosomes, in MKN45 and AsPC-1 cells treated with PM-8 and PM-17 [9]. This study on LC3, a protein conjugation system required for the autophagosome formation as well as its transport and maturation showed that in AsPC-1 cells the LC3 expression increased proportionally with time, confirming the hypothesis of an autophagic pathway [9]. To our knowledge, studies with POVs and autophagy are scarce or inexistent. Nevertheless, a plausible mechanism for the prevention of autophagy by decavanadate, for instance, would be the inhibition of the endoplasmic reticulum (ER) calcium pump

(Ca^{2+}-ATPase), thus inducing a depletion of calcium in the ER. Decavanadate was also described to prevent bone mineralization by activation of both MAPK and PI-3K pathways, besides inhibiting enzymatic activity of alkaline phosphatase (ALP) [58, 59]. It is suggested that by acting through these signaling pathways, decavanadate will inhibit the autophagy process.

8.3.3 Inhibition of Ecto-Nucleosidases and Histone Deacetylases (HDACs)

Several studies point out that POMs are strong and selective inhibitors of ecto-ATPases [18, 32]. The ecto-ATPases, also known as ecto-nucleoside triphosphate diphosphohydrolase (E-NTPDase), are able to hydrolyze extracellular nucleotide, such as ATP, ADP, UTP, and UDP. These nucleotides regulate several tissue functions via specific receptors called P2. The activation of such receptors triggers multiple processes affecting, among other mechanisms, the cellular metabolism. The inhibition of E-NTPD leads to the inactivation of P2 receptors, disturbing the cellular transport and signaling. Müller *et al.* have demonstrated that six different POTs, presenting different sizes, shapes, and charge, could differently inhibit three kinds of NTDPases. Among all these POMs, two have proven antitumor activity [32], suggesting correlation with the inhibition of ecto-nucleosides.

In eukaryotic cells, histone molecules form a complex with cellular DNA called chromatin. Chromatin has a compact organization in which most DNA sequences are structurally inaccessible and functionally inactive. Histones are submitted to different kinds of epigenetic modifications (acetylation, methylation, and phosphorylation). The acetylation (induced by histone acetylase) of lysine in the N-terminal tails of histones is related with a looser chromatin state and a gene transcription activation, whereas the deacetylation (induced by HDAC) of the same lysine residues correspond to a more condensed state and gene silencing.

Tumor progression can be induced by epigenetics modifications. Studies have reported that an aberrant recruitment and altered expression of HDAC are responsible for the repression of tumor-suppressor genes leading to malignancies. HDAC inhibitors are able to inhibit tumor cell growth causing a transcriptional change in genes that regulate biological processes in tumor proliferation [18].

Dong *et al.* have discovered POM-based HDACs inhibitors enhancing strong antitumor activity [33]. Three different germanotungstate Keggin-types and one phosphotungstate Dawson-type were used against several tumor cell lines (A549, SW620, HepG2, MM-231, and MGC-803). It was shown that the Dawson-type exhibits higher inhibitory effect against HDAC than others POTs [33]. Besides being the stronger HDAC inhibitor, the Dawson-type has demonstrated good antitumor activity (IC_{50} of 19.2 μg/mL on HepG2) and low cytotoxicity (IC_{50} of 42.8 μg/mL on L-02). Besides, tests *in vivo* have proved that the POT phosphotungstate could induce hyperacetylation of histone H3 in a dose-dependent manner [33].

8.3.4 Inhibition of ALPs, Kinases, P-type ATPases and Aquaporins (AQPs)

Decavanadate species and POMs were described as strong inhibitors of phosphatases, such as ALP [18, 23, 24, 34]. Decavanadate and POMs were assessed for their inhibitory effect on ALP and as putative antitumor agent [34]. Seven POTs were assessed for their inhibitory effect on ALP and as putative antitumor agent [34]. Abnormal levels of ALP in the serum are detected in cancer patients since tumors are abnormal cellular growth proliferating faster than normal cells, and thus, the inhibition of ALP will affect tumor cell metabolism and function. For a concentration of 100 μM, all compounds demonstrated more than 70% inhibition on intestinal alkaline phosphatase (IAP) and tissue nonspecific alkaline phosphatase (TNAP). Two of these POTs, induced higher inhibition on IAP and TNAP, respectively, with Ki values of 313 and 135 nM. More recently, three different POMs, P_5W_{30}, V_{10}, and TeW_6, with chitosan-encapsulated nanoassemblies were tested as anticancer agents on HeLa cells [19]. The maximum cytotoxicity against HeLa cells was observed for the compound CTS-P_5W_{30} which also has higher phosphatase inhibition. It was suggested, in both studies, that the POT with the largest number of tungsten and phosphorus atoms may provide most optimum interaction with the phosphatase [19, 34].

P-type ATPases play a crucial role in cellular ion homeostasis and have been described as potential molecular targets for several types of compounds used as anti-ulcers, anticancer, and heart ischemic failure, among other diseases. Among these compounds, several POMs have been described as P-type inhibitors [60, 61] and the effects compare with therapeutic drugs [62]. Recent insights into decavanadate interactions with the Ca^{2+}-ATPase can be very useful to understand the mechanism of action of POMs as antitumoral agents. As described recently, decavanadate non-competitive interaction with the Ca^{2+}-ATPase strongly inhibits the ATPase activity (IC_{50} = 15 μM) [60]. Further studies referred to Ca^{2+}-ATPase cysteine oxidation upon V_{10} incubation is not prevented by antioxidants. It was suggested the involvement of cysteines at the V_{10} binding site as well as the participating of vanadyl species on the process of enzymatic inhibition [61]. It was also previously suggested that the Ca^{2+}-ATPase V_{10} binding site, which is formed by three protein domains [63], is located at the cellular cytoplasmic side. V_{10} can interact with the protein by electrostatic interaction or by hydrogen bonding. However, the specific residues involved in the V_{10}-Ca^{2+}-ATPase interaction, perhaps with a cysteine residue, are yet to be totally clarified [61]. Furthermore, because it was verified that V_{10} can bind to the two conformations E1 and E2, (contrary to vanadate), being phosphorylated or not, it can target both E1 and E2 ATPase conformations [60]. Taking this observation into consideration, it was suggested that decavanadate could interact with the ion pumps without the necessity to cross the membrane from the extracellular to the intracellular side [62]. Therefore, decavanadate interactions with membrane proteins can more rapidly induce changes in calcium homeostasis with implications in, for example, muscle contraction, calcium accumulation in mitochondria, and concomitantly ROS production and cell death.

Besides decavanadate, others POMs, such as the POT P_2W_{18}, have been observed to inhibit Ca^{2+}-ATPase with IC_{50} values of inhibition about 30 times lower (IC_{50} = 0.6 μM) than decavanadate, and with a mixed type of inhibition [64], on contrary to the non-competitive mode of inhibition observed for V_{10} and decaniobate (Nb_{10}) [60]. Recently, AQPs were also described to be potential protein membrane targets for POTs [65]. AQPs were found to be overexpressed in tumors making their inhibitors of particular interest as anticancer drugs [66]. It was suggested that POTs strongly affect AQP3 activity and induce inhibition of melanoma cancer cell migration and growth, unveiling their potential as anticancer drugs against tumors, opening a new window in this field of research [65]. In sum, POTs present high affinity for both P-type ATPases [67] and AQPs [65] targets (Fig. 8.6) that might be responsible for the observed inhibitory effect on melanoma cell migration and proliferation.

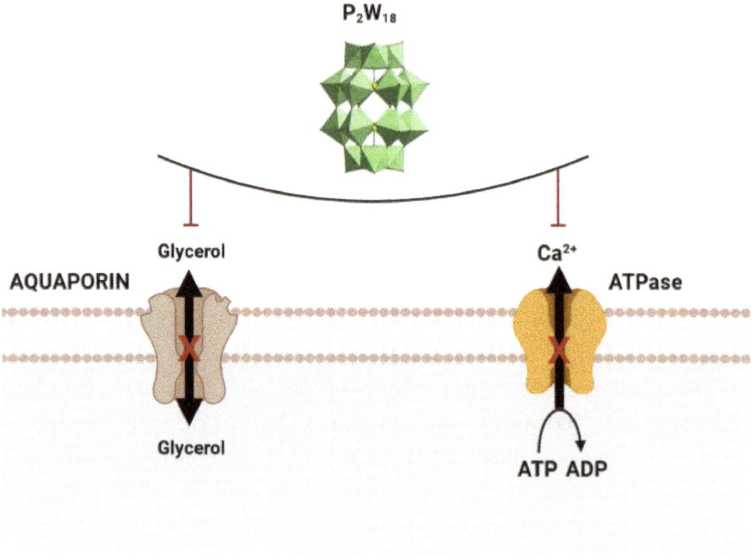

Figure 8.6 Ca^{2+}-ATPase activity and AQP3 glycerol permeability inhibition by P_2W_{18} as a putative mode of action for melanoma anticancer activity [65, 67].

Decavanadate protein interactions, for instance with myosin, actin, and calcium ATPase, are mostly due to electrostatic and hydrogen bonds, although covalent bonds could also occur [68, 69]. Similarly, other POMs and protein interactions were described recently [70–72]. Besides P-type ATPases, it was observed that $V_{10}O_{28}^{6-}$ also interacts with actin [73, 74], which can lead to actin cytoskeleton damage and cellular death processes. Actin is one of the most abundant proteins in cells, being involved in many cellular and biological processes. However, studies between POMs and actin are scarce and only V_{10}–actin studies have been described so far [73, 74].

Biomedical research is looking for new non-ATP competitive inhibitors of protein kinases, like the human protein kinase CK2 inhibitors that have already been designated as promising drugs targets in cancers [75, 76]. POMs, such as P_2Mo_{18}, have been described as non-competitive and potent CK2 inhibitors (IC_{50} = 5 nM) [76], although, due to its instability, it was not possible to know if this POT was responsible for the observed effects. Nevertheless, POMos derivatives exhibit strong efficiency and selectivity, thus representing non-classical kinase inhibitors with increasing interest.

8.4 Antibacterial Activity of POMs

As referred before, the antibiotic resistance represents a real threat to the world public health since a high proportion of bacteria are resistant to antibiotics. Combination between different antibiotics is commonly adopted to solve the antibiotherapy gap necessary to combat infections by multiresistant bacteria. This gap paves the way for the exploration of new combinations as well as new type of antibiotics for future adjuvant strategies [77].

In addition to other recent antibacterial therapies, such as the discovery of teixobactin produced by the Gram negative *Eleftheria terrae* [78, 79], new inorganic compounds, such as metal oxides, have been attracting interest by offering mechanisms of action different from the usual [1, 80, 81]. Thus, some of the studies described in the present chapter, reporting the antibacterial activity of POMs, have been conducted against resistant bacteria [22, 31, 57, 82–86]. Similarly, as described in the previous section, 75 studies were found since 1996 for POMs and antibacterial effects, with an increasing number in the last decade (Fig. 8.7A) making them putative future drugs in the control against pathogenic bacteria. Among the studies found, the POVs (including V_{10}), POTs, and POMos represent the major contribution in this field (Fig. 8.7B).

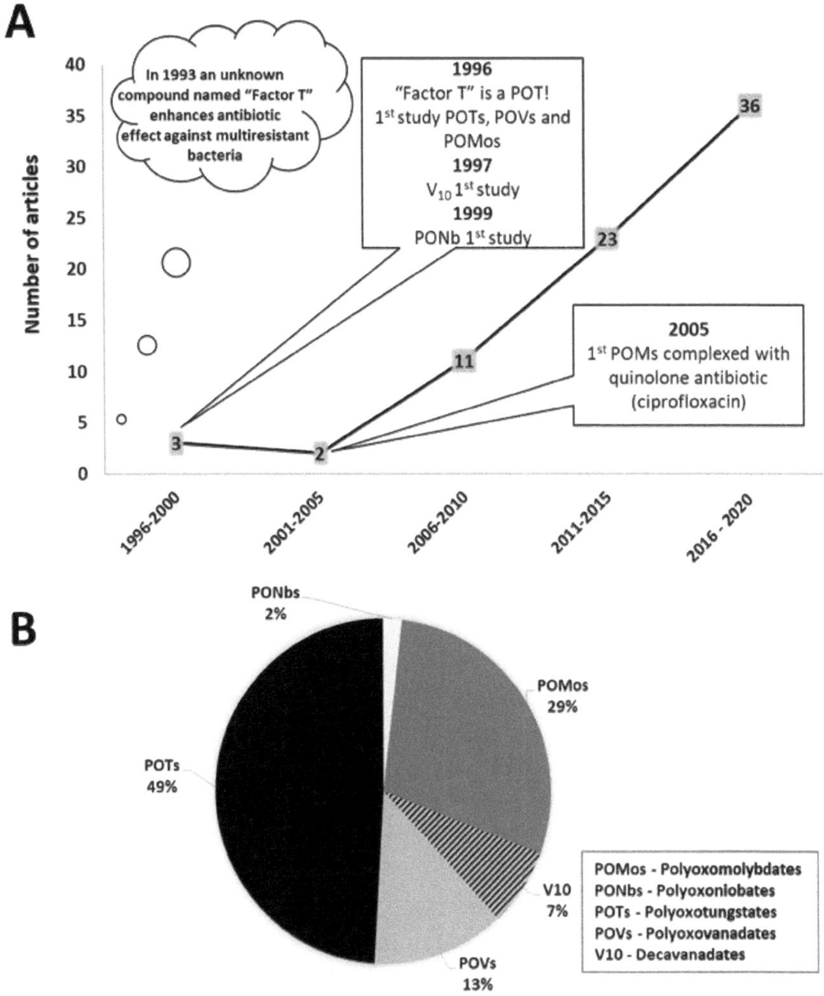

Figure 8.7 Studies for POMs tested against bacteria. (A) Number of articles related with POMs and bacteria since 1996. The first studies described for POTs, POMos, POV (including V_{10}) are highlighted as well as for POMs complexed with antibiotics; (B) distribution in percentage of the different POMs tested for antibacterial activity since 1996.

History of POMs as antibacterial agent started serendipitously as other recognized antibiotics known to date, such as penicillin. In 1993, an unknown compound named initially as "factor T" was observed to enhance the antibacterial effect of *β*-lactams inhibiting methicillin-resistant *Staphylococcus aureus* (MRSA) strains, one of the first antibiotic menaces in the combat of infections worldwide [87]. The aged mixture of tungstate and phosphate named "Factor T" was later found to be a POT: PW_{11}, the first of many other POMs investigated as antibacterial agent [31].

In fact POMs have more than synergistic effects. Yamase *et al.* [83] reported firstly a minimum inhibitory concentration (MIC) of 3.2 mg/mL for POT PW_{12} against MRSA SR 605 and ATCC 43300. The MIC values of POTs against a clinical MRSA strain and the reference *Staphylococcus aureus* ATCC 6538 reached 1.6 mg/mL for both bacteria [84].

8.4.1 Decavanadate and POVs

A recent review on the structure-antibacterial activity relationship of POMs pointed to POVs, and especially decavanadate (V_{10}), as the most potent against certain bacteria, such as *Streptococcus pneumoniae* [1]. Also, recently, it was described that decavanadate inhibited the growth of *Mycobacterium smegmatis* and *Mycobacterium tuberculosis* more potently than metavanadate [88]. In addition, association of V_{10} with (iso) nicotinamide compounds also increased the toxicity toward *Escherichia coli* at concentrations below 1 mM [89].

It was also shown that besides V_{10}, two manganese-polyoxovanadates, namely, MnV_{11} and MnV_{13}, inhibit the growth of *E. coli* [90]. MnV_{11}, MnV_{13}, and V_{10} were all more potent than simple vanadate, with 50% maximal growth inhibition concentrations (GI_{50}) of 0.21, 0.27, 0.58, and 1.1 mM, respectively. Strikingly, the V_{10} isostructural species decaniobate (Nb_{10}) revealed only residual effects on *E. coli* growth [90].

In previous studies, the MIC value of POVs for *Streptococcus pneumoniae* reached 4–32 μg/mL, in contrast for POTs and POMos the MIC value is much higher (128–1000 μg/mL) [91]. In this study, the morphology of *Streptococcus pneumoniae* changed after the exposure to POVs and the effect of POVs is assumed to be related with the interaction between POVs and the cell membrane. For concentrations higher than the MIC value, cell lysis was observed displaying a bactericidal effect.

Chen *et al.* [57] have studied the antibacterial activity of a chitosan-V_{10} complex against *E. coli* and *S. aureus*. Chitosan is famous for its antimicrobial activity as it inhibits the mRNA synthesis after penetration into the nuclei of microorganisms. The polyanion used in this interesting study, $V_{10}O_{28}(NH_4)_6$, can assemble with the cationic chitosan through non-covalent interactions. This POV compound ($V_{10}O_{28}^{6-}$), seems to agglomerate on the membrane surface of the chitosan polymer, presenting some shortcomings, which are overcome by the addition of an inorganic cation, such as Ca^{2+}. The complex demonstrated the same antibacterial activity with an MIC value of 12.5 μM against both bacteria [57]. On the other hand, $V_{10}O_{28}^{6-}$ is known for its inhibition of P-type ATPases, such as Ca^{2+}-ATPase [60, 61, 69], leading to a disturbance in the molecular ion transport across the membrane, thus devastating the bacteria metabolism, whereas chitosan can inhibit the mRNA synthesis [57].

8.4.2 POTs and POMos

In order to elucidate the role of the monomeric tungstate (WO_4^{2-}) in the antibacterial action, the study conducted by Yamase (2006) showed that the tungstate itself does not contribute to the antibacterial action of different POTs [83]. In their study, the MIC value for PW_{12} and As_4W_{40} for MRSA SR 3605 and MRSA ATCC 43300 was 3.2 mg/mL and 0.4 mg/mL, respectively. In contrast, the tungstate itself achieves an MIC value of 102 mg/mL for both strains [83]. In another study, it was also observed that the tungsten, up to the studied concentration of 0.256 mg/mL, showed no antibacterial effect against *Helicobacter pylori* strains resistant to the antibiotic metronidazole [85]. However, these strains of *Helicobacter pylori* showed for the POTs As_4W_{40} and P_2W_{18} MIC values that varied between 8–32 μg/mL and 128–256 μg/mL, (mM concentrations), respectively.

Staphylococcus aureus is a Gram-positive bacterium that can easily develop antibiotic resistance. In 2006, Yamase group

[92] examined the antibacterial effect of the *β*-lactam antibiotic, oxacillin, and the glycopeptide antibiotic vancomycin against MRSA and vancomycin-resistant *S. aureus* (VRSA) in the presence of three different POMs. Molybdenum- and tungsten-based POMs, such as $K_6[P_2W_{18}O_{62}]$, $K_4[SiMo_{12}O_{40}]$, and $K_7[PTi_2W_{10}O_{40}]$ (abbreviated P_2W_{18}, $SiMo_{12}$, and PTi_2W_{10}, respectively). The synergistic effect of oxacillin with POMs was evaluated against three strains of MRSA (ATCC 43300, SR3605, MRS394-1) and against two strains of VRSA (Mu_3 and Mu_{50}) using a disc diffusion method. It was reported that P_2W_{18} and PTi_2W_{10} were able to change the *β*-lactam resistance to a *β*-lactam susceptible profile for all strains, except for Mu_3. $SiMo_{12}$ displayed the highest synergistic action by changing the resistance profile to oxacillin of all strains to a susceptible profile.

In a previous study of the same group [81], the $Na_3[PMo_{12}O_{40}]$, with an identical Keggin structure as $SiMo_{12}$, did not show any antibacterial activity against MRSA. The difference that the authors found between the two studies might be due to a structural instability at physiological pH of the compound $SiMo_{12}$ [83, 92]. Regarding the mechanism of action of $SiMo_{12}$ and P_2W_{18}, it was showed that the reduction of $SiMo_{12}$ and P_2W_{18} allow their penetration and uptake into the bacterial cell, going right after the electron transfer system. Due to the redox potential of these two POMs, the oxidation of some compounds involved in the respiratory mechanism, leading to reduction of others, could direct to a possible inhibition of ATP synthesis, thus inducing cell death. The improvement of the antibacterial activity of the antibiotics oxacillin and vancomycin by the POMs P_2W_{18}, $SiMo_{12}$, and PTi_2W_{10} was associated with the inhibition of the transcription level of the genes *mecA* and *pbp genes* (the *mecA* gene codifies for PBP2a that has low affinity to *β*-lactam antibiotics and the *pbp* genes [pbpA- and pbp2–4] codify for different penicillin binding proteins) [92].

Recently, the activity of several POMs against the Gram negative human mucosal pathogen *Moraxella catarrhalis* was compared and the Preyssler-type POT $[NaP_5W_{30}O_{110}]^{14-}$ (abbreviated P_5W_{30}) showed the highest activity [93]. This compound presented the lower MIC value against *M. catarrhalis* (1 μg/mL) among all tested POMs, as well as against *S. aureus*

(16 μg/mL) and *E. faecalis* (8 μg/mL), but was much less potent against *E. coli* (MIC > 256 μg/mL). It is to be noticed that the MIC value observed for P_5W_{30} was much lower (64 times lower) in comparison with other POT with identical charge and size, although with Se instead of P in the composition (Se_2W_{29}), reinforcing that the antibacterial activity depends on the structure, shape, and/or composition of POTs. For *H. pylori*, POMs exhibiting the highest activity were mostly Keggin-type POTs, polyoxovanadotungstates, and large highly negatively charged POMs [1].

8.5 Mechanisms of Action of POMs as Antibacterial Agents

The mechanism of action of POMs as antibacterial agents is not yet fully understood. Several possible mechanisms were anticipated but further studies are required [1, 57, 85, 86, 92]. Recently, the antibacterial activity of POMs, structures, antibiotic effects, modes of action and future perspectives were reviewed [1]. In this section, we resume some of these modes, and in Table 8.2, We describe some of the effects induced by the exposure of bacteria to POMs in the studies mentioned in the above section.

8.5.1 Modulation of Gene Expression

It is known that the bacterial β-lactam resistance is associated with the peptidoglycan-synthetic enzyme that possesses a low affinity to β-lactams, called PBP2a. As mentioned above, this protein is encoded by the *mecA* gene. Yamase group has investigated the PBP2a expression and the data showed that all POMs studied (except for $SiMo_{12}$) diminished the *mecA*-induced mRNA band intensity in the electrophoresis, thus suggesting that POMs were affecting the transcription process [92]. As for the POMo compound, it seemed that the decrease of PBP2a expression was related to another mechanism, such as the inhibition of the translation from *mecA*-induced mRNA to PBP2a or of the folding of the polypeptide chain [85].

Table 8.2 Comparison of the antibacterial activities and effects of some POMs

POMs	Bacteria	MIC [μM]	Effects	Refs
$V_{15}O_{36}$ V_4O_{12}	*Staphylococcus pneumoniae*	4–32	Interference with the transport system of ions and substrates	[57]
Chitosan-$Ca_3V_{10}O_{28}$	*Staphylococcus aureus* *Escherichia coli*	12.5	Inhibition of mRNA synthesis interfering with protein's production and inhibition of Na^+/K^+-ATPase destroying bacteria's metabolism	[57]
P_2W_{18} $SiMo_{12}$ PTi_2W_{10}	*Staphylococcus aureus* MRSA	100–25 600	Suppression of *mecA* induced mRNA expression; inhibition of translation process; inhibition of translational process	[12, 88]
KAs_4W_{40} KSb_9W_{21}	*Helicobacter pylori*	<256	High negative charge of POMs induces a morphological change from bacillary to coccoid form, inducing a bactericidal action	[85]

8.5.2 POMs Interactions with Proteases, Phosphatases, P-Type ATPases and Actin

Inhibition of sulfotransferases and phosphatases has been reported as a mechanism of antibacterial activity of POMs [57, 94]. Moreover, the disturbance in the molecular transport across the membrane, that devastates the bacteria metabolism, was reported by Chen *et al.* [57]. As described above, $V_{10}O_{28}^{6-}$ is known for its inhibition of Na^+/K^+-ATPase [95] and Ca^{2+}-ATPase [60, 69], possibly causing a disturbance in the molecular transport across the membrane and consequently disturbing the bacterial metabolism. P-type ATPases play a crucial role in cellular ion homeostasis and have been described as potential molecular targets of POMs [64]. As described before, decavanadate non-competitive interaction with Ca^{2+}-ATPase strongly inhibits the ATPase activity (IC_{50} = 15 μM) [60]. It is well-established that the V_{10} binding site, which is formed by three proteins domains [63], is located at the intracellular cytoplasmic side.

POMs can also affect bacterial cells by changing their morphological structure and leading them to death. Inoue *et al.* have reported that POTs, such as As_4W_{40} and Sb_9W_{21}, could enable morphological changes of *H. pylori* from bacillary to a U-shaped or coccoid form [85]. Fiorani *et al.* have also revealed extensive degradation of the *E. coli* cell disrupting their rod-shaped morphology by a complex formed of a chitosan compound with V_2Mo_{10} [96]. Regarding the morphology and shape of the cells, it was described that V_{10} interacts with G-actin polymerization/depolymerization dynamics in skeletal muscle cells [73, 74], which can lead to actin cytoskeleton damage and cellular death processes.

Recently, simultaneous XANES and EXAFS studies concluded that decavanadate interacts in a different mode with G-actin and F-actin [97]. Regarding the V_{10}–G-actin interaction, besides decavanadate species, an oxidovanadium (IV) signal, identical to that observed for V(IV)–$S_{Cys,}$ was detected, evidencing V_{10} reduction upon protein interaction, which is prevented by ATP. This might suggest an interaction with ATP binding site that leads to vanadate reduction upon POMs interaction. It is believed that the detailed knowledge of the molecular basis of decavanadate–actin interplay and its involvement in the G-actin polymerization process allows a better understanding of putative POM activities and pharmacological applications of metal-based drugs.

8.5.3 Effects on Sialyl- and Sulfotransferase

The difference between a sialyltransferase and a sulfotransferase is that a glycosyltransferase catalyzes the transfer of a sialic acid residue to the terminal position of an oligosaccharide chain of glycoproteins and glycolipids [98], whereas the sulfotransferase catalyzes the transfer reaction of a sulfate group to an acceptor sugar chain on the surface of cells [99]. These modifications of carbohydrate chains play a role in cell–cell recognition, serving as a target for bacterial and also viral infections. Therefore, it is conceivable that the inhibition of these catalytic reactions could lead to the inhibition of biological activities of these sugar chains. Seko *et al.* [94] have evidenced the effective non-competitive inhibition of sialyl and sulfotransferases by POMs in their antibacterial activity. Among 20 POMs, 3 tungstate-based POMs, $[SiNiW_{11}O_{40}]$ $[Cu_3(PW_9O_{34})_2]$, and $[SiVW_{11}O_{40}]$, were shown to have a higher inhibition activity, with IC_{50} values as low as 0.2 nM for the α-2,3-sialyltransferase (ST3Gal-I). For three vanadium-based POMs, $[KV_{13}O_{31}(MePO_3)_3]$ $[V_{18}O_{42}]$, and $[V_{18}O_{44}(N_3)]$, it was proven that the latter had a higher inhibition for the galactose-3-O-sulfotransferase (Gal3ST-2), with an IC_{50} value of 3 nM.

8.5.4 Bioenergetic and Redox Disturbance

The reduction of some POMs allows their penetration and uptake into the bacterial cell, going right after the electron transfer system. Due to the redox potential of POMs $SiMo_{12}$ and P_2W_{18}, the oxidation of some compounds involved in the respiratory mechanism, leading to reduction of others, could direct to a possible inhibition of ATP synthesis, thus inducing bacterial death of resistant bacteria [92].

Fukuda and Yamase (1997) reported changes in cell morphology when *S. pneumoniae* was exposed to POVs [91]. With highly negative-charged POTs, *H. pylori* changed from bacillary to coccoid morphology, and fluorescent X-ray analysis suggest that As_4W_{40}, Sb_9W_{21} and $SiVW_{11}$, or at least some of the elements, were taken into the bacteria cells [85].

In another study with different POMs, *S. aureus* strains were able to reduce to blue color forms the P_2W_{18} and $SiMo_{12}$ compounds with more positive redox potentials (cyclic voltammetry) [92]. In the bacterial cells treated with P_2W_{18}, energy dispersive X-ray analysis indicated the localization of W atoms at the periphery of the cells. Similar blue staining of *E. coli* cells was observed in presence of vanadium (V) compounds, probably by reduced forms containing oxidovanadium (IV) [90]. Although it is not clear the molecular mechanisms responsible for the reduction of these POMs by bacteria, it is reasonable to anticipate that it diverts electrons from cell redox systems and hinders bacterial growth.

POMs that can catalyze the production of ROS, as the phosphomolybdates and V_{10}-copper(II) tris derivatives recently described, can further disrupt bacterial cell integrity showing promising antibacterial activities [100–102].

8.6 Antiviral Activity of POMs

As it was mentioned, the number of studies using POMs that address viral infection is comparatively lower than the ones found for cancer and bacteria. Nevertheless, this number is increasing, and in the past 10 years represents almost 40% of the total (Fig. 8.8A). Among the studies described so far, the ones using POTs represent the major contribution in this field reaching 75% (Fig. 8.8B).

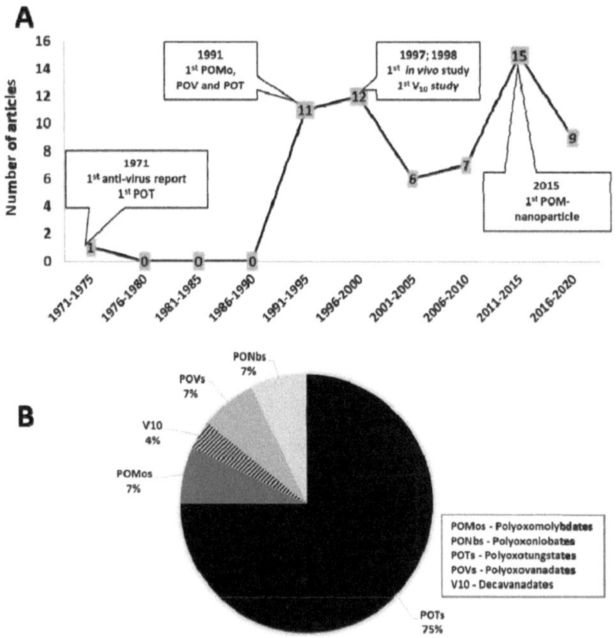

Figure 8.8 Studies using POMs as antiviral agents. (A) Number of articles relating virus and POMs since 1971. The first antiviral studies described for POTs, POMos, POV, and V_{10} are highlighted as well as the first *in vivo* studies and the first study using POM-nanoparticle. (B) Distribution, in percentage of tested POMs, namely, V_{10}, POVs, POTs, and POMos on the antiviral studies over the covered periods.

8.6.1 Decavanadate and POVs

Decavanadate ($V_{10}O_{28}^{6-}$), an isoPOM with 10 octahedral vanadium (V) centers (Fig. 8.1G), is known to interact with a wide range of biomolecules and exerts several biological activities, *in vitro* and *in vivo*, by affecting mitochondrial function, increasing glucose uptake in adipocytes, and inhibiting the process of muscle contraction as well as the actin polymerization [103–105]. Note that the most cited paper regarding decavanadate (V_{10}) in biology is the interaction of V_{10} in a spatially selective manner within the protein cages of virions [13]. Besides preventing the formation of virions, decavanadate is also able to inhibit viral activities by preventing the virus-cell host binding [13].

8.6.2 POTs and POMos

The antiviral activity of POMs was initially studied and tested *in vivo* in France in 1972 by Raynaud and Jasmin group [106]. They demonstrated that POTs, such as $[NH_4]_{17}Na[Na(SbW_7O_{24})_3(SbO_7)_2]$, could exhibit inhibition of the RNA-dependent DNA polymerases of retroviruses. This discovery led to the hypothesis of the activity against the human immunodeficiency virus (HIV) that was later demonstrated for some POTs [13, 107, 108]. Some studies demonstrated that POMs exhibited their antiviral effect by inhibiting the binding of virus to the host cell and/or its penetration [13, 25, 50, 85, 92, 107–109].

It is known that HIV specially targets CD4 molecules present in T lymphocytes, monocytes, and macrophage lineage. A glycoprotein, denominated gp120, allows its binding on CD4 cells and, consequently, the injection of viral material into the host cell [50]. In 2000, Witvrouw *et al.* [109] have demonstrated that the anti-HIV activity of heteropolytungstates had a structure–activity relationship and could inhibit the replication of HIV-1 and HIV-2. In fact it was observed that among the 28 types of POMs used, all of them showed activity against retrovirus with a significant variability in efficiency depending on their structure. The single Wells–Dawson structure of the compound 13 (α_2-$[NMe_3H]_7[CH_3C_5H_4TiP_2W_{17}O_{61}]$) and the double Wells–Dawson of the structure compound 17 ($Na_{16}[Mn_4(H_2O)_2(P_2W_{15}O_{56})_2]$) both inhibited the binding of HIV particles to CD4 cells by blocking the binding of gp120 to SUP-T1 cells, in a concentration dependent manner. The former compound 17 showed higher activities once its inhibition for CD4 and gp120 was 83 and 95%, respectively, at 100 μg/mL compared to 61 and 78% for the compound 13. It was suggested that some conformations and structures are better suited for inhibition than others, pointing out the possibility of target molecules that have marked structural and conformational differences.

In 2005, Yamase group have demonstrated that V/W and Ti/W mixed Keggin derivatives exhibit a wide-ranging of anti-RNA virus activities against MDCK, Vero, Hep-2, and MT-4 cells. All the compounds showed a toxicity at concentrations above 200 μM (except for MT-4 cells, which was between 46 and 100 μM) [22]. The studies described are summarized in Table 8.3.

Table 8.3 Types of virus and POMs effects

Virus	POMs	Effect	References
Zika	POT	Targets the entry process of ZIKV infection	[110]
Hepatitis C	POT	Disrupts HCV particles	[108]
FluV-A	POT POV	Cell surface location of POT Inhibits the fusion of virus	[111–115]
FluV-B	POT POV	Cell surface location	[111–113, 115]
SR-RSV-A	POT	Decreased pp60 (v-rsc) activity	[116]
RSV-A	POT PONb	Inhibits adsorption; inhibits fusion	[106, 114]
RSV-B	POT	n.d.	[111]
PfluV 2	POT	n.d.	[111]
PfluV 3	POT	n.d.	[111]
MLSV	POT	n.d.	[111]
MPSV	POT	n.d.	[111]
Herpes type 1	POT	Cell surface location	[112, 118]

Herpes type 2	POT	Cell surface location	[112, 118]
HIV-1	POT POMo	Blocks the gp120-binding Domain of the CD4 receptor	[109, 111, 112, 119–121]
HIV-2	POT	n.d.	[109]
Hepatitis B	POT	Anti-HBV activity	[122, 123]
HCMV	POT	n.d.	[118]
VZV	POT	n.d.	[118]
Vaccinia virus	POT	n.d.	[118]
Adenovirus type 1	POT	n.d.	[118]
Adenovirus type 5	POT	n.d.	[118]
TGEV	POT, PONb, POV	n.d.	[111]
FIPV	POT, PONb, POV	n.d.	[111]
SARS-V	POT, PONb, POV	n.d.	[111]
SARS-CoV	POT, POMo	Interaction with $3CL^{pro}$ protein	[124, 125]

Note: n.d., not determined.

Another study with one Keggin-type POM, ($K_5[SiVW_{11}O_{40}]$) and two double Keggin-type POMs, ($K_{10}Na[(VO)_3(SbW_9O_{33})_2]$) and ($K_{11}H[(VO)_3(SbW_9O_{33})_2]$) showed activity, by inhibition, against dengue virus (DFV), influenza virus (FluV-A), respiratory syncytial virus (RSV), parainfluenza virus (PfluV 2), distemper virus (CDV), and HIV [111]. Both compounds showed higher toxicity, as their effective concentration (EC) values were lower than 1 μM, in comparison with others (EC_{50} >10–60 μM). It was also demonstrated that ($K_{10}Na[(VO)_3(SbW_9O_{33})_2]$) strongly inhibits the binding of the viral gp120 antibodies. Figure 8.9 summarizes the types of virus studied for all POMs.

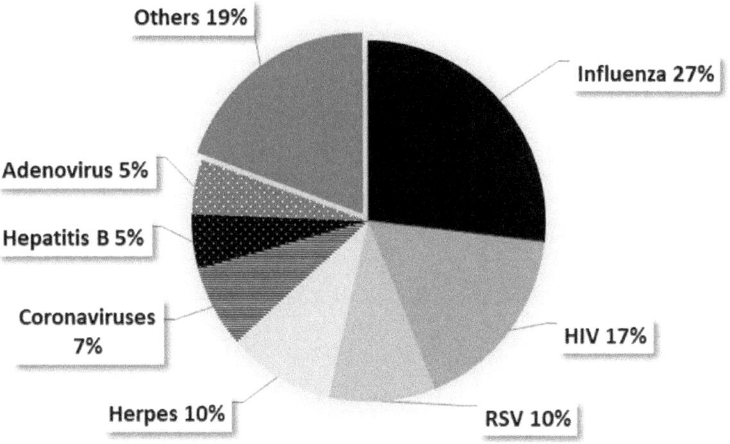

Figure 8.9 Studies published with all types of POMs on different types of viruses (HIV, Adeno, Corona, Herpes, Influenza, others).

Hosseini *et al.* evaluated the effect of a novel POM derivate $[P_2W_{18}O_{62}]^{6-}$ (abbreviated P_2W_{18}) on influenza virus (FluV) in MDCK cell line [126]. It was showed that this P_2W_{18} derivative had an EC_{50} of 100 μM, whereas no effective cytotoxic effect on MDCK cells was observed at a concentration up to 50 μM. It was also demonstrated that P_2W_{18} could inhibit the virus hemagglutinin A (HA), responsible for the first stage of viral attachment. Thus, the Wells–Dawson type POM P_2W_{18} is likely to have a dual mechanism of action in the inhibition of FluV replication: it reduces the binding of HA trimer to the cell receptors and inhibits the fusion of viral particles into the cell.

8.7 POMs Mechanisms of Antiviral Activity

As referred above, Hosseini *et al.* confirmed that P_2W_{18} is likely to have a dual mechanism of action in the inhibition of FluV replication: it interacts with HA responsible for the first stage of viral attachment and inhibits fusion of viral particles into the cell [126]. As mentioned before, inhibition of catalytic reactions promoted by sialyltransferases and sulfotransferases would affect the carbohydrate chains in glycoproteins that play a major part in cell-viral recognition serving as a target for viral infections. Thus, by targeting virus membrane proteins POMs would affect the early stage of viral infection (Fig. 8.10).

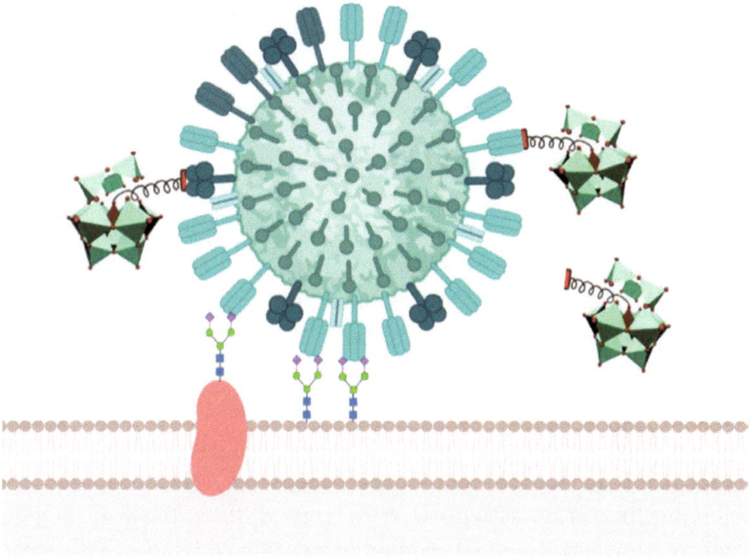

Figure 8.10 POMs putative interactions with viral membrane proteins as hemagglutinin preventing the early stage of infection. Besides, putative inhibition of neuraminidase would prevent a later stage of infection. Color code: glycoprotein, dark pink; hemagglutinin, dry green; neuraminidase, dark green; POM, green.

Certain POMs, such as $[SiW_9Nb_3O_{40}]$, as described by Qi *et al.*, can act directly on hepatitis C virus (HCV) virion particles and destabilize the integrity of its structure [108]. Their data suggested that it could specifically inhibit HCV infection at an early stage of its life cycle. The inhibition of the HCVcc and HCVpp surface proteins of the HCV envelope, with an EC_{50} of 0.8 and 1.2 μM, respectively, could explain the specificity of $SiW_9Nb_3O_{40}$ to HCV envelope proteins. Furthermore, it seemed that POM-12 could endorse the degradation of HCV genomic RNA.

HIV is known to specially target CD4 molecules present in T lymphocytes, monocytes, and macrophage lineage. A glycoprotein, denominated gp120 which is located on the surface of the virus, is the principal weapon of HIV because it allows its binding on CD4 cells and the injection of the viral material into the cell [50].

Other studies reported that POMs could inhibit proteases in a non-competitive manner at low micro molar concentrations [23, 107, 109]. The HIV-1 protease is important for the maturation of protein components of an infectious HIV virion, thus, its inhibition could be responsible for the anti-HIV effect of POMs.

8.8 Conclusion and Perspective

The anticancer, antibacterial, and antiviral properties of POMs are promising. POMs are interesting compounds that possess a rich diversity of structures which applications are yet to be discovered. POMs have been studied and used in a large variety of fields, and have become of great interest in medicine. Indeed many POMs have proved to be efficient against viruses, bacteria, and also tumor cells. POTs are among those POMs that have been the subject of the largest number of studies on anticancer, antiviral, and antibacterial activities. With regard to anticancer activity, cervix cancer is the most studied in relation to studies with POMs, in antibacterial activity a greater number of studies have focused on Gram + bacteria and the antiviral activity of POMs has been prevalent in respiratory tract viruses, mainly influenza viruses. Thus, 52% of the antitumor studies carried out with POMs were with tungsten metallates; 49% in antibacterial studies and 75% in anti-viral studies. POMos are the second most studied type of POM in the three mentioned biological activities, with 36, 29, and 7%, respectively.

The effect of each POM depends on its structure, metal nature, and biomolecular target, pointing out to new POMs for future biological applications. Several molecular targets are unveiling their potential as anticancer, antibacterial, and antiviral drugs of the future, thus, opening new windows for future research in these fields. Despite promising results against cancer, virus, and bacteria, POMs were not yet tested in clinical trials. This may be due to the lack of *in vivo* studies, which determine its behavior in terms of absorption, distribution, metabolism, elimination, and toxicity. To be approved as a drug, the POM must show higher activity against its biological target and very low toxicity toward normal cells in comparison to the approved drug. Further studies involving interdisciplinary teams must be conducted to understand which POM is better tuned against a particular disease or infection.

8.8.1 Anticancer Activities

POMs present antitumor activity against a large number of tumor cells. Their mechanisms of action are still difficult to be determined. Several studies have investigated possible pathways leading to this cellular death. In general, POMs are able to inhibit tumor growth by inducing apoptosis. Some suggested that POMs, once entered into the mitochondrion, can form a complex with the FMN leading to the inhibition of ATP-synthesis. Generation of oxidative stress, activation of caspase-3 and presence of DNA ladder were also factors that contribute to apoptosis. Others studies suggested that the antitumor activity of POMs is correlated with the inhibition of the ecto-nucleosidases, HDACs or, as recently referred, to the inhibition of AQP's function. A synergy between apoptosis and autophagy was also proposed by some authors. Nevertheless, based on these putative mechanisms of action, the antitumor activity of a POM might depend on its structure, metal nature, size, and/or its target. For example, decavanadate and other POMs exhibited antitumor activities at different concentrations and with different IC_{50} values through the inhibition of ALP and kinases, whereas so far no decavanadates were found to inhibit HDAC or ecto-ATPases. In sum, the cytotoxic effect on different types of cancer cells was also found to be different for several POMs.

8.8.2 Antibacterial Activities

Several POMs have demonstrated good activity against antibiotic resistant bacteria. Some molybdenum- and tungsten-based POMs can disable the β-lactam resistance of MRSA and VRSA by inhibiting the transcription of the *mecA* and *pbp* genes. In other cases, POMs seem to dispatch their antibacterial activity by inducing morphological changes through cytoskeleton interactions or interfering with the ionic transport systems, leading bacteria to death. In fact, decavanadate and other POMs were described as potent P-type ATPase inhibitors whereas V_{10} prevention of actin polymerization points out a putative mechanism of action at the cytoskeleton level. The majority of the studies have addressed the potential of POMs to control the bacterial growth and some explored the mechanism of action of these compounds, however, several aspects of the virulence and community life of bacteria were not yet explored.

8.8.3 Antiviral Activities

Finally, antiviral activity is one of the most known effects of POTs. SbW_7 was the first and the only POM to have been tested in clinical trials in 1972, but was also the most toxic one. Despite its noxious effect, SbW7 gave hope for further studies regarding POMs and their antiviral activity against HIV and several other viruses. It seems that POMs are able to inhibit the DNA- or RNA-virus activities by preventing the virus-cell host binding at an early stage of the viral infection by targeting viral membrane proteins. It seems that some conformations and structures are better suited for inhibition of antiviral activities than others. For example, in the study of POMs against HIV, the single and double Dawson structures were the ones who enhanced higher inhibition whereas decavanadate is linked within the protein cages of virions.

Acknowledgments

This study received national funds through FCT – Foundation for Science and Technology through project UIDB/04326/2020 (MA, GF, and CF).

Abbreviations

A549	Human lung cancer cells
ALP	Alkaline phosphatase

AsPC-1	Pancreatic cancer cells
BWCN	$(Himi)_2\,[Bi_2W_{20}O_{66}(OH)_4CO_2(H_2O)_6Na_4(H_2O)_{14}]\bullet 17H_2O$
Ca^{2+}-ATPase	Calcium adenosine triphosphatase
CTS-$Ca_3V_{10}O_{28}$	Chitosan-$Ca_3V_{10}O_{28}(NH_4)_6$
E-NTPDase	Ecto-nucleotide triphosphate diphosphohydrolase
FluV FIPV 5-FU GI_{50}	Influenza virus Feline infectious peritonitis virus fluorouracil 50% maximal growth inhibition concentration
gp120 HCB HCMV HCV	Glycoprotein expressed by HIV Hepatitis B virus Human cytomegalovirus Hepatitis C virus
HDAC	Histone deacetylase
HEK-293	Human embryonic cells
HeLa	Human cervical cancer cells
Hep-2	Human Larynx carcinoma cell line
HepG2	Human hepato-cellular carcinoma
HIV	Human immunodeficiency virus
HPA-23	$[NH_4]_{17}Na[Na(SbW_7O_{24})_3(Sb_3O_7)_2]^{18-}$
HT-29	Colon cancer cells
HUVEC	Human umbilical vein endothelial cells
IC_{50}	Half inhibitory concentration
K-562	Human myelogenous leukemia
L-02	Human normal hepatocellular cells
MCF-7	Human breast cancer cells
MDA-MB-231	Human breast adenocarcinoma cell line
MDCK	Madin–Darby canine kidney
mecA	gene that code for PBP2a
MIC	Minimum inhibitory concentration
MKN45 MLSV	Gastric cancer cells Measles virus

MPSV MT-4 MRSA	Mumps virus T-cell acute lymphoblastic leukemia cells methicillin-resistant *Staphylococcus aureus*
P-388	Murine leukemia cells
PBP	penincilin-binding proteins
POMos	Polyoxomolybdates
POMs	Polyoxometalates
PONbs	Polyoxoniobates
POTs	Polyoxotungstates
POVs	Polyoxovanadates
PM-8	$[NH_3Pr]_6[Mo_7O_{24}] \bullet 3H_2O$
PM-17 pp60(v-rsc) RSV	$[Me_3NH]_6[H_2Mo_{12}O_{28}(OH)_{12}(MoO_3)_4]2H_2O$ Rous sarcoma virus gene that encodes a tyrosine kinase that causes a type of cancer in chickens Respiratory syncytial virus
P_2W_{18}	$K_6[P_2W_{18}O_{62}] \bullet 14H_2O$
S180 SARS-V SARS-CoV	Murine sarcoma cells Severe acute respiratory syndrome virus Severe acute respiratory syndrome coronavirus
SGC-7901	Human gastric cancer cells
SKOV-3	Human ovarian cancer cells
SMMC-7721 SR-RSV	Human hepatocellular cancer cells Schmidt–Ruppin strain of Rous sarcoma virus
SUP-T1 TGEV	Expresses high levels of surface CD4 and are useful in studies of cell fusion and cytopathic effects of HIV-1 as well as cytopathic strains of HIV-2 Transmissible gastroenteritis virus
U-87	Human brain-like glioblastoma cells
U251 VRSA	Human malignant glioma cells vancomycin-resistant *Staphylococcus aureus*
V_{10}	Decavanadate
Vero VZV ZIKV	Lineages of cells used in cell cultures as host cells for growing virus Varicella zoster virus Zika virus

References

1. Bijelic, A., Aureliano, M. and Rompel, A. (2018). The antibacterial activity of polyoxometalates: structures, antibiotic effects and future perspectives., *Chem. Commun. (Camb).*, **54**, pp. 1153–1169.
2. Bijelic, A., Aureliano, M. and Rompel, A. (2019). Polyoxometalates as potential next-generation metallodrugs in the combat against cancer, *Angew. Chemie – Int. Ed.*, **58**, pp. 2980–2999.
3. Hayashi, Y. (2011). Hetero and lacunary polyoxovanadate chemistry: synthesis, reactivity and structural aspects., *Coord. Chem. Rev.*, **255**, pp. 2270–2280.
4. Chen, X., Yan, S., Wang, H., Hu, Z., Wang, X. and Huo, M. (2015). Aerobic oxidation of starch catalyzed by isopolyoxovanadate $Na_4Co(H_2O)_6V_{10}O_{28}$, *Carbohydr. Polym.*, **117**, pp. 673–680.
5. Mohapatra, L. and Parida, K.M. (2014). Dramatic activities of vanadate intercalated bismuth doped LDH for solar light photocatalysis, *Phys. Chem. Chem. Phys.*, **16**, pp. 16985–16996.
6. Bijelic, A. and Rompel, A. (2015). The use of polyoxometalates in protein crystallography – an attempt to widen a well-known bottleneck, *Coord. Chem. Rev.*, **299**, pp. 22–38.
7. Rhule, J., Hill, C., Judd, D. and Schinazi, R. (1998). Polyoxometalates in medicine, *Chem. Rev.*, **98**, pp. 327–358.
8. Mitsui, S., Ogata, A., Yanagie, H., Kasano, H., Hisa, T., Yamase, T. and Eriguchi, M. (2006). Antitumor activity of polyoxomolybdate, $[NH_3Pri]_6[Mo_7O_{24}]\cdot 3H_2O$, against, human gastric cancer model, *Biomed. Pharmacother.*, **60**, pp. 353–358.
9. Ogata, A., Yanagie, H., Ishikawa, E., Morishita, Y., Mitsui, S., Yamashita, A., Hasumi, K., Takamoto, S., Yamase, T. and Eriguchi, M. (2007). Antitumour effect of polyoxomolybdates: induction of apoptotic cell death and autophagy in *in vitro* and *in vivo* models, *Br. J. Cancer*, **98**, pp. 399–409.
10. Zhao, W., Wang, C., Dong, S., Li, Y., Zhang, D. and Han, L. (2011). *in vitro* study on the antitumor activity of several new polyoxometalates. In *Proceedings 2011 International Conference on Human Health and Biomedical Engineering*, pp. 594–597, August 19–22, 2011, Jilin, China.
11. Wang, L., Yu, K., Zhou, B.-B., Su, Z.-H., Gao, S., Chu, L.-L. and Liu, J.-R. (2014). The inhibitory effects of a new cobalt-based polyoxometalate on the growth of human cancer cells, *Dalton Trans.*, **43**, p. 6070.

12. Li, Y.-T., Zhu, C.-Y., Wu, Z.-Y., Jiang, M. and Yan, C.-W. (2010). Synthesis, crystal structures and anticancer activities of two decavanadate compounds, *Transit. Met. Chem.*, **35**, pp. 597–603.

13. Douglas, T. and Young, M. (1998). Host-guest encapsulation of materials by assembled virus protein cages, *Nature*, **393**, pp. 152–155.

14. Zhang, B., Qiu, J., Wu, C., Li, Y. and Liu, Z. (2015). Anti-tumor and immunomodulatory activity of iron hepta-tungsten phosphate oxygen clusters complex, *Int. Immunopharmacol.*, **29**, pp. 293–301.

15. Dianat, S., Bordbar, A.K., Tangestaninejad, S., Yadollahi, B., Zarkesh-Esfahani, S.H. and Habibi, P. (2013). ctDNA binding affinity and *in vitro* antitumor activity of three Keggin type polyoxotungestates, *J. Photochem. Photobiol. B Biol.*, **124**, pp. 27–33.

16. Fu, L., Gao, H., Yan, M., Li, S., Li, X., Dai, Z. and Liu, S. (2015). Polyoxometalate-based organic-inorganic hybrids as antitumor drugs, *Small*, **11**, pp. 2938–2945.

17. Bâlici, Ş., Şuşman, S., Rusu, D., Zsolt Nicula, G., Soritau, O., Rusu, M., Biris, A.S., and Matei, H. (2016). Differentiation of stem cells into insulin-producing cells under the influence of nanostructural polyoxometalates, *J. Appl. Toxicol.*, **36**, pp. 373–84.

18. Lee, S.Y., Fiene, A., Li, W., Hanck, T., Brylev, K.A., Fedorov, V.E., Lecka, J., Haider, A., Pietzsch, H.-J., Zimmermann, H., Sévigny, J., Kortz, U., Stephan, H., and Müller, C.E. (2015). Polyoxometalates – potent and selective ecto-nucleotidase inhibitors, *Biochem Pharmacol.*, **93**, pp. 171–81.

19. Saeed, S.H., Al-Oweini, R., Haider, A., Kortz, U. and Iqbal, J. (2014). Cytotoxicity and enzyme inhibition studies of polyoxometalates and their chitosan nanoassemblies, *Toxicol. Reports*, **1**, pp. 341–352.

20. Iqbal, J., Barsukova-Stuckart, M., Ibrahim, M., Ali, S.U., Khan, A.A. and Kortz, U. (2012). Polyoxometalates as potent inhibitors for acetyl and butyrylcholinesterases and as potential drugs for the treatment of Alzheimer's disease, *Med. Chem. Res.*, **22**, pp. 1224–1228.

21. Hasenknopf, B. (2005). Polyoxometalates: introduction to a class of inorganic compounds and their biomedical applications, *Front. Biosci.*, **10**, p. 275.

22. Yamase, T. (2005). Anti-tumor, -viral, and -bacterial activities of polyoxometalates for realizing an inorganic drug, *J. Mater. Chem.*, **15**, p. 4773.

23. Stephan, H., Kubeil, M., Emmerling, F. and Müller, C.E. (2012). Polyoxometalates as versatile enzyme inhibitors, *Eur. J. Inorg. Chem.*, **2013**, pp. 1585–1594.

24. Turner, T.L., Nguyen, V.H., McLauchlan, C.C., Dymon, Z., Dorsey, B.M., Hooker, J.D. and Jones, M.A. (2012). Inhibitory effects of decavanadate on several enzymes and *Leishmania tarentolae in vitro, J. Inorg. Biochem.*, **108**, pp. 96–104.

25. Ogata, A., Mitsui, S., Yanagie, H., Kasano, H., Hisa, T., Yamase, T. and Eriguchi, M. (2005). A novel anti-tumor agent, polyoxomolybdate induces apoptotic cell death in AsPC-1 human pancreatic cancer cells, *Biomed. Pharmacother.*, **59**, pp. 240–244.

26. De Luyart Lubice, J.J.; De Luyart Lubice, F.C. (1783). *Extr. las Juntas Gen. Celebr. por la Real Soc. Bascon.*, pp. 46–88.

27. Berzelius, J.J. (1826). Beitrag zur näheren Kenntniss des Molybdäns, *Ann. Phys.*, **82**, pp. 369–392.

28. Rosenheim, A. and Liebknecht, O. (1899). Zur Kenntniss der Jodsäure und Ueberjodsäure, *Justus Liebig's Ann. der Chemie*, **308**, pp. 40–67.

29. Mukherjee, H.N. (1965). Treatment of cancer of the intestinal tract with a complex compound of phosphotungstic phosphomolybdic acids and caffeine, *J. Indian Med. Assoc.*, **44**, pp. 477–479.

30. Raynaud, M., Chermann, J.C., Plata, F., Jasmin, C. and Mathé, G. (1971). Inhibitors of viruses of the leukemia and murine sarcoma group. Biological inhibitor CJMR, *C. R. Acad. Hebd. Seances Acad. Sci. D.*, **272**, pp. 2038–2040.

31. Tajima, Y., Nagasawa, Z., Tanabe, I., Kusaba, K. and Tadano, J. (1996). Anionic properties of *β*-lactam-enhancing factor on methicillin-resistan *Staphylococcus aureus, Res. Microbiol.*, **147**, pp. 279–286.

32. Müller, C.E., Iqbal, J., Baqi, Y., Zimmermann, H., Röllich, A. and Stephan, H. (2006). Polyoxometalates – a new class of potent ecto-nucleoside triphosphate diphosphohydrolase (NTPDase) inhibitors, *Bioorg. Med. Chem. Lett.*, **16**, pp. 5943–5947.

33. Dong, Z., Tan, R., Cao, J., Yang, Y., Kong, C., Du, J., Zhu, S., Zhang, Y., Lu, J., Huang, B. and Liu, S. (2011). Discovery of polyoxometalate-based HDAC inhibitors with profound anticancer activity *in vitro* and *in vivo*, *Eur. J. Med. Chem.*, **46**, pp. 2477–2484.

34. Raza, R., Matin, A., Sarwar, S., Barsukova-Stuckart, M., Ibrahim, M., Kortz, U. and Iqbal, J. (2012). Polyoxometalates as potent and selective inhibitors of alkaline phosphatases with profound anticancer and amoebicidal activities, *Dalton Trans.*, **41**, p. 14329.

35. Zhai, F., Wang, X., Li, D., Zhang, H., Li, R. and Song, L. (2008). Synthesis and biological evaluation of decavanadate $Na_4Co(H_2O)_6V_{10}O_{28}\cdot 18H_2O$, *Biomed. Pharmacother.*, **63**, pp. 51–55.

36. Fujita, H., Fujita, T., Sakurai, T., Yamase, T. and Seto, Y. (1992). Antitumor activity of new antitumor substance, polyoxomolybdate, against several human cancers in athymic nude mice, *Tohoku J. Exp. Med.*, **168**, pp. 421–426.

37. Galani, A., Tsitsias, V., Stellas, D., Psycharis, V., Raptopoulo, C.P. and Karaliota, A. (2015). Two novel compounds of vanadium and molybdenum with carnitine exhibiting potential pharmacological use, *J. Inorg. Biochem.*, **142**, pp. 109–117.

38. León, I.E., Porro, V., Astrada, S., Egusquiza, M.G., Cabello, C.I., Bollati-Fogolin, M. and Etcheverry, S.B. (2014). Polyoxometalates as antitumor agents: bioactivity of a new polyoxometalate with copper on a human osteosarcoma model, *Chem. Biol. Interact.*, **222**, pp. 87–96.

39. She, S., Bian, S., Huo, R., Chen, K., Huang, Z., Zhang, J., Hao, J. and Wei, Y. (2016). Degradable organically-derivatized polyoxometalate with enhanced activity against glioblastoma cell line, *Sci. Rep.*, **6**, p. 33529.

40. Zhu, N., Zhang, D., Wang, W., Li, X., Yang, B., Song, J., Zhao, X., Huang, B., Shi, W., Lu, R., Niu, P., Zhan, F., Ma, X., Wang, D., Xu, W., Wu, G., Gao, G. F. and Tan, W. (2020). A novel coronavirus from patients with pneumonia in China, 2019, *N. Engl. J. Med.*, **382**, pp. 727–733.

41. Cheng, M., Li, N., Wang, N., Hu, K., Xiao, Z., Wu, P., Wei, Y. (2018). Synthesis, structure and antitumor studies of a novel decavanadate complex with a wavelike two-dimensional network, *Polyhedron*, **155**, pp. 313–319.

42. Qi, W., Zhang, B., Qi, Y., Guo, S., Tian, R., Sun, J. and Zhao, M. (2017). The anti-proliferation activity and mechanism of action of $K_{12}[V_{18}O_{42}(H_2O)]\cdot 6H_2O$ on breast cancer cell lines, *Molecules*, **22**, p. 1535.

43. Karimian, D., Yadollahi, B. and Mirkhani, V. (2017). Dual functional hybrid-polyoxometalate as a new approach for multidrug delivery, *Microporous Mesoporous Mater.*, **247**, pp. 23–30.

44. Zheng, Y., Gan, H., Zhao, Y., Li, W., Wu, Y., Yan, X., Wang, Y., Li, J. and Wang, X. (2019). Self-assembly and antitumor activity of a polyoxovanadate-based coordination nanocage. *Chem. Eur. J.*, **25**, pp. 15326–15332.

45. Yamase, T., Fujita, H. and Fukushima, K. (1988). Medical chemistry of polyoxometalates. Part 1. Potent antitumor activity of polyoxomolybdates on animal transplantable tumors and human cancer xenograft, *Inorganica Chim. Acta*, **151**, pp. 15–18.

46. Sun, T., Cui, W., Yan, M., Qin, G., Guo, W., Gu, H., Liu, S. and Wu, Q. (2016). Target delivery of a novel antitumor organoplatinum (IV)-substituted polyoxometalate complex for safer and more effective colorectal cancer therapy *in vivo*, *Adv. Mater.*, **28**, pp. 7397–7404.

47. Dong, Z., Yang, Y., Liu, S., Lu, J., Huang, B. and Zhang, Y. (2018). HDAC inhibitor PAC-320 induces G2/M cell cycle arrest and apoptosis in human prostate cancer, *Oncotarget*, **9**, pp. 512–523.

48. Yong, Y., Zhang, C., Gu, Z., Du, J., Guo, Z., Dong, X., Xie, J., Zhang, G., Liu, X. and Zhao, Y. (2017). Polyoxometalate-based radiosensitization platform for treating hypoxic tumors by attenuating radioresistance and enhancing radiation response, *ACS Nano*, **11**, pp. 7164–7176.

49. Boulmier, A., Feng, X., Oms, O., Mialane, P., Rivière, E., Shin, C.J., Yao, J., Kubo, T., Furuta, T., Oldfield, E., Dolbecq, A. (2017). Anticancer activity of polyoxometalate-bisphosphonate complexes: synthesis, characterization, *in vitro* and *in vivo* results, *Inorg. Chem.*, **56**, pp. 7558–7565.

50. Chanh, T.C., Dreesman, G.R. and Kennedy, R.C. (1987). Monoclonal anti-idiotypic antibody mimics the CD4 receptor and binds human immunodeficiency virus, *Proc. Natl. Acad. Sci.*, **84**, pp. 3891–3895.

51. Gândara, R.M.C., Soares, S.S., Martins, H., Gutiérrez-Merino, C. and Aureliano, M. (2005). Vanadate oligomers: *in vivo* effects in hepatic vanadium accumulation and stress markers, *J. Inorg. Biochem.*, **99**, pp. 1238–1244.

52. Soares, S.S., Martins, H. and Aureliano, M. (2005). Vanadium distribution following decavanadate administration, *Arch. Environ. Contam. Toxicol.*, **50**, pp. 60–64.

53. Soares, S.S., Martins, H., Duarte, R.O., Moura, J.J.G., Coucelo, J., Gutiérrez-Merino, C. and Aureliano, M. (2007). Vanadium distribution, lipid peroxidation and oxidative stress markers upon decavanadate *in vivo* administration, *J. Inorg. Biochem.*, **101**, pp. 80–88.

54. Soares, S.S., Henao, F., Aureliano, M. and Gutiérrez-Merino, C. (2008). Vanadate-induced necrotic death in neonatal rat cardiomyocytes through mitochondrial membrane depolarization, *Chem. Res. Toxicol.*, **21**, pp. 607–618.

55. Soares, S.S., Gutiérrez-Merino, C. and Aureliano, M. (2007). Decavanadate induces mitochondrial membrane depolarization and inhibits oxygen consumption, *J. Inorg. Biochem.*, **101**, pp. 789–796.

56. Soares, S.S., Gutiérrez-Merino, C. and Aureliano, M. (2007). Mitochondria as a target for decavanadate toxicity in *Sparus aurata* heart, *Aquat. Toxicol.*, **83**, pp. 1–9.

57. Chen, S., Wu, G., Long, D. and Liu, Y. (2006). Preparation, characterization and antibacterial activity of chitosan-Ca3V10O28 complex membrane, *Carbohydr. Polym.*, **64**, pp. 92–97.

58. Tiago, D.M., Laizé, V., Cancela, M.L., and Aureliano, M. (2007). Impairment of mineralization by metavanadate and decavanadate solutions in a fish bone-derived cell line, *Cell Biol. Toxicol.*, **24**, pp. 253–263.

59. Tiago, D.M., Laizé, V., Aureliano, M. and Cancela, M.L. (2008). Vanadate proliferative and anti-mineralogenic effects are mediated by MAPK and PI-3K/Ras/Erk pathways in a fish chondrocyte cell line, *FEBS Lett.*, **582**, pp. 1381–1385.

60. Fraqueza, G., Ohlin, C.A., Casey, W.H. and Aureliano, M. (2012). Sarcoplasmic reticulum calcium ATPase interactions with decaniobate, decavanadate, vanadate, tungstate and molybdate, *J. Inorg. Biochem.*, **107**, pp. 82–89.

61. Fraqueza, G., Batista de Carvalho, L.A.E., Marques, M.P.M., Maia, L., Ohlin, C.A., Casey, W.H. and Aureliano, M. (2012). Decavanadate, decaniobate, tungstate and molybdate interactions with sarcoplasmic reticulum Ca^{2+}-ATPase: quercetin prevents cysteine oxidation by vanadate but does not reverse ATPase inhibition, *Dalton Trans.*, **41**, p. 12749.

62. Aureliano, M., Fraqueza, G. and Ohlin, C.A. (2013). Ion pumps as biological targets for decavanadate, *Dalton Trans.*, **42**, p. 11770.

63. Hua, S., Inesi, G. and Toyoshima, C. (2000). Distinct topologies of mono- and decavanadate binding and photo-oxidative cleavage in the sarcoplasmic reticulum ATPase, *J. Biol. Chem.*, **275**, pp. 30546–30550.

64. Fraqueza, G., Fonseca, C., Al-Sayed, E., Rompel, A. and Aureliano, M. (2017). Polyoxometalates as inhibitors of P-Type ATPases and the role of polyphenols, *Arch. Int. Soc. Antioxidants Nutr. Heal.*, **5**, pp. 17–20.

65. Pimpão, C., da Silva, I.V., Mósca, A.F., Pinho, J.O., Gaspar, M.M., Gumerova, N.I., Rompel, A., Aureliano, M. and Soveral, G. (2020). The aquaporin-3-inhibiting potential of polyoxotungstates, *Int. J. Mol. Sci.*, **21**, p. pii: E2467.

66. Soveral, G. and Casini, A. (2016). Aquaporin modulators: a patent review (2010–2015), *Expert Opin. Ther. Pat.*, **27**, pp. 49–62.

67. Gumerova, N., Krivosudský, L., Fraqueza, G., Breibeck, J., Al-Sayed, E., Tanuhadi, E., Bijelic, A., Fuentes, J., Aureliano, M. and Rompel, A.

(2018). The P-type ATPase inhibiting potential of polyoxotungstates. *Metallomics*, **10**, pp. 287–295.

68. Aureliano, M. and Ohlin, C.A. (2014). Decavanadate *in vitro* and *in vivo* effects: facts and opinions, *J. Inorg. Biochem.*, **137**, pp. 123–130.

69. Aureliano, M. (2016). Decavanadate toxicology and pharmacological activities: V10 or V1, both or none?, *Oxid. Med. Cell. Longev.*, **2016**, pp. 1–6.

70. Solé-Daura, A., Goovaerts, V., Stroobantsm K., Absillis, G., Jiménez-Lobato, P., Poblet, J.M., Hirst, J.D., Parac-Vogt, T.N. and Carbó, J.J. (2016). Probing polyoxometalate-protein interactions using molecular dynamics simulations, *Chem. Eur. J.*, **22**, pp. 15280–15289.

71. Blazevic, A. and Rompel, A. (2016). The Anderson–Evans polyoxometalate: from inorganic building blocks via hybrid organic-inorganic structures to tomorrows "Bio-POM", *Coord. Chem. Rev.*, **307**, pp. 42–64.

72. Arefian, M., Mirzaei, M., Eshtiagh-Hosseini, H. and Frontera, A. (2017). A survey of the different roles of polyoxometalates in their interaction with amino acids, peptides and proteins, *Dalt. Trans.*, **46**, pp. 6812–6829.

73. Ramos, S., Duarte, R.O., Moura, J.J.G. and Aureliano, M. (2009). Decavanadate interactions with actin: cysteine oxidation and vanadyl formation, *Dalt. Trans.*, p. 7985.

74. Ramos, S., Duarte, R.O., Moura, J.J.G. and Aureliano, M. (2012). Recent advances into vanadyl, vanadate and decavanadate interactions with actin, *Metallomics*, **4**, pp. 16–22.

75. Nienberg, C., Garmann, C., Gratz, A., Bollacke, A., Götz, C. and Jose, J. (2017). Identification of a potent allosteric inhibitor of human protein kinase CK2 by bacterial surface display library screening, *Pharmaceuticals*, **10**, p. 6.

76. Prudent, R., Moucadel, V., Laudet, B., Barette, C., Lafanechère, L., Hasenknopf, B., Li, J., Bareyt, S., Lacôte, E., Thorimbert, S., Malacria, M, Gouzerh, P. and Cochet, C. (2008). Identification of polyoxometalates as nanomolar noncompetitive inhibitors of protein kinase CK2, *Chem. Biol.*, **15**, pp. 683–692.

77. Worthington, R.J. and Melander, C. (2013). Combination approaches to combat multidrug-resistant bacteria, *Trends Biotechnol.*, **31**, pp. 177–184.

78. Ling, L.L., Schneider, T., Peoples, A.J., Spoering, A.L., Engels, I., Conlon, B.P., Mueller, A., Schäberle, T.F., Hughes, D.E., Epstein, E., Jones, M.,

Lazarides, L., Steadman, V.A., Cohen, D.R., Felix, C.R., Fetterman, K.A., Millett, W.P., Nitti, A.G., Zullo, A.M., Chen, C. and Lewis, K. (2015). A new antibiotic kills pathogens without detectable resistance, *Nature*, **517**, pp. 455–459.

79. Fiers, W.D., Craighead, M. and Singh, I. (2017). Teixobactin and its analogues: a new hope in antibiotic discovery, *ACS Infect. Dis.*, **3**, pp. 688–690.

80. Dizaj, S.M., Lotfipour, F., Barzegar-Jalali, M., Zarrintan, M.H. and Adibkia, K. (2014). Antimicrobial activity of the metals and metal oxide nanoparticles, *Mater. Sci. Eng. C*, **44**, pp. 278–284.

81. Farzana, R., Iqra, P. and Hunaiza, T. (2018). Antioxidant and antimicrobial effects of polyoxometalates., *Microbiol. Curr. Res.*, **2**, pp. 7–11.

82. Tajima, Y., Nagasawa, Z. and Tadano, J. (1993). A factor found in aged tungstate solution enhanced the antibacterial effect of beta-lactams on methicillin-resistant *Staphylococcus aureus.*, *Microbiol. Immunol.*, **37**, pp. 695–703.

83. Yamase, T., Fukuda, N. and Tajima, Y. (1996). Synergistic effect of polyoxotungstates in combination with beta-lactam antibiotics on antibacterial activity against methicillin-resistant *Staphylococcus aureus.*, *Biol. Pharm. Bull.*, **19**, pp. 459–65.

84. Balici, S., Niculae, M., Pall, E., Rusu, M., Rusu, D. and Matei, H. (2016). Antibiotic-like behaviour of polyoxometalates. *in vitro* comparative study: seven polyoxotungstates-nine antibiotics against gram-positive and gram-negative bacteria, *Rev. Chim.*, **67**, pp. 485–490.

85. Inoue, M., Segawa, K., Matsugana, S., Matsumoto, N., Oda, M. and Yamase, T. (2005). Antibacterial activity of highly negative charged polyoxotungstates, $K_{27}[KAs_4W_{40}O_{140}]$ and $K_{18}[KSb_9W_{21}O_{86}]$, and Keggin-structural polyoxotungstates against *Helicobacter pylori*, *J. Inorg. Biochem.*, **99**, pp. 1023–1031.

86. Fukuda, N., Yamase, T. and Tajima, Y. (1999). Inhibitory effect of polyoxotungstates on the production of penicillin- binding proteins and *β*-lactamase against methicillin-resistant *Staphylococcus aureus*, *Biol. Pharm. Bull.*, **22**, pp. 463–470.

87. Tajima, Y., Nagasawa, Z. and Tadano, J. (1993). A factor found in aged tungstate solution enhanced the antibacterial effect of *β*-lactams on methicillin-resistant *Staphylococcus aureus*, *Microbiol. Immunol.*, **37**, pp. 695–703.

88. Samart, N., Arhouma, Z., Kumar, S., Murakami, H.A., Crick, D.C. and Crans, D.C. (2018). Decavanadate inhibits mycobacterial growth more potently than other oxovanadates, *Front. Chem.*, **6**, pp. 1–16.

89. Missina, J.M., Gavinho, B., Postal, K., Santana, F.S., Valdameri, G., de Souza, E.M., Hughes, D.L., Ramirez, M.I., Soares, J.F. and Nunes, G.G. (2018). Effects of decavanadate salts with organic and inorganic cations on *Escherichia coli*, *Giardia intestinalis*, and vero cells, *Inorg. Chem.*, **57**, pp. 11930–11941.

90. Marques-Da-Silva, D., Fraqueza, G., Lagoa, R., Vannathan, A.A., Mal, S.S. and Aureliano, M. (2019). Polyoxovanadate inhibition of: *Escherichia coli* growth shows a reverse correlation with Ca^{2+}-ATPase inhibition, *New J. Chem.*, **43**, pp. 17577–17587.

91. Fukuda, N. and Yamase, T. (1997). *in vitro* antibacterial activity of vanadate and vanadyl compounds against *Streptococcus pneumoniae*, *Biol. Pharm. Bull.*, **20**, pp. 927–930.

92. Inoue, M., Suzuki, T., Fujita, Y., Oda, M., Matsumoto, N. and Yamase, T. (2006). Enhancement of antibacterial activity of beta-lactam antibiotics by $[P_2W_{18}O_{62}]^{6-}$, $[SiMo_{12}O_{40}]^{4-}$, and $[PTi_2W_{10}O_{40}]^{7-}$ against methicillin-resistant and vancomycin-resistant *Staphylococcus aureus*, *J. Inorg. Biochem.*, **100**, pp. 1225–1233.

93. Gumerova, N., Al-Sayed, E., Krivosudský, L., Cipcic-Paljetak, H., Verbanac, D. and Rompell, A. (2018). Antibacterial activity of polyoxometalates against *Moraxella catarrhalis*, *Front. Chem.*, **6**, pp. 336–345.

94. Seko, A., Yamase, T. and Yamashita, K. (2009). Polyoxometalates as effective inhibitors for sialyl- and sulfotransferases, *J. Inorg. Biochem.*, **103**, pp. 1061–1066.

95. Krstić, D., Colović, M., Bosnjaković-Pavlović, N., Spasojević-De Bire, A. and Vasić, V. (2009). Influence of decavanadate on rat synaptic plasma membrane ATPases activity, *Gen. Physiol. Biophys.*, **28**, pp. 302–306.

96. Fiorani, G., Saoncella, O., Kaner, P., Atinkaya, S.A., Figoli, A., Bonchio, M. and Carraro, M. (2014). Chitosan-polyoxometalate nanocomposites: synthesis, characterization and application as antimicrobial agents, *J. Clust. Sci.*, **25**, pp. 839–854.

97. Marques, M.P.M., Gianolio, D., Ramos, S., Batista de Carvalho, L.A.E. and Aureliano, M. (2017). An EXAFS approach to the study of polyoxometalate-protein interactions: the case of decavanadate-actin, *Inorg. Chem.*, **56**, pp. 10893–10903.

98. Harduin-Lepers, A., Mollicone, R., Delannoy, P. and Oriol, R. (2005). The animal sialyltransferases and sialyltransferase-related genes: a phylogenetic approach, *Glycobiology*, **15**, pp. 805–817.

99. Negishi, M., Pedersen, L.G., Petrochenko, E., Shevtsov, S., Gorokhov, A., Kakuta, Y. and Pedersen, L.C. (2001). Structure and function of sulfotransferases, *Arch. Biochem. Biophys.*, **390**, pp. 149–157.

100. Fang, Y., Xing, C., Zhan, S., Zhao, M., Li, M. and Liu, H. (2019). A polyoxometalate-modified magnetic nanocomposite: a promising antibacterial material for water treatment, *J. Mater. Chem. B*, **7**, pp. 1933–1944.

101. Ma, X., Zhou, F., Yue, Hua, J. and Ma, P. (2019). A nano-linear zinc-substituted phosphomolybdate with reactive oxygen species catalytic ability and antibacterial activity, *J. Mol. Struct.*, **1198**, p. 126865.

102. Missina, J.M., Leme, L.B.P., Postal, K., Santana, F.S., Hughes, D.L., de Sá, E.L., Ribeiro, R.R. and Nunes, G.G. (2020). Accessing decavanadate chemistry with tris(hydroxymethyl)aminomethane, and evaluation of methylene blue bleaching, *Polyhedron*, **180**, p. 114414.

103. Aureliano, M. and Crans, D. C. (2009). Decavanadate ($V_{10}O_{28}^{6-}$) and oxovanadates: Oxometalates with many biological activities, *J. Inorg. Biochem.*, **103**, pp. 536–546.

104. Pereira, M. J., Carvalho, E., Eriksson, J. W., Crans, D. C., and Aureliano, M. (2009). Effects of decavanadate and insulin enhancing vanadium compounds on glucose uptake in isolated rat adipocytes, *J. Inorg. Biochem.*, **103**, pp. 1687–1692.

105. Aureliano, M. (2009). Decavanadate: a journey in a search of a role, *Dalton Trans.*, **42**, pp. 9093–9100.

106. Haapala, D.K., Jasmin, C., Sinoussi, F., Chermann, J.C. and Raynaud, M. (1973). Inhibition of tumour virus RNA-dependent DNA polymerase by the heteropolyanion, silicotungstate, *Biomedicine*, **19**, pp. 7–11.

107. Flütsch, A., Schroeder, T., Grütter, M.G. and Patzke, G.R. (2011). HIV-1 protease inhibition potential of functionalized polyoxometalates, *Bioorg. Med. Chem. Lett.*, **21**, pp. 1162–1166.

108. Qi, Y., Xiang, Y., Wang, J., Qi, Y., Li, J., Niu, J. and Zhong, J. (2013). Inhibition of hepatitis C virus infection by polyoxometalates, *Antiviral Res.*, **100**, pp. 392–398.

109. Witvrouw, M., Weigold, H., Pannecouque, C., Schols, D., De Clercq, E., and Holan, G. (2000). Potent anti-HIV (type 1 and type 2) activity of

polyoxometalates: structure-activity relationship and mechanism of action, *J. Med. Chem.*, **43**, pp. 778–783.

110. Francese, R., Civra, A., Rittà, M., Donalisio, M., Argenziano, M., Cavalli, R., Mougharbel, A.S., Kortz, U. and Lembo, D. (2019). Anti-zika virus activity of polyoxometalates, *Antiviral Res.*, **163**, pp. 29–33.

111. Shigeta, S., Mori, S., Yamase, T., Yamamoto, N. and Yamamoto, N. (2006). Anti-RNA virus activity of polyoxometalates, *Biomed. Pharmacother.*, **60**, pp. 211–219.

112. Wang, J., Liu, Y., Xu, K., Qi, Y., Zhong, J., Zahng, K., Li, J., Wang, E., Wu, Z. and Kang, Z. (2014). Broad-spectrum antiviral property of polyoxometalate localized on a cell surface, *ACS Appl. Mater. Interfaces*, **6**, pp. 9785–9789.

113. Liu, S.X., Wang, C.L., Yu, M., Li, Y.X. and Wang, E.B. (2005). Synthesis and anti-influenza virus activity of polyoxometalates containing amantadine, *Acta Chim. Sin.*, **63**, pp. 1069–1074.

114. Shigeta, S., Mori, S., Watanabe, J., Soeda, S., Takahashi, K. and Yamase, T. (1997). Synergistic anti-influenza virus A (H1N1) activities of PM-523 (polyoxometalate) and ribavirin *in vitro* and *in vivo*, *Antimicrob. Agents Chemother.*, **41**, pp. 1423–1427.

115. Huffman, J.H., Sidwell, R.W., Barnard, D.L., Morrison, A., Otto, M.J., Hill, C.L. and Schinazi, R.F. (1997). Influenza virus-inhibitory effects of a series of germanium- and silicon-centred polyoxometalates, *Antivir. Chem. Chemother.*, **8**, pp. 75–83.

116. Schoeberl, C., Boehner, R., Krebs, B., Mueller, C. and Barnekow, A. (1998). A new polyoxometalate complex inhibits retrovirus encoded reverse transcriptase activity *in vitro* and *in vivo*, *Int. J. Oncol.*, **12**, pp. 153–159.

117. Barnard, D.L., Hill, C.L., Gage, T., Matheson, J.E., Huffman, J.H., Sidwell, R.W., Otto, M.J. and Schinazi, R.F. (1997). Potent inhibition of respiratory syncytial virus by polyoxometalates of several structural classes, *Antiviral Res.*, **34**, pp. 27–37.

118. Fukuma, M., Seto, Y. and Yamase, T. (1991). *In vitro* antiviral activity of polyoxotungstate (PM-19) and other polyoxometalates against herpes simplex virus, *Antiviral Res.*, **16**, pp. 327–339.

119. Wang, X., Wang, J., Zahng, W., Li, B., Zhu, Y., Hu, Q., Yang, Y., Zhang, X., Yan, H. and Zeng, Y. (2018). Inhibition of human immunodeficiency virus type 1 entry by a Keggin polyoxometalate, *Viruses*, **10**, pp. 265.

120. Liu, Y., Liu, S., Wang, E., Zeng, Y. and Li, Z. (2006). Synthesis and *in vitro* anti-HIV-1 activity of electrochemically reduced praseodymium salts of 12-tungstophosphate acid, *Int. J. Virol.*, **2**, pp. 25–34.

121. Inouye, Y., Tokutake, Y., Kunihara, J., Yoshida, T., Yamase, T., Nakata, A. and Nakamura, S. (1992). Suppressive effect of polyoxometalates on the cytopathogenicity of human immunodeficiency virus Type-1 (HIV-1) *in vitro* and their inhibitory activity against HIV-1 reverse transcriptase, *Chem. Pharm. Bull. (Tokyo).*, **40**, pp. 805–807.

122. Li, Q., Zhang, H., Qi, Y., Wang, J., Li, J. and Niu, J. (2019). Antiviral effects of a niobium-substituted heteropolytungstate on hepatitis B virus-transgenic mice, *Drug Dev. Res.*, **80**, pp. 1062–1070.

123. Zhang, H., Qi, Y., Ding, Y., Wang, J., Li, Q., Zahng, J., Jiang, Y., Chi, X., Li, J. and Niu, J. (2012). Synthesis, characterization and biological activity of a niobium-substituted-heteropolytungstate on hepatitis B virus, *Bioorg. Med. Chem. Lett.*, **22**, pp. 1664–1669.

124. Shao, C., Wang, J.-P., Yang, G.-C., Su, Z.-M., Hu, D.-H. and Sun, C.-C. (2008). Interactions of $[Mo_6O_{19}]^{2-}$ and its derivatives substituted with organic groups inhibitor with SARS-CoV 3CLpro by molecular modeling, *Gaodeng Xuexiao Huaxue Xuebao/Chemical J. Chinese Univ.*, **29**, pp. 165–169.

125. Hu, D., Shao, C., Guan, W., Su, Z. and Sun, J. (2007). Studies on the interactions of Ti-containing polyoxometalates (POMs) with SARS-CoV 3CLpro by molecular modeling, *J. Inorg. Biochem.*, **101**, pp. 89–94.

126. Hosseini, S.M., Amini, E., Kheiri, M.T., Mehrbod, P., Shahidi, M. and Zabihi, E. (2012). Anti-influenza activity of a novel polyoxometalate derivative (POM-4960), *Int. J. Mol. Cell. Med.*, **1**, pp. 21–29.

Index